Analysis and Control of Finite-Value Systems

Analysis and Control of Finite-Value Systems

Haitao Li
School of Mathematics and Statistics
Shandong Normal University
Jinan, People's Republic of China

Guodong Zhao
School of Mathematics and Statistics
Shandong Normal University
Jinan, People's Republic of China

Peilian Guo
School of Information Science and Engineering
Shandong Normal University
Jinan, People's Republic of China

Zhenbin Liu
Science and Information College
Qingdao Agricultural University
Qingdao, People's Republic of China

CRC Press is an imprint of the
Taylor & Francis Group, an **informa** business
A SCIENCE PUBLISHERS BOOK

CRC Press
Taylor & Francis Group
6000 Broken Sound Parkway NW, Suite 300
Boca Raton, FL 33487-2742

First issued in paperback 2020

ISBN-13: 978-1-138-55650-8 (hbk)
ISBN-13: 978-0-367-78125-5 (pbk)

Visit the Taylor & Francis Web site at
http://www.taylorandfrancis.com

and the CRC Press Web site at
http://www.crcpress.com

Dedication

To our Ph.D. supervisor Prof. Yuzhen Wang for his guidance

Preface

In the last few decades, the importance of finite-value systems has been recognized. Scholars in different research fields often resort to the variation of finite-value systems as mathematical tools to investigate problems. For example, Boolean network has become a powerful tool in describing and analyzing cellular networks. However, there very few books related to finite-value systems available.

The purpose of this book is to present recent developments and applications in finite-value systems. The book covers various branches of finite-value systems: Boolean networks, mix-valued logical networks, delayed logical networks, switched logical networks, and probabilistic logical networks. Some applications of finite-value systems, which include networked evolutionary games, nonlinear feedback shift register, graph theory, and finite-field networks, are discussed. For every kind of finite-value system covered in the book, we have selected the most representative results. These results can represent the most recent topics in the research of finite-value systems.

The fundamental tool in this book is a new matrix product, called semi-tensor product (STP) of matrices. The STP of matrices is a generalization of the conventional matrix product to the case that the dimension matching condition is not satisfied. This generalization keeps all the major properties of the conventional matrix product unchanged. Using the STP, a logical function can be converted into a multilinear mapping, called the matrix expression of logic. From this construction, the dynamics of finite-value systems can be expressed as a conventional discrete-time multi-linear system. With the help of linear expression and classic control theory, many important issues about finite-value systems, such as stability analysis, controllability analysis, feedback stabilization, output tracking, disturbance decoupling, and optimal control, can be well studied. Furthermore, using the STP method, the theory of finite-value systems can be applied to networked evolutionary games, nonlinear feedback shift register, graph theory, and finite-field networks.

Chapter 1 provides a brief introduction to finite-value systems and STP. In Chapter 2, we present the recent developments on Boolean (control) networks, including Lyapunov-based stability analysis, function perturbation, feedback stabilization and output tracking of Boolean networks. Chapter 3 is devoted to mix-valued logical networks. Chapter 4 studies the topological structure, trajectory controllability and output tracking of delayed logical networks. The stability, controllability and disturbance decoupling of switched logical networks are investigated in Chapter 5. Chapter 6 studies the controllability, optimal control and output tracking control of probabilistic logical control networks. In Chapter 7, we present the application of finite-value systems in networked evolutionary games. Chapter 8 is devoted to the investigation of nonlinear feedback shift registers. Chapter 9 applies the finite-value systems and STP method to graph theory. Finite-field networks are discussed in Chapter 10.

This book is self-contained. The prerequisites for its use are linear algebra and some basic knowledge of classic control theory. Almost all the research works in this book were supervised by Professor Yuzhen Wang, who is the Ph.D. supervisor of all four authors. With his hard work, the theory of finite-value systems has made rapid progress in recent years. The authors are also in debt to Professor Daizhan Cheng for his warmhearted support and a few postgraduate students, including Yating Zheng, Yalu Li, Xueying Ding, Xiaojing Xu and Xiaodong Li, for the editorial support.

The research works involved in this book were supported by the National Natural Science Foundation of China under grants 61374065, 61503225 and 61403223, the Research Fund for the Taishan Scholar Project of Shandong Province, the Natural Science Fund for Distinguished Young Scholars of Shandong Province under grant JQ201613, and the Natural Science Foundation of Shandong Province under grant ZR2015FQ003.

Contents

List of Figures

List of Tables

Part I

Theory

1 Introduction and Preliminaries

1.1 FINITE-VALUE NETWORKS

Finite-value network (FVN) is a type of system whose state, input and output take values from a finite set. FVNs have wide applications in the research fields such as systems biology [4], game theory [3], digital circuits [5], finite automata [13], graph theory [12], fuzzy control [2] and nonlinear feedback shift register [6]. The main characterization of FVNs is parameter-free, and hence one can use FVNs to model large scale systems.

As a special class of FVNs, k-valued logical networks (KVLNs) take values from the finite set $\mathscr{D}_k = \{0, 1, \cdots, k-1\}$. When k is a prime number, KVLNs become finite-field networks [9]. Especially, when $k = 2$, KVLNs are called Boolean networks. If the state, input and output of FVNs take values from different finite sets, we call this kind of FVNs mix-valued logical networks (MVLNs). In the following section, we present some practical examples for FVNs.

Example 1.1: Boolean network model of a sub-network of signal transduction networks

Consider the following Boolean network, which is a sub-network of signal transduction networks [11]:

$$\begin{cases} x_1(t+1) = x_8(t), \\ x_2(t+1) = x_1(t), \\ x_3(t+1) = x_2(t), \\ x_4(t+1) = x_8(t), \\ x_5(t+1) = x_4(t), \\ x_6(t+1) = x_3(t) \vee x_5(t), \\ x_7(t+1) = x_8(t), \\ x_8(t+1) = x_6(t) \wedge \neg x_7(t), \end{cases} \tag{1.1}$$

where x_1 stands for the nitric oxide synthase (NOS), x_2 represents the nitric oxide (NO), x_3 is the guanyl cyclase (GC), x_4 is the phospholipase C (PLC), x_5 represents the inositol-1,4,5-trisphosphate (InsP3), x_6 is the Ca^{2+} influx to the cytosol from intracellular stores (CIS), x_7 stands for the Ca^{2+}ATPases and Ca^{2+}/H^+ antiporters responsible for Ca^{2+} efflux from the cytosol (Ca^{2+}ATPase), and x_8 is the cytosolic Ca^{2+} increase (Ca_c^{2+}).

Example 1.2: Boolean network model for the lactose operon in Escherichia coli

Consider the following Boolean network, which is a five-variable model of the lactose operon in Escherichia coli [10]:

$$\begin{cases} x_1(t+1) = x_3(t), \\ x_2(t+1) = x_1(t), \\ x_3(t+1) = x_3(t) \vee (x_4(t) \wedge x_2(t)), \\ x_4(t+1) = x_5(t) \vee (x_4(t) \wedge \neg x_2(t)), \\ x_5(t+1) = x_1(t), \end{cases} \tag{1.2}$$

where $x_1 \in \mathscr{D}$ denotes the mRNA, $x_2 \in \mathscr{D}$ the β-galactosidase, $x_3 \in \mathscr{D}$ the allolactose, $x_4 \in \mathscr{D}$ the intracellular lactose, and $x_5 \in \mathscr{D}$ the lac permease.

The "minimal" Boolean model for the lactose operon in Escherichia coli is given as follows [10]:

$$\begin{cases} x_1(t+1) = \neg u_1(t) \wedge (x_3(t) \vee u_2(t)), \\ x_2(t+1) = x_1(t), \\ x_3(t+1) = \neg u_1(t) \wedge [(x_2(t) \wedge u_2(t)) \\ \qquad\qquad \vee (x_3(t) \wedge \neg x_2(t))], \end{cases} \tag{1.3}$$

where $x_1 \in \mathscr{D}$ denotes the mRNA, $x_2 \in \mathscr{D}$ the lacZ polypeptide, $x_3 \in \mathscr{D}$ the intracellular lactose, $u_1 \in \mathscr{D}$ the external glucose, and $u_2 \in \mathscr{D}$ the external lactose.

Example 1.3: Rock-Scissors-Paper game

Consider a networked evolutionary game (NEG) consisting of four players, in which the set of players are denoted by $N = \{P_1, P_2, P_3, P_4\}$ and the network graph of the game is string. The neighborhood of each P_i is denoted by $U(i)$. The basic game of this NEG is the Rock-Scissors-Paper game [3], whose payoff matrix is given in Table 1, where "Rock", "Scissors" and "Paper" are denoted by "1", "2" and "3", respectively. Hence, all the players have the same set of strategies: $S = \{1,2,3\}$.

Table 1: Payoff Matrix.

$P_1 \setminus P_2$	1	2	3
1	(0, 0)	(1, –1)	(–1, 1)
2	(–1, 1)	(0, 0)	(1, –1)
3	(1, –1)	(–1, 1)	(0, 0)

Suppose that the game can be repeated infinitely. At each time, P_i only plays the Rock-Scissors-Paper game with its neighbors in $U(i)$, and its aggregate payoff $c_i : S^{|U(i)|} \to \mathbb{R}$ is the sum of payoffs gained by playing with all its neighbors in $U(i)$, that is,

$$c_i(P_i, P_j | j \in U(i)) = \sum_{j \in U(i)} c_{ij}(P_i, P_j), \tag{1.4}$$

where $c_{ij} : S \times S \to \mathbb{R}$ denotes the payoff of P_i playing with its neighbor P_j, $j \in U(i)$.

The strategy updating rule is: for each $i = 1,2$, $P_i(t+1)$ is updated by the best strategy from strategies of its neighbors in $U(i)$ at time t. Precisely, if $j^* = \arg\max_{j \in U(i)} c_j(P_j, P_k | k \in U(j))$, then $P_i(t+1) = P_{j^*}(t)$. When the neighbors with maximum payoff are not unique, say, $\arg\max_{j \in U(i)} c_j(P_j, P_k | k \in U(j)) := \{j_1^*, \cdots, j_r^*\}$, we choose $j^* = \min\{j_1^*, \cdots, j_r^*\}$.

According to the strategy updating rule, we obtain the following 3-valued logical network:

$$P_i(t+1) = f_i(P_1(t), P_2(t), P_3(t), P_4(t)), \tag{1.5}$$

where $f_i, i = 1,2,3,4$ are 3-valued logical functions, which can be uniquely determined by the strategy updating rule.

From the above examples, one can see that the dynamics of FVNs can be expressed in the following form:

$$\begin{cases} x_i(t+1) = f_i(X(t), U(t), \Xi(t)), i = 1, \cdots, n; \\ y_j(t) = h_j(X(t)), j = 1, \cdots, p, \end{cases} \tag{1.6}$$

where $X(t)$, $U(t)$ and $\Xi(t)$ denote the state, control input and disturbance input, respectively, and $y_j(t), j = 1, \cdots, p$ is the output. There exist several tools for the study of FVNs, including computer simulation method, polynomial theory over finite-field, and semi-tensor product (STP) of matrices [1, 7, 8]. Compared with other methods, STP method is a powerful mathematical tool which can convert any FVN into a (bi)linear form. This conversion forms a bridge between classic control theory and FVNs. We will give a detailed introduction on STP in the following sections.

1.2 NOTATIONS

The following notations will be used in the whole book.

- "$\neg$", "$\wedge$" and "$\vee$" represents "Negation", "Conjunction" and "Disjunction", respectively.
- $\mathbb{R}$, $\mathbb{N}$ and $\mathbb{Z}_+$ denote the sets of real numbers, natural numbers and positive integers, respectively.
- $\mathscr{D}_k := \{0, 1, \cdots, k-1\}$, and $\mathscr{D}_k^n := \underbrace{\mathscr{D}_k \times \cdots \times \mathscr{D}_k}_{n}$. When $k = 2$, we denote $\mathscr{D}_2$ by $\mathscr{D}$.
- $\Delta_n := \{\delta_n^k : 1 \leq k \leq n\}$, where δ_n^k represents the k-th column of the identity matrix I_n. When $k = 2$, we denote Δ_2 by Δ.
- An $n \times t$ logical matrix $M = [\delta_n^{i_1}\ \delta_n^{i_2}\ \cdots\ \delta_n^{i_t}]$ is denoted by $M = \delta_n[i_1\ i_2\ \cdots\ i_t]$. $\mathscr{L}_{n \times t}$ represents the set of $n \times t$ logical matrices.
- $Blk_i(A)$ denotes the i-th $n \times n$ block of an $n \times mn$ matrix A.
- For a real matrix $A \in \mathbb{R}^{n \times m}$, $(A)_{i,j}$, $Col_i(A)$ and $Row_i(A)$ denote the (i, j)-th element of A, the i-th column of A, and the i-th row of A, respectively. We call $A > 0$, if $(A)_{i,j} > 0$ holds for any i and j.
- "$\otimes$" represents the Kronecker product of matrices.

- The Khatri-Rao product of $A \in \mathbb{R}^{p\times n}$ and $B \in \mathbb{R}^{q\times n}$ is:

$$A * B = [Col_1(A)\otimes Col_1(B)\ \ Col_2(A)\otimes Col_2(B)\ \ \cdots\ \ Col_n(A)\otimes Col_n(B)].$$

- An $n\times t$ matrix $A=(a_{ij})$ is called a Boolean matrix, if $a_{ij}\in\mathscr{D}$, $\forall\ i=1,\cdots n,\ j=1,\cdots,t$. Denote the set of $n\times t$ Boolean matrices by $\mathscr{B}_{n\times t}$.

1.3 SEMI-TENSOR PRODUCT OF MATRICES

In this section, we state the definition and some basic properties of STP. For details, please refer to [1].

DEFINITION 1.1

The semi-tensor product of two matrices $A\in\mathbb{R}^{m\times n}$ and $B\in\mathbb{R}^{p\times q}$ is defined as:

$$A\ltimes B=(A\otimes I_{\frac{\alpha}{n}})(B\otimes I_{\frac{\alpha}{p}}), \tag{1.7}$$

where $\alpha = lcm(n,p)$ is the least common multiple of n and p.

It should be pointed out that the semi-tensor product of matrices is a generalization of the conventional matrix product. Thus, we omit the symbol "$\ltimes$" if no confusion arises in the following.

PROPOSITION 1.1

STP has the following properties:

(i) Let $A\in\mathbb{R}^{m\times n}$, $B\in\mathbb{R}^{p\times q}$ and $C\in\mathbb{R}^{r\times s}$. Then, $(A\ltimes B)\ltimes C = A\ltimes(B\ltimes C)$.
(ii) Let $X\in\mathbb{R}^{t\times 1}$ be a column vector and $A\in\mathbb{R}^{m\times n}$. Then,

$$X\ltimes A=(I_t\otimes A)\ltimes X. \tag{1.8}$$

(iii) Let $X\in\mathbb{R}^{m\times 1}$ and $Y\in\mathbb{R}^{n\times 1}$ be two column vectors. Then,

$$Y\ltimes X = W_{[m,n]}\ltimes X\ltimes Y, \tag{1.9}$$

where $W_{[m,n]}\in\mathscr{L}_{mn\times mn}$ is called the swap matrix, which is given as:

$$\begin{aligned} W_{[m,n]} \ = \ & \delta_{mn}[1\ \ m+1\ \ \cdots\ \ (n-1)m+1 \\ & 2\ \ m+2\ \ \cdots\ \ (n-1)m+2 \\ & \cdots \\ & m\ \ m+m\ \ \cdots\ \ (n-1)m+m]. \end{aligned}$$

(iv) Let $X \in \Delta_m$ and $Y \in \Delta_n$. Define two dummy matrices $D_f[m,n]$ and $D_r[m,n]$ as follows:

$$D_f[m,n] = \delta_m[\underbrace{1\ 1 \cdots 1}_{n}\ \underbrace{2\ 2 \cdots 2}_{n} \cdots \underbrace{m\ m \cdots m}_{n}], \tag{1.10}$$

$$D_r[m,n] = \delta_n[\underbrace{\underbrace{1\ 2 \cdots n}\ \underbrace{1\ 2 \cdots n} \cdots \underbrace{1\ 2 \cdots n}}_{m}]. \tag{1.11}$$

Then, $D_f[m,n] \ltimes X \ltimes Y = X$ and $D_r[m,n] \ltimes X \ltimes Y = Y$.

(v) Let $X \in \mathbb{R}^{n\times 1}$ be a column vector. Then,

$$M_{r,n}X = X^2,$$

where $M_{r,n} = diag\{\delta_n^1, \delta_n^2, \cdots, \delta_n^n\}$.

In the following section, some useful results on the matrix expression of Boolean functions are stated.

Identifying $i \sim \delta_k^{i+1}$, $i \in \mathscr{D}_k$, we have $\mathscr{D}_k \sim \Delta_k$, where "$\sim$" denotes two different forms of the same object. We call δ_k^{i+1} the vector form of logical values.

PROPOSITION 1.2

Let $f(x_1,x_2,\cdots,x_s) : \mathscr{D}_k^s \mapsto \mathscr{D}_k$ be a logical function. Then, there exists a unique matrix $M_f \in \mathscr{L}_{k\times k^s}$, called the structural matrix of f, such that,

$$f(x_1,x_2,\cdots,x_s) = M_f \ltimes_{i=1}^s x_i, \quad x_i \in \Delta_k, \tag{1.12}$$

where $\ltimes_{i=1}^s x_i := x_1 \ltimes \cdots \ltimes x_s$.

For example, the structural matrices of Negation ($\neg$), Conjunction ($\wedge$) and Disjunction ($\vee$) are $\delta_2[2\ 1]$, $\delta_2[1\ 2\ 2\ 2]$ and $\delta_2[1\ 1\ 1\ 2]$, respectively.

Using Proposition 1.2, one can convert (1.6) into the following component-wise algebraic form:

$$\begin{cases} x_i(t+1) = M_i u(t)x(t)\xi(t), i = 1,\cdots,n; \\ y_j(t) = H_j x(t), j = 1,\cdots,p, \end{cases} \tag{1.13}$$

where M_i and H_j are structural matrices of f_i and h_j, respectively. Multiplying the equations in (1.13) together yields the following algebraic form:

$$\begin{cases} x(t+1) = Lu(t)x(t)\xi(t), \\ y(t) = Hx(t), \end{cases} \tag{1.14}$$

where $L = M_1 * M_2 * \cdots * M_n$ and $H = H_1 * \cdots * H_p$. Obviously, (1.14) has a bilinear form.

For example, the algebraic form of (1.1) is

$$x(t+1) = Lx(t),$$

where:

$$\begin{array}{rl} L = \delta_{256}[& 2\ 148\ \ 1\ 147\ \ 2\ 148\ \ 2\ 148\ \ 2\ 148\ \ 1\ 147\ \ 2\ 148\ \ 2\ 148 \\ & 10\ 156\ \ 9\ 155\ \ 10\ 156\ \ 10\ 156\ \ 10\ 156\ \ 9\ 155\ \ 10\ 156\ \ 10\ 156 \\ & 2\ 148\ \ 1\ 147\ \ 2\ 148\ \ 2\ 148\ \ 6\ 152\ \ 5\ 151\ \ 6\ 152\ \ 6\ 152 \\ & 10\ 156\ \ 9\ 155\ \ 10\ 156\ \ 10\ 156\ \ 14\ 160\ \ 13\ 159\ \ 14\ 160\ \ 14\ 160 \\ & 34\ 180\ \ 33\ 179\ \ 34\ 180\ \ 34\ 180\ \ 34\ 180\ \ 33\ 179\ \ 34\ 180\ \ 34\ 180 \\ & 42\ 188\ \ 41\ 187\ \ 42\ 188\ \ 42\ 188\ \ 42\ 188\ \ 41\ 187\ \ 42\ 188\ \ 42\ 188 \\ & 34\ 180\ \ 33\ 179\ \ 34\ 180\ \ 34\ 180\ \ 38\ 184\ \ 37\ 183\ \ 38\ 184\ \ 38\ 184 \\ & 42\ 188\ \ 41\ 187\ \ 42\ 188\ \ 42\ 188\ \ 46\ 192\ \ 45\ 191\ \ 46\ 192\ \ 46\ 192 \\ & 66\ 212\ \ 65\ 211\ \ 66\ 212\ \ 66\ 212\ \ 66\ 212\ \ 65\ 211\ \ 66\ 212\ \ 66\ 212 \\ & 74\ 220\ \ 73\ 219\ \ 74\ 220\ \ 74\ 220\ \ 74\ 220\ \ 73\ 219\ \ 74\ 220\ \ 74\ 220 \\ & 66\ 212\ \ 65\ 211\ \ 66\ 212\ \ 66\ 212\ \ 70\ 216\ \ 69\ 215\ \ 70\ 216\ \ 70\ 216 \\ & 74\ 220\ \ 73\ 219\ \ 74\ 220\ \ 74\ 220\ \ 78\ 224\ \ 77\ 223\ \ 78\ 224\ \ 78\ 224 \\ & 98\ 244\ \ 97\ 243\ \ 98\ 244\ \ 98\ 244\ \ 98\ 244\ \ 97\ 243\ \ 98\ 244\ \ 98\ 244 \\ & 106\ 252\ \ 105\ 251\ \ 106\ 252\ \ 106\ 252\ \ 106\ 252\ \ 105\ 251\ \ 106\ 252\ \ 106\ 252 \\ & 98\ 244\ \ 97\ 243\ \ 98\ 244\ \ 98\ 244\ \ 102\ 248\ \ 101\ 247\ \ 102\ 248\ \ 102\ 248 \\ & 106\ 252\ \ 105\ 251\ \ 106\ 252\ \ 106\ 252\ \ 110\ 256\ \ 109\ 255\ \ 110\ 256\ \ 110\ 256]. \end{array}$$

The algebraic form of (1.3) is:

$$x(t+1) = Lu(t)x(t),$$

where,

$$\begin{array}{rl} L = \delta_8[& 6\ 6\ 6\ 6\ 8\ 8\ 8\ 8\ 6\ 6\ 6\ 6\ 8\ 8\ 8\ 8 \\ & 1\ 1\ 1\ 2\ 3\ 3\ 3\ 4\ 2\ 6\ 1\ 6\ 4\ 8\ 3\ 8]. \end{array}$$

REFERENCES

1. Cheng, D., Qi, H. and Li, Z. (2011). Analysis and Control of Boolean Networks: A Semi-Tensor Product Approach. London, Springer.
2. Cheng, D., Feng, J. and Lv, H. (2012). Solving fuzzy relational equations via semitensor product. IEEE Transactions on Fuzzy Systems, 20(2): 390–396.
3. Cheng, D., He, F., Qi, H. and Xu, T. (2015). Modeling, analysis and control of networked evolutionary games. IEEE Trans. Aut. Contr., 60(9): 2402–2415.
4. Kauffman, S. (1969). Metabolic stability and epigenesis in randomly constructed genetic nets. Journal of Theoretical Biology, 22(3): 437–467.

5. Liu, Z., Wang, Y. and Li, H. (2014). A new approach to derivative calculation of multi-valued logical functions with application to fault detection of digital circuits. IET Control Theory and Applications, 8: 554–560.
6. Liu, Z., Wang, Y. and Cheng, D. (2015). Nonsingularity of feedback shift registers. Automatica, 55: 247–253.
7. Lu, J., Li, H., Liu, Y. and Li, F. (2017). Survey on semi-tensor product method with its applications in logical networks and other finite-value systems. IET Control Theory and Applications, 11(13): 2040–2047.
8. Li, H., Zhao, G., Meng, M. and Feng, J. (2017). A survey on applications of semi-tensor product method in engineering. Science China Information Sciences, DOI: 10.1007/s11432-017-9238-1.
9. Pasqualetti, F., Borra, D. and Bullo, F. (2014). Consensus networks over finite fields. Automatica, 50(2): 349–358.
10. Robeva, R. and Hodge, T. (2013). Mathematical Concepts and Methods in Modern Biology: Using Modern Discrete Models. Academic Press.
11. Saadatpour, A., Albert, I. and Albert, R. (2010). Attractor analysis of asynchronous Boolean models of signal transduction networks. J. Theoretical Biology, 266: 641–656.
12. Wang, Y., Zhang, C. and Liu, Z. (2012). A matrix approach to graph maximum stable set and coloring problems with application to multi-agent systems. Automatica, 48: 1227–1236.
13. Xu, X. and Hong, Y. (2013). Matrix approach to model matching of asynchronous sequential machines. IEEE Trans. Aut. Contr., 58(11): 2974–2979.

2 Boolean Networks

2.1 INTRODUCTION TO BOOLEAN NETWORKS

Boolean network is a special kind of finite-value system whose state, input and output take values from $\mathscr{D} := \{0, 1\}$. Boolean networks were first introduced by Jacob and Monod in 1960s, and then were used by Kauffman to describe, analyze and simulate gene regulatory networks [1].

In a Boolean network, the gene state can be determined by Boolean difference equations and the activation of other genes. In order to manipulate Boolean networks, one needs to introduce binary control inputs and outputs to the network dynamics, which yields Boolean control networks (BCNs). Akutsu et al. pointed out that the control problems of Boolean networks are NP-hard [2].

Example 2.1: Boolean network model of apoptosis networks

Consider the following apoptosis network [3]:

$$\begin{cases} x_1(t+1) = \neg x_2(t) \wedge u(t), \\ x_2(t+1) = \neg x_1(t) \wedge x_3(t), \\ x_3(t+1) = x_2(t) \vee u(t), \end{cases} \tag{2.1}$$

where the concentration level (high or low) of the inhibitor of apoptosis proteins (IAP) is denoted by x_1, the concentration level of the active caspase 3 (C3a) by x_2, and the concentration level of the active caspase 8 (C8a) by x_3; the concentration level of the tumor necrosis factor (TNF, a stimulus) is regarded as the control input u.

From the above example, one can see that the dynamics of Boolean networks can be expressed in the following form:

$$\begin{cases} x_1(t+1) = f_1(x_1(t), \cdots, x_n(t)), \\ \quad \vdots \\ x_n(t+1) = f_n(x_1(t), \cdots, x_n(t)), \end{cases} \tag{2.2}$$

where $x_i(t) \in \mathscr{D}, i = 1, \cdots, n$ denotes the state variable, and $f_i : \mathscr{D}^n \to \mathscr{D}, i = 1, \cdots, n$ are Boolean functions.

Daizhan Cheng [4] proposed an algebraic state space representation approach (ASSR) to the analysis and control of Boolean networks. In the ASSR framework, 1 and 0 are identified as $\delta_2^1 := \begin{bmatrix} 1 \\ 0 \end{bmatrix}$ and $\delta_2^2 := \begin{bmatrix} 0 \\ 1 \end{bmatrix}$, respectively. System (2.2) has the following component-wise algebraic form:

$$\begin{cases} x_1(t+1) &= K_1 x(t), \\ &\vdots \\ x_n(t+1) &= K_n x(t), \end{cases} \tag{2.3}$$

where $K_i \in \mathcal{L}_{2\times 2^n}$ is the structural matrix of f_i, $i = 1, \cdots, n$, and $x(t) = \ltimes_{i=1}^n x_i(t)$. Then, system (2.2) can be converted into the following algebraic form:

$$x(t+1) = Lx(t), \tag{2.4}$$

where $L = K_1 * \cdots * K_n \in \mathcal{L}_{2^n \times 2^n}$. For example, the algebraic form of system (2.1) is

$$x(t+1) = \delta_8[7\ 7\ 3\ 3\ 5\ 7\ 1\ 3\ 7\ 7\ 8\ 8\ 5\ 7\ 6\ 8]x(t).$$

It was proved in [4] that (2.2) is equivalent to (2.4). In the following section we explain how to obtain (2.2) from (2.4).

PROPOSITION 2.1

Split K_i into two equal parts as $K_i = [K_i^1,\ K_i^2]$, then,

$$\begin{aligned} f_i(x_1, \cdots, x_n) &= (x_1 \wedge f_i^1(x_2, \cdots, x_n)) \vee \\ &\quad (\neg x_1 \wedge f_i^2(x_2, \cdots, x_n)), \end{aligned} \tag{2.5}$$

where f_i^j has its structural matrix as K_i^j, $j = 1,\ 2$.

2.2 TOPOLOGICAL STRUCTURE OF BOOLEAN NETWORKS

Consider system (2.4). we state the definitions of fixed point and cycle for Boolean networks.

DEFINITION 2.1

1. A state $x_0 \in \Delta_{2^n}$ is called a fixed point of the system (2.4), if $Lx_0 = x_0$.
2. $\{x_0, Lx_0, \cdots, L^k x_0\}$ is called a cycle of the system (2.4), if $L^k x_0 = x_0$, and the elements in the set $\{x_0, \cdots, L^{k-1} x_0\}$ are pairwise distinct.
3. S is called the basin of the attractor (fixed point or cycle) C, if S is the set of points which converge to C.

Based on the algebraic form (2.4), we have the following result on the calculation of fixed point and cycle.

Theorem 2.1

The number of cycles with length d for system (2.4), denoted by N_d, is inductively determined by:

$$\begin{cases} N_1 = tr(L), \\ N_d = \frac{tr(L^d) - \sum_{k \in P(d)} kN_k}{d}, 2 \leq d \leq 2^n, \end{cases} \tag{2.6}$$

where $tr(L^d)$ denotes the trace of matrix L^d, and $P(d)$ denotes the set of proper factors of d. ■

Theorem 2.2

The basin of attractor C, denoted by S, can be calculated as follows:

$$S = C \cup L^{-1}(C) \cup L^{-2}(C) \cup \cdots \cup L^{-T_t}(C), \tag{2.7}$$

where $L^{-k}(C) = \{q | L^k q \in C\}$, and T_t denotes the transient period of system (2.4). ■

For example, when $u = 1$, system (2.1) has two fixed points δ_8^3, δ_8^5 and a cycle with length 2, that is, $\{\delta_8^1, \delta_8^7\}$. Moreover, the basin of $\{\delta_8^1, \delta_8^7\}$ is $\{\delta_8^1, \delta_8^2, \delta_8^6, \delta_8^7\}$. When $u = 0$, system (2.1) has two fixed points δ_8^5, δ_8^8 and a cycle with length 2, that is, $\{\delta_8^6, \delta_8^7\}$.

One can see that the dimension of the state transition matrix obtained by the semi-tensor product method grows exponentially. A natural question arises: given a Boolean network, how to reduce the dimension of the state transition matrix without changing the topological structure (that is, all the fixed points and cycles)? We study this problem by using the logical matrix factorization approach.

Consider system (2.4). For $L = \delta_{2^n}[i_1 \ i_2 \ \cdots \ i_{2^n}] \in \mathscr{L}_{2^n \times 2^n}$, let $\Gamma = \{\widehat{i_j} : j = 1, \cdots, r; \widehat{i_j} < \widehat{i_k}, j < k\}$ denote the set of distinct indices in the set $\{i_1, i_2, \cdots, i_{2^n}\}$, and $r := |\Gamma|$. Let s_j be the number of indices in $(i_1, i_2, \cdots, i_{2^n})$ coinciding with $\widehat{i_j}$. Using a permutation matrix $Q \in \mathscr{L}_{2^n \times 2^n}$, one can permute the columns of L as:

$$LQ = \delta_{2^n}\Big[\underbrace{\widehat{i_1} \cdots \widehat{i_1}}_{s_1} \cdots \underbrace{\widehat{i_j} \cdots \widehat{i_j}}_{s_j} \cdots \underbrace{\widehat{i_r} \cdots \widehat{i_r}}_{s_r}\Big]. \tag{2.8}$$

Then, we can factorize L into:

$$L = L_1 L_2, \tag{2.9}$$

where,

$$L_1 = \delta_{2^n}[\widehat{i}_1 \ \cdots \ \widehat{i}_r] \in \mathcal{L}_{2^n \times r},$$

and

$$L_2 = \delta_r\Big[\underbrace{1 \cdots 1}_{s_1} \cdots \underbrace{r \cdots r}_{s_r}\Big]Q^{-1} \in \mathcal{L}_{r\times 2^n}.$$

Now, we define a bijective map from $\{\delta_{2^n}^{\widehat{i}_j} : \widehat{i}_j \in \Gamma\}$ to Δ_r as $\varphi(\delta_{2^n}^{\widehat{i}_j}) = \delta_r^j$, $\forall\ j = 1, \cdots, r$. Setting $\widehat{L} = L_2 L_1 \in \mathcal{L}_{r\times r}$, we obtain a new system with $\widehat{L}$ as:

$$z(t+1) = \widehat{L}z(t), \tag{2.10}$$

where $z(t) \in \Delta_r$. Then, we have the following two theorems.

Theorem 2.3

$x_e = \delta_{2^n}^{\widehat{i}_\mu}$ is a fixed point of system (2.4), if and only if $z_e = \delta_r^\mu$ is a fixed point of system (2.10). ■

Proof. (Necessity) Assume that $x_e = \delta_{2^n}^{\widehat{i}_\mu}$ is a fixed point of system (2.4). Then,

$$\begin{aligned}\delta_{2^n}^{\widehat{i}_\mu} &= L\delta_{2^n}^{\widehat{i}_\mu} = L_1 L_2 \delta_{2^n}^{\widehat{i}_\mu} \\ &= L_1 Col_{\widehat{i}_\mu}(L_2) = L_1 \delta_r^\mu,\end{aligned}$$

which together with the fact that all the columns of L_1 are distinct imply that $L_2\delta_{2^n}^{\widehat{i}_\mu} = Col_{\widehat{i}_\mu}(L_2) = \delta_r^\mu$. Combining $L_2\delta_{2^n}^{\widehat{i}_\mu} = \delta_r^\mu$ with $L_1\delta_r^\mu = \delta_{2^n}^{\widehat{i}_\mu}$, we have,

$$\widehat{L}\delta_r^\mu = L_2 L_1 \delta_r^\mu = L_2 \delta_{2^n}^{\widehat{i}_\mu} = \delta_r^\mu.$$

Thus, $z_e = \delta_r^\mu$ is a fixed point of system (2.10).

(Sufficiency) Suppose that $z_e = \delta_r^\mu$ is a fixed point of system (2.10). Then $\widehat{L}\delta_r^\mu = \delta_r^\mu$. Since $L_1\delta_r^\mu = \delta_{2^n}^{\widehat{i}_\mu}$, one can see that:

$$\delta_r^\mu = \widehat{L}\delta_r^\mu = L_2 L_1 \delta_r^\mu = L_2 \delta_{2^n}^{\widehat{i}_\mu}.$$

Thus,

$$L\delta_{2^n}^{\widehat{i}_\mu} = L_1(L_2\delta_{2^n}^{\widehat{i}_\mu}) = L_1\delta_r^\mu = \delta_{2^n}^{\widehat{i}_\mu},$$

which implies that $x_e = \delta_{2^n}^{\widehat{i}_\mu}$ is a fixed point of system (2.4).

Theorem 2.4

$\{\delta_{2^n}^{\widehat{i_\mu}}, L\delta_{2^n}^{\widehat{i_\mu}}, \cdots, L^k\delta_{2^n}^{\widehat{i_\mu}}\}$ is a cycle of system (2.4) with length k, if and only if $\{\delta_r^{\mu}, \widehat{L}\delta_r^{\mu}, \cdots, \widehat{L}^k\delta_r^{\mu}\}$ is a cycle of the system (2.10) with length k. ■

Proof. The proof of this theorem is similar to that of Theorem 2.3, and thus we omit it.

For the number of cycles of system (2.4), we have the following new result.

Theorem 2.5

The number of cycles with length s for system (2.4), denoted by N_s, is inductively determined by,

$$\begin{cases} N_1 = Trace(\widehat{L}), \\ N_s = \frac{Trace(\widehat{L}^s) - \sum_{k \in \mathscr{P}(s)} kN_k}{s}, 2 \leq s \leq r, \end{cases} \tag{2.11}$$

where $\widehat{L} = L_2 L_1 \in \mathscr{L}_{r \times r}$. ■

Using the same technique as in (2.10), one can factorize $\widehat{L}$ as $\widehat{L} = \widehat{L}_1 \widehat{L}_2$, and obtain a new logical matrix $\widehat{\widehat{L}} = \widehat{L}_2 \widehat{L}_1$, where $\widehat{L}_1 \in \mathscr{L}_{r \times r_1}$, $\widehat{L}_2 \in \mathscr{L}_{r_1 \times r}$, $\widehat{\widehat{L}} \in \mathscr{L}_{r_1 \times r_1}$ and $r_1 \leq r$. Keep this procedure going, we finally obtain the smallest logical matrix $L^* = \delta_{r^*}[\alpha_1 \ \cdots \ \alpha_{r^*}] \in \mathscr{L}_{r^* \times r^*}$ with $\alpha_i \neq \alpha_j$, $\forall\, i \neq j$. It should be pointed out that Theorems 2.3 and 2.4 also hold when replacing $\widehat{L}$ by L^*. In this case, we obtain the smallest-size system:

$$z(t+1) = L^* z(t) \tag{2.12}$$

whose topological structure is the same as that of the original Boolean network (2.4). We call r^* and L^* the attractor index and the attractor characteristic matrix of system (2.4), respectively.

2.3 LYAPUNOV-BASED STABILITY ANALYSIS

As a special kind of discrete-time dynamic systems, Boolean networks should have Lyapunov functions. In this section, we define a Lyapunov function and establish a Lyapunov theory for Boolean networks.

First we state the definition and properties of pseudo-Boolean functions. For details, please refer to [5].

DEFINITION 2.2

A n-ary pseudo-Boolean function $f(x_1,x_2,\cdots,x_n)$ is a mapping from $\mathscr{D}^n$ to $\mathbb{R}$, where $\mathscr{D}^n := \underbrace{\mathscr{D}\times\cdots\times\mathscr{D}}_{n}$.

PROPOSITION 2.2

Every pseudo-Boolean function $f(x_1,x_2,\cdots,x_n)$ can be uniquely represented in the multi-linear polynomial form of:

$$f(x_1,x_2,\cdots,x_n) = c_0 + \sum_{k=1}^{m} c_k \prod_{i\in A_k} x_i, \tag{2.13}$$

where $c_0, c_1, \cdots, c_m$ are real coefficients, $A_1, A_2, \cdots, A_m$ are nonempty subsets of $N = \{1,2,\cdots,n\}$, and the product is the conventional one.

PROPOSITION 2.3

Assume that $f(x_1,x_2,\cdots,x_n) : \Delta^n \mapsto \mathbb{R}$ is a pseudo-Boolean function. Then, there exists a unique matrix $M_f \in \mathbb{R}^{2\times 2^n}$ such that:

$$f(x_1,x_2,\cdots,x_n) = J_1 M_f \ltimes_{i=1}^{n} x_i, \quad x_i \in \Delta, \tag{2.14}$$

where $J_1 = [1\ 0]$ is called the selection matrix to be used to obtain the first row of M_f.

Proof. From Proposition 2.2, we have,

$$\begin{aligned} & f(x_1,x_2,\cdots,x_n) \\ = \ & c_0 + \sum_{k=1}^{m} c_k \prod_{i\in A_k} x_i \\ = \ & J_1\left(c_0(E_d)^n(I_{2^n}\otimes\delta_2^1) + \sum_{k=1}^{m} c_k M_k\right) \ltimes_{i=1}^{n} x_i \\ := \ & J_1 M_f \ltimes_{i=1}^{n} x_i, \end{aligned}$$

where M_k is the structural matrix of term $(\prod_{i\in A_k} x_i)$. Thus, the proof is completed.

In the following section, we define a Lyapunov function for Boolean networks, and present some Lyapunov-based stability results.

System (2.2) is said to be asymptotically stable, if it has a fixed point x_e as its only attractor. Without loss of generality, we always assume that if system (2.2) is asymptotically stable, then the unique fixed point is $x_e = (0,0,\cdots,0)$ (a coordinate transformation can guarantee this assumption). In general, system (2.2) has a few attractors including fixed points and/or cycles. We denote by $\mathcal{O}_e$ the set of the fixed points, and $\mathcal{S}$ the set of both, the fixed points and the points in all cycles. Obviously, $\mathcal{O}_e \subseteq \mathcal{S}$.

Notice that a Lyapunov function of Boolean network (2.2) must be a real-valued function of $x_i \in \mathcal{D}$, $i = 1,2,\cdots,n$; otherwise, it only takes two values, that is, 1 and 0, and surely cannot be used as an energy function for the system. Based on this analysis, a Lyapunov function of system (2.2) should be a pseudo-Boolean function in the form of:

$$V(x_1,x_2,\cdots,x_n) = c_0 + c_1x_1 + c_2x_2 + \cdots + c_nx_n \\ + c_{n+1}x_1x_2 + \cdots + c_{2^n-1}x_1x_2\cdots x_n, \tag{2.15}$$

where the total number of terms is $C_n^0 + C_n^1 + \cdots + C_n^n = 2^n$; c_i, $i = 0,1,\cdots,2^n-1$ are real coefficients; and the similar terms are arranged in the order of dictionary, that is, for any two terms $c_sx_{i_1}x_{i_2}\cdots x_{i_r}$ and $c_tx_{j_1}x_{j_2}\cdots x_{j_r}$, $s < t$ if and only if $i_k \leqslant j_k$ holds for $k = 1,2,\cdots,r$.

Theorem 2.6

Boolean network (2.2) is asymptotically stable at x_e, if there exists a pseudo-Boolean function in the form of (2.15) satisfying:

(i) $V(x_1,x_2,\cdots,x_n) > 0$ for $\forall\ (x_1,x_2,\cdots,x_n) \neq x_e \in \mathcal{D}^n$, and $V(x_e) = 0$;
(ii) along the trajectories of system (2.2), $\Delta V(x_1(t),\cdots,x_n(t)) := V(x_1(t+1),\cdots,x_n(t+1)) - V(x_1(t),\cdots,x_n(t)) < 0$ holds for $(x_1(t),\cdots,x_n(t)) \neq x_e$, and $\Delta V(x_1(t),\cdots,x_n(t)) = 0$ holds for $(x_1(t),\cdots,x_n(t)) = x_e$. ■

Proof. Assume that $V(x_1,x_2,\cdots,x_n)$ in the form of (2.15) satisfies (i) and (ii). Using the vector form of logical variables and setting $x = \ltimes_{i=1}^n x_i$, it can be seen from Theorem 2.2 that $V(x_1,x_2,\cdots,x_n)$ can be expressed as:

$$V(x) = J_1M_V \ltimes_{i=1}^n x_i. \tag{2.16}$$

Thus, along the trajectories of system (2.4), we have,

$$\begin{aligned} \Delta V(x(t)) &= V(x_1(t+1),\cdots,x_n(t+1)) - \\ &\quad\ V(x_1(t),\cdots,x_n(t)) \\ &= J_1M_Vx(t+1) - J_1M_Vx(t) \\ &= J_1M_VLx(t) - J_1M_Vx(t) \\ &= J_1M_V(L - I_{2^n})x(t), \end{aligned}$$

From the above equation and Conditions (i) and (ii) we obtain:

$$\begin{cases} c_0 = Col_{2^n}(J_1 M_V) = 0; \\ Col_i(J_1 M_V) > 0, \ \forall \ 1 \leq i \leq 2^n - 1; \\ Col_i(J_1 M_V(L - I_{2^n})) < 0, \ \forall \ 1 \leq i \leq 2^n - 1; \\ Col_{2^n}(J_1 M_V(L - I_{2^n})) = 0. \end{cases} \tag{2.17}$$

Now, we prove that system (2.4) is globally convergent to x_e. We divide the proof into the following two steps.

Step 1. Let us show that x_e in the vector form of $\delta_{2^n}^{2^n}$ is a fixed point of the system (2.4).

In fact, if $\delta_{2^n}^{2^n}$ is not a fixed point of the system, then $L\delta_{2^n}^{2^n} = \delta_{2^n}^i$, $i \neq 2^n$. Thus,

$$Col_{2^n}(L - I_{2^n}) = [0, \cdots, 0, \underbrace{1}_{i-th}, 0, \cdots, 0, -1]^T.$$

Using (2.17), we obtain:

$$\begin{aligned} & Col_{2^n}(J_1 M_V(L - I_{2^n})) = J_1 M_V Col_{2^n}(L - I_{2^n}) \\ & = Col_i(J_1 M_V) - Col_{2^n}(J_1 M_V) = Col_i(J_1 M_V) > 0, \end{aligned}$$

which is a contradiction with $Col_{2^n}(J_1 M_V(L - I_{2^n})) = 0$. Therefore, $\delta_{2^n}^{2^n}$ is a fixed point of the system.

Step 2. We prove that system (2.4) is asymptotically stable at x_e.

Let $x(0) = \delta_{2^n}^{i_0}$ be any initial point. If $\delta_{2^n}^{i_0} \neq \delta_{2^n}^{2^n}$, then we obtain $x(1) = Lx(0) = \delta_{2^n}^{i_1}$. If $\delta_{2^n}^{i_1} \neq \delta_{2^n}^{2^n}$, then we have $x(2) = Lx(1) = \delta_{2^n}^{i_2}$. $\cdots$ Keep going, we obtain $x(k) = Lx(k-1) = \delta_{2^n}^{i_k}$ $\cdots$. Thus, we have the sequence,

$$x(0) \rightarrow x(1) \rightarrow x(2) \rightarrow \cdots \rightarrow x(k) \rightarrow \cdots. \tag{2.18}$$

From Conditions (i) and (ii), we know that $V(x(k)) = Col_{i_k}(J_1 M_V) > 0$, $k = 0, 1, 2, \cdots$, and

$$V(x(0)) > V(x(1)) > \cdots > V(x(k)) > \cdots. \tag{2.19}$$

Since $x(k) = \delta_{2^n}^{i_k} \in \Delta_{2^n}$, $k = 0, 1, \cdots$ and Δ_{2^n} is a finite set, we conclude that there exists an integer k_0 $(0 \leqslant k_0 \leqslant 2^n - 1)$ such that $x(k_0) = \delta_{2^n}^{2^n}$. If not, the sequence (2.18) is an infinite one, and there exist j_1 and j_2 such that $j_1 < j_2$ and $x(j_1) = x(j_2)$, which implies that $V(x(j_1)) = V(x(j_2))$. On the other hand, from (2.19) we have $V(x(j_1)) > V(x(j_2))$. This is a contradiction. Thus, there exists k_0 such that $x(k_0) = \delta_{2^n}^{2^n}$, which implies the sequence (2.18) converges to x_e.

From the arbitrariness of $x(0)$, system (2.4) is globally asymptotically stable at x_e.

Now, based on Theorem 2.6, we give the definition of Lyapunov function for Boolean networks.

DEFINITION 2.3

A pseudo-Boolean function $V(x_1,\cdots,x_n)$: $\mathscr{D}^n \mapsto \mathbb{R}$ in the form of (2.15) is called a strict-Lyapunov function of Boolean network (2.2), if,

(i) $V(x_1,\cdots,x_n) > 0$ for $\forall$ $(x_1,\cdots,x_n) \neq x_e \in \mathscr{D}^n$, and $V(x_e) = 0$;
(ii) along the trajectories of Boolean network (2.2), $\Delta V(x_1(t),\cdots,x_n(t)) < 0$ holds for $(x_1(t),\cdots, x_n(t)) \neq x_e$, and $\Delta V(x_1(t),\cdots,x_n(t)) = 0$ holds for $(x_1(t),\cdots,x_n(t)) = x_e$.

For the general case that Boolean network (2.4) has a few attractors including fixed points and/or cycles, we give the following definition.

DEFINITION 2.4

A pseudo-Boolean function $V(x_1,\cdots,x_n)$: $\mathscr{D}^n \mapsto \mathbb{R}$ in the form of (2.15) is called a Lyapunov function of Boolean network (2.2), if,

(i) $V(x_1,\cdots,x_n) > 0$, $\forall (x_1,\cdots,x_n) \in \mathscr{D}^n \setminus \mathscr{O}_e$, and $V(x_1,\cdots,x_n) = 0$ holds $\forall (x_1,\cdots,x_n) \in \mathscr{O}_e$;
(ii) along the trajectories of Boolean network (2.2), $\Delta V(x_1(t),\cdots,x_n(t)) < 0$ holds for $(x_1(t),\cdots, x_n(t)) \notin \mathscr{S}$, and $\Delta V(x_1(t),\cdots,x_n(t)) = 0$ holds for $(x_1(t),\cdots,x_n(t)) \in \mathscr{S}$.

We give an illustrative example.

Example 2.2:

Consider the following Boolean network:

$$\begin{cases} x_1(t+1) = x_1(t)\bar{\vee}x_2(t)), \\ x_2(t+1) = \neg(x_1(t) \rightarrow x_2(t)), \end{cases} \tag{2.20}$$

where $x_i \in \mathscr{D}$, $i = 1,2$.

Choose $V(x_1,x_2) = 2x_1 + 3x_2 - 4x_1x_2$, then it is easy to check that $V(x_1,x_2) > 0$ for $(x_1,x_2) \neq (0,0) \in \mathscr{D}^2$, and $V(0,0) = 0$. On the other hand, we can easily check that along the trajectories of Boolean network (2.20), $\Delta V(x_1(t),x_2(t)) < 0$ holds for $(x_1(t),x_2(t)) \neq (0,0)$, and $\Delta V(x_1(t),x_2(t)) = 0$ holds for $(x_1(t),x_2(t)) = (0,0)$. Thus, $V(x_1,x_2)$ is a strict-Lyapunov function of system (2.20). By Theorem 2.6, system (2.20) is globally convergent to $(0,0)$.

Finally, we present a method to construct a Lyapunov function for a given Boolean network.

Consider Boolean network (2.4). Assume that a Lyapunov function $V(x_1,x_2,\cdots,x_n)$ of the system is given in the form of (2.15), where c_i, $i=0,1,\cdots,2^n-1$ are real coefficients to be determined.

Set,

$$J_1M_V = [a_1, a_2, \cdots, a_{2^n}]. \tag{2.21}$$

Since,

$$\begin{cases} V(1,1,\cdots,1,1) &= c_0+\cdots+c_{2^n-1}=a_1, \\ V(1,1,\cdots,1,0) &= c_0+\cdots+c_{n-1}+c_{n+1} \\ &+ c_{2^n-n-1}+\cdots=a_2, \\ &\vdots \\ V(0,0,\cdots,0,0) &= c_0=a_{2^n}, \end{cases} \tag{2.22}$$

there exists a $2^n\times 2^n$ matrix P_n such that:

$$P_n[c_0, c_1, \cdots, c_{2^n-1}]^T = [a_1, a_2, \cdots, a_{2^n}]^T.$$

It is easy to see that P_n is nonsingular, with which we have:

$$[c_0, c_1, \cdots, c_{2^n-1}]^T = P_n^{-1}[a_1, a_2, \cdots, a_{2^n}]^T. \tag{2.23}$$

J_1M_V should satisfy:

$$\begin{aligned} &Col_i(J_1M_V)>0,\ \ i\in\{1,2,\cdots,2^n\}\setminus\mathscr{I}_e; \\ &Col_i(J_1M_V)=0,\ \ i\in\mathscr{I}_e; \\ &Col_i(J_1M_V(L-I_{2^n}))<0, i\in\{1,2,\cdots,2^n\}\setminus\mathscr{I}_{\mathscr{S}}; \\ &Col_i(J_1M_V(L-I_{2^n}))=0,\ \ i\in\mathscr{I}_{\mathscr{S}}, \end{aligned} \tag{2.24}$$

which are equivalent to:

$$\begin{cases} a_i=0,\ i\in\mathscr{I}_e, \\ a_i>0,\ i\in\{1,2,\cdots,2^n\}\setminus\mathscr{I}_e, \\ [a_1, a_2,\cdots, a_{2^n}]Col_i(L-I_{2^n})=0,\ \ i\in\mathscr{I}_{\mathscr{S}}, \\ [a_1, a_2,\cdots, a_{2^n}]Col_i(L-I_{2^n})<0,\ i\in\{1,2, \\ \quad\cdots,2^n\}\setminus\mathscr{I}_{\mathscr{S}}. \end{cases} \tag{2.25}$$

We have the following result on the solvability of (2.25).

PROPOSITION 2.4

The set of inequalities/equations (2.25) is always solvable.

We have the following algorithm to construct the desired Lyapunov function $V(x_1,x_2,\cdots,x_n)$ for Boolean network (2.4).

ALGORITHM 2.1

Consider Boolean network (2.4) and assume that $V(x_1,\cdots,x_n)$ in the form of (2.15) is a Lyapunov function of the system to be found. To construct $V(x_1,\cdots,x_n)$, we follow the following steps:

1). Compute the matrix L, and find out the index sets $\mathscr{I}_e$ and $\mathscr{I}_{\mathscr{S}}$ from L;
2). Solve the set of inequalities/equations (2.25) and obtain a solution $(a_1, a_2, \cdots, a_{2^n})$;
3). Compute the matrix P_n, and then find out all the c_i by the equation (2.23) with the obtained solution $(a_1, a_2, \cdots, a_{2^n})$. Then, the desired Lyapunov function is given as:

$$V(x_1,\cdots,x_n) = c_0 + c_1x_1 + c_2x_2 + \cdots + c_nx_n$$
$$+c_{n+1}x_1x_2 + \cdots + c_{2^n-1}x_1x_2\cdots x_n,$$

where $x_i \in \mathscr{D}$, $i = 1,2,\cdots,n$.

Example 2.3:

Construct a Lyapunov function for the following Boolean network:

$$\begin{cases} x_1(t+1) = [x_1(t) \wedge (x_2(t) \rightarrow x_3(t))] \\ \qquad\qquad \vee(\neg x_1(t) \wedge x_3(t)), \\ x_2(t+1) = [x_1(t) \wedge (x_2(t) \vee x_3(t))] \vee \neg x_1(t), \\ x_3(t+1) = [x_3(t) \wedge (x_1(t) \leftrightarrow x_2(t))] \vee \neg x_3(t), \end{cases} \tag{2.26}$$

where $x_i \in \mathscr{D}$, $i = 1,2,3$.

Assume that the pseudo-Boolean function,

$$V(x_1,x_2,x_3) = c_0 + c_1x_1 + c_2x_2 + c_3x_3 + c_4x_1x_2$$
$$+c_5x_1x_3 + c_6x_2x_3 + c_7x_1x_2x_3 \tag{2.27}$$

is a Lyapunov function of system (2.26) to be found, where c_i, $i = 0,1,\cdots,7$ are real coefficients to be determined. We now use Algorithm 2.1 to calculate all the c_i.

Using the vector form of logical variables and letting $x(t) = \ltimes_{i=1}^{3} x_i(t)$, one can easily obtain the algebraic form of system (2.26) as $x(t+1) = Lx(t)$, where $L = \delta_8[1\ 5\ 2\ 3\ 2\ 5\ 1\ 5]$. Moreover, it is easy to see from L that $\mathscr{I}_e = \{1\}$ and $\mathscr{I}_{\mathscr{S}} = \{1,2,5\}$.

For this example, the set of ineqalities/equations (2.25) reduces to:

$$\begin{cases} a_1 = 0,\ a_i > 0,\ i = 2,\cdots,8, \\ a_2 = a_5, \\ a_2 - a_3 < 0,\ a_3 - a_4 < 0, \\ a_5 - a_6 < 0,\ a_1 - a_7 < 0,\ a_5 - a_8 < 0, \end{cases} \tag{2.28}$$

which has an infinite number of solutions. For example, $(a_1,a_2,a_3,a_4,a_5,a_6,a_7,a_8) = (0,1,2,3,\ 1,2,1,2)$, $(0,2,3,4,2,6,7,8)$, and so on.

Choose a solution, say, $(0,1,2,3,1,2,1,2)$. It is easy to obtain $[c_0,c_1,c_2,c_3,c_4,c_5,c_6,c_7]^T = P_3^{-1}[0,1,2,3,1,2,1,2]^T = [2,1,0,-1,-2,0,0,0]^T$. Thus, a Lyapunov function of system (2.26) is given as:

$$V(x_1,x_2,x_3) = 2 + x_1 - x_3 - 2x_1x_2, \tag{2.29}$$

where $x_i \in \mathscr{D}$, $i = 1,2,3$.

2.4 FUNCTION PERTURBATION OF BOOLEAN NETWORKS

In Boolean networks, there may exist measurement error, immeasurable variables, or gene mutation. Due to this, the perturbation impact analysis of Boolean networks, including the state and function perturbation, has been studied. Consequently, lots of results have been established for intervention and long-run behavior analysis of Boolean networks [6]. In this section, we study the function perturbation of Boolean networks by using the semi-tensor product of matrices.

Consider system (2.2) with its component-wise algebraic form (2.3) and algebraic form (2.4). We first investigate how do K_i and L change when some values in the truth table of the Boolean function f_i alter. It is well known that the first row of the structural matrix K_i corresponds to the truth value of f_i. Thus, the truth value change is equivalent to the change of K_i. Next, we study this problem based on the change of K_i.

Suppose that the j-th column of K_i alters. Then, from the fact that $Col_k(L) = \ltimes_{j=1}^n Col_k(K_j)$, $k = 1,\cdots,2^n$, we conclude that $Col_k(L)$, $k \neq j$ do not change, and $Col_j(L)$ changes.

We further suppose that $L = \delta_{2^n}[i_1,i_2,\cdots,i_{2^n}]$, and after the j-th column of K_i alters, L becomes $L' = \delta_{2^n}[i'_1,i'_2,\cdots,i'_{2^n}]$. Then, we have the following proposition.

PROPOSITION 2.5

Consider system (2.2) with its component-wise algebraic form (2.3) and algebraic form (2.4). Suppose that the j-th column of K_i alters. Then, $i_k = i'_k$, $k \neq j$, and,

$$i'_j = \begin{cases} i_j + 2^{n-i}, Col_j(K_i) \text{ changes from } \delta_2^1 \text{ to } \delta_2^2, \\ i_j - 2^{n-i}, Col_j(K_i) \text{ changes from } \delta_2^2 \text{ to } \delta_2^1. \end{cases} \tag{2.30}$$

Proof. We first prove that when $Col_j(K_i)$ changes from δ_2^1 to δ_2^2, $i'_j = i_j + 2^{n-i}$.

Denote $Col_j(K_p) = \delta_2^{i_j^p}$, $i_j^p \in \{1,2\}$, $p = 1,\cdots,n$. We have:

$$i_j = \sum_{p=1}^{n} (i_j^p - 1)2^{n-p} + 1.$$

Thus:

$$\begin{aligned} i'_j &= \sum_{p=1}^{n} (i'^{p}_j - 1)2^{n-p} + 1 \\ &= \sum_{p \neq i} (i^p_j - 1)2^{n-p} + (i'^{i}_j - 1)2^{n-i} + 1 \\ &= \sum_{p=1}^{n} (i^p_j - 1)2^{n-p} + 1 + (i'^{i}_j - i^i_j)2^{n-i} \\ &= i_j + (2-1)2^{n-i} - (1-1)2^{n-i} = i_j + 2^{n-i}. \end{aligned}$$

Similarly, one can prove that when $Col_j(K_i)$ changes from δ_2^2 to δ_2^1, $i'_j = i_j - 2^{n-i}$. We omit it.

Suppose that after the function perturbation, there are m ($m \leq 2^n$) columns of $L = \delta_{2^n}[i_1, i_2, \cdots, i_{2^n}]$ to be altered, denoted by $j_1, \cdots, j_m$. Furthermore, assume that $Col_{j_q}(L) = \delta_{2^n}^{i_{j_q}}$, $q = 1, \cdots, m$ change to $\delta_{2^n}^{i'_{j_q}}$, and L changes to L'. Our goal is stated as: For the attractor (fixed point or cycle) of system (2.4), denoted by $C = \{\delta_{2^n}^{k_1}, \cdots, \delta_{2^n}^{k_r}\}$, judge whether or not C is invariant after the function perturbation.

We have the following result about the function perturbation in Boolean networks.

Theorem 2.7

Consider system (2.2) with its component-wise algebraic form (2.3) and algebraic form (2.4). Suppose that after the function perturbation, there are m ($m \leq 2^n$) columns of L to be altered, denoted by $j_1, \cdots, j_m$. Furthermore, assume that $C = \{\delta_{2^n}^{k_1}, \cdots, \delta_{2^n}^{k_r}\}$ is an attractor of system (2.4). Then, C is invariant after the function perturbation, if and only if:

$$\{j_1, \cdots, j_m\} \cap \{k_1, \cdots, k_r\} = \emptyset. \tag{2.31}$$

■

Proof. (Necessity) Suppose that C is invariant after the function perturbation, but (2.31) is not satisfied. That is, there exists a $j_l \in \{k_1, \cdots, k_r\}$, say $j_l = k_v$. Then, on one hand,

$$L'\delta_{2^n}^{k_v} = L'\delta_{2^n}^{j_l} = \delta_{2^n}^{i'_{j_l}} \neq \delta_{2^n}^{i_{j_l}},$$

and on the other hand,

$$L'\delta_{2^n}^{k_v} = L\delta_{2^n}^{k_v} = L\delta_{2^n}^{j_l} = \delta_{2^n}^{i_{j_l}},$$

which implies a contradiction. Therefore, (2.31) holds true.

(Sufficiency) Supposing that (2.31) is satisfied, then $Col_{k_v}(L),\ v = 1,\cdots,r$ are invariant. Thus $L'\delta_{2^n}^{k_v} = L\delta_{2^n}^{k_v},\ v = 1,\cdots,r$, which implies that C is invariant.

Finally, we focus our attention on how does an attractor change when just one column of L alters. Given an attractor (fixed point or cycle) of system (2.2), denoted by $C = \{\delta_{2^n}^{k_1},\cdots,\delta_{2^n}^{k_r}\}$. Suppose that after the function perturbation, the j-th column of $L = \delta_{2^n}[i_1,i_2,\cdots,i_{2^n}]$ is altered. Furthermore, assume that $\delta_{2^n}^{i_j}$ changes to $\delta_{2^n}^{i'_j}$, and L changes to L'. According to Theorem 2.7, C is invariant if and only if $j\overline{\in}\{k_1,\cdots,k_r\}$. Thus, to study how does C change when the j-th column of L is altered, we suppose $j = k_v$, then, $\delta_{2^n}^{i_j} = L\ltimes\delta_{2^n}^{k_v}$ and $\delta_{2^n}^{i'_j} \neq L\ltimes\delta_{2^n}^{k_v}$.

Intuitively, there are 3 possible changes of the attractor C when the j-th column of L is altered. 1. If $\delta_{2^n}^{i'_j}\overline{\in}S$, then $L'\ltimes\delta_{2^n}^{k_v} = \delta_{2^n}^{i'_j}\overline{\in}S$, which implies that C disappears with no new attractors generated, and vice versa. 2. If $\delta_{2^n}^{i'_j} \in S/C := \{x|x\in S,\ \text{but}\ x\overline{\in}C\}$, then $L'\ltimes\delta_{2^n}^{k_v} = \delta_{2^n}^{i'_j} \in S/C$, that is, C disappears with a new attractor generated, and vice versa. 3. If $r > 1$, and $\delta_{2^n}^{i'_j} \in C/\{L\ltimes\delta_{2^n}^{k_v}\}$, then $L'\ltimes\delta_{2^n}^{k_v} = \delta_{2^n}^{i'_j} \in C/\{L\ltimes\delta_{2^n}^{k_v}\}$, that is, C shrinks, and vice versa.

Based on the above analysis, we have the following result.

Theorem 2.8

Consider system (2.2) with its component-wise algebraic form (2.3) and algebraic form (2.4). Suppose that after the function perturbation, the j-th column of L is altered, and $\delta_{2^n}^{i_j}$ changes to $\delta_{2^n}^{i'_j}$. Furthermore, assume that $C = \{\delta_{2^n}^{k_1},\cdots,\delta_{2^n}^{k_r}\}$ is an attractor of system (2.4), and $j = k_v,\ 1\leq v\leq r$. Then,

(i). C disappears with no new attractors generated, if and only if $\delta_{2^n}^{i'_j}\overline{\in}S$.

(ii). C disappears with a new attractor generated, if and only if $\delta_{2^n}^{i'_j} \in S/C$.

(iii). for $r > 1$, C shrinks, if and only if $\delta_{2^n}^{i'_j} \in C/\{L\ltimes\delta_{2^n}^{k_v}\}$. ■

An application of the above results is given below.

Example 2.4:

Consider the following WNT5A Boolean network introduced in Table 3 of [6]:

$$\begin{cases} x_1(t+1) = f_1 = \neg x_6(t), \\ x_2(t+1) = f_2 = [x_2(t) \wedge (x_4(t) \vee x_6(t))] \\ \qquad\qquad\qquad \vee(\neg x_2(t) \wedge x_4(t) \wedge x_6(t)), \\ x_3(t+1) = f_3 = \neg x_7(t), \\ x_4(t+1) = f_4 = x_4(t), \\ x_5(t+1) = f_5 = x_2(t) \vee (\neg x_2(t) \wedge \neg x_7(t)), \\ x_6(t+1) = f_6 = x_3(t) \vee x_4(t), \\ x_7(t+1) = f_7 = x_2(t) \rightarrow x_7(t). \end{cases} \tag{2.32}$$

Using the vector form of logical variables and setting $x(t) = \ltimes_{i=1}^{7} x_i(t)$, by the semi-tensor product method we have,

$$x(t+1) = Lx(t), \tag{2.33}$$

where $L =$

$$\begin{aligned} \delta_{128}[&81\ 66\ 17\ 2\ 81\ 66\ 17\ 2\ 93\ 74\ 61\ 42\ 93\ 74\ 61\ 42 \\ &81\ 66\ 17\ 2\ 81\ 66\ 17\ 2\ 95\ 76\ 63\ 44\ 95\ 76\ 63\ 44 \\ &81\ 65\ 49\ 33\ 81\ 65\ 49\ 33\ 125\ 105\ 61\ 41\ 125\ 105\ 61\ 41 \\ &81\ 65\ 49\ 33\ 81\ 65\ 49\ 33\ 127\ 107\ 63\ 43\ 127\ 107\ 63\ 43 \\ &81\ 66\ 17\ 2\ 81\ 66\ 17\ 2\ 93\ 74\ 61\ 42\ 93\ 74\ 61\ 42 \\ &81\ 66\ 17\ 2\ 81\ 66\ 17\ 2\ 95\ 76\ 63\ 44\ 95\ 76\ 63\ 44 \\ &81\ 65\ 49\ 33\ 81\ 65\ 49\ 33\ 125\ 105\ 61\ 41\ 125\ 105\ 61\ 41 \\ &81\ 65\ 49\ 33\ 81\ 65\ 49\ 33\ 127\ 107\ 63\ 43\ 127\ 107\ 63\ 43]. \end{aligned}$$

One can easily obtain the entire attractors of (2.32) as follows: $\delta_{128}^{63} \sim (1,0,0,0,0,0,1)$, $\delta_{128}^{66} \sim (0,1,1,1,1,1,0)$, $\delta_{128}^{74} \sim (0,1,1,0,1,1,0)$, $\delta_{128}^{81} \sim (0,1,0,1,1,1,1)$. Since the WNT5A gene is up-regulated [6], the attractor δ_{128}^{63} is undesirable.

The goal of this example is to eliminate the attractor δ_{128}^{63} by some function perturbations in the WNT5A network (2.32) with two constraints: (i) to perturb as few truth values of the Boolean functions as possible; (ii) to affect as few states as possible.

For the constraint (i), the best choice is to perturb the 63-th column of the structural matrix of some f_i, $i = 1, \cdots, 7$. Then we have seven choices, that is, to perturb the 63-th column of the structural matrix of some f_i, $i = 1, \cdots, 7$. By Proposition 2.5, through the perturbation of the 63-th column of the structural matrix of f_1, $Col_{63}(L)$ changes from δ_{128}^{63} to δ_{128}^{127}. Similarly, by the perturbation of the 63-th column of the structural matrix of f_2, f_3, f_4, f_5, f_6 and f_7, $Col_{63}(L)$ changes from δ_{128}^{63} to δ_{128}^{31}, δ_{128}^{47}, δ_{128}^{55}, δ_{128}^{59}, δ_{128}^{61} and δ_{128}^{64} respectively.

Noting that the number of affected states by one column perturbation of f_i equals 2^{n-k_i} (here, k_i denotes the number of variables of f_i), thus to achieve the constraint (ii), we first choose f_2 as our candidate perturbed function. Then $Col_{63}(L)$ changes from δ_{128}^{63} to δ_{128}^{31}. Since δ_{128}^{31} is in the basin of the attractor δ_{128}^{63}, by Theorem 2.8, a new attractor $\{\delta_{128}^{31}, \delta_{128}^{63}\}$ is generated, which is undesirable.

Now to achieve the constraint (ii), we choose f_5, f_6 and f_7 as our candidate perturbed function, respectively. By Theorem 2.8 again, one can easily see that the perturbation of f_5 or f_6 or f_7 is also undesirable since a new attractor is generated.

Next, we choose f_1, f_3 and f_4 as our candidate perturbed function, respectively. By Theorem 2.8, we conclude that the perturbation of f_1 or f_3 is undesirable since a new attractor is generated, while the perturbation of f_4 is desirable since no new attractor is generated.

Based on the above analysis, we obtain the best choice of the perturbed function to achieve our goal: to perturb the 63-th column of the structural matrix of f_4.

2.5 FEEDBACK STABILIZATION OF BOOLEAN CONTROL NETWORKS

In this section, we study the feedback stabilization of the following Boolean control network. Consider the following Boolean control network:

$$\begin{cases} x_1(t+1) = f_1(X(t),U(t)), \\ x_2(t+1) = f_2(X(t),U(t)), \\ \vdots \\ x_n(t+1) = f_n(X(t),U(t)); \\ y_j(t) = g_j(X(t)), \quad j = 1,\cdots,p, \end{cases} \tag{2.34}$$

where $X(t) = (x_1(t),x_2(t),\cdots,x_n(t)) \in \mathscr{D}^n$, $U(t) = (u_1(t),\cdots,u_m(t)) \in \mathscr{D}^m$ and $Y(t) = (y_1(t),\cdots,y_p(t)) \in \mathscr{D}^p$ are the state, the control input and the output of system (2.34), respectively, and $f_i : \mathscr{D}^{m+n} \mapsto \mathscr{D}$, $i = 1,\cdots,n$ and $g_j : \mathscr{D}^n \mapsto \mathscr{D}$, $j = 1,\cdots,p$ are logical functions.

We now state the definition of stabilization for BCNs.

DEFINITION 2.5

For a given state $X_e = (x_1^e,x_2^e,\cdots,x_n^e) \in \mathscr{D}^n$, system (2.34) is said to be globally stabilizable to X_e, if there exist a control sequence $\{U(t), t \in \mathbb{N}\}$ and a positive integer τ such that $X(t;X_0;U) = X_e$, $\forall\, X_0 \in \mathscr{D}^n$ and $\forall\, t \geq \tau$.

We consider the following two kinds of feedback controls:

- State feedback stabilizer in the form of:

$$\begin{cases} u_1(t) = k_1(x_1(t),x_2(t),\cdots,x_n(t)), \\ \vdots \\ u_m(t) = k_m(x_1(t),x_2(t),\cdots,x_n(t)), \end{cases} \tag{2.35}$$

where $k_i : \mathscr{D}^n \mapsto \mathscr{D}$, $i = 1,\cdots,m$ are logical functions to be determined.

- Output feedback stabilizer in the form of:

$$\begin{cases} u_1(t) = h_1(y_1(t), \cdots, y_p(t)), \\ \vdots \\ u_m(t) = h_m(y_1(t), \cdots, y_p(t)), \end{cases} \tag{2.36}$$

where $h_i : \mathscr{D}^p \mapsto \mathscr{D}$, $i = 1, \cdots, m$ are logical functions to be determined.

In the following section, we convert system (2.34) and feedback controls (2.35) and (2.36) into equivalent algebraic forms, respectively.

Using the vector form of logical variables and setting $x(t) = \ltimes_{i=1}^n x_i(t) \in \Delta_{2^n}$, $u(t) = \ltimes_{i=1}^m u_i(t) \in \Delta_{2^m}$ and $y(t) = \ltimes_{i=1}^p y_i(t) \in \Delta_{2^p}$, by the semi-tensor product method, one can convert (2.34), (2.35) and (2.36) into:

$$\begin{cases} x(t+1) = Lu(t)x(t), \\ y(t) = Hx(t), \end{cases} \tag{2.37}$$

$$u(t) = Gx(t), \tag{2.38}$$

and

$$u(t) = Ky(t), \tag{2.39}$$

respectively, where $L \in \mathscr{L}_{2^n \times 2^{m+n}}$, $H \in \mathscr{L}_{2^p \times 2^n}$, $G \in \mathscr{L}_{2^m \times 2^n}$ and $K \in \mathscr{L}_{2^m \times 2^p}$. Moreover, the equilibrium $x_e = \ltimes_{i=1}^n x_i^e = \delta_{2^n}^{\alpha}$.

For the state feedback stabilization control design, we have the following result.

For $k = 1, 2, \cdots$, if,

$$E_k(\alpha) = \{x_0 \in \Delta_{2^n} : \text{ there exist } u(0), \cdots, u(k-1) \in \Delta_{2^m} \\ \text{such that } x(k; x_0; u(0), \cdots, u(k-1)) = \delta_{2^n}^{\alpha}\} \tag{2.40}$$

then the state feedback stabilizer of system (2.34) can be designed by the following result.

Theorem 2.9

Consider system (2.37) with $L = \delta_{2^n}[\alpha_1 \ \alpha_2 \ \cdots \ \alpha_{2^{m+n}}]$. Suppose that $\delta_{2^n}^{\alpha} \in E_1(\alpha)$ and that there exists an integer $1 \leq N \leq 2^n - 1$ such that $E_N(\alpha) = \Delta_{2^n}$. For each $1 \leq i \leq 2^n$ which corresponds to a unique integer $1 \leq l_i \leq N$ such that $\delta_{2^n}^i \in E_{l_i}(\alpha) \setminus E_{l_i - 1}(\alpha)$, where $E_0(\alpha) := \emptyset$, let $1 \leq p_i \leq 2^m$ be such that:

$$\begin{cases} \alpha_{(p_i-1)2^n+i} = \alpha, & l_i = 1, \\ \delta_{2^n}^{\alpha_{(p_i-1)2^n+i}} \in E_{l_i-1}(\alpha), & l_i \geq 2. \end{cases} \tag{2.41}$$

Then, the state feedback control $u(t) = Gx(t)$ with:

$$G = \delta_{2^m}[p_1 \ \cdots \ p_{2^n}] \tag{2.42}$$

globally stabilizes system (2.34) to $x_e = \delta_{2^n}^{\alpha}$. ■

Now, based on Theorem 2.9, we design output feedback stabilizers for BCNs.

From Theorem 2.9, for each integer $1 \le i \le 2^n$, one can find all integers $1 \le p_i \le 2^m$ satisfying (2.41), which form a set, denoted by P_i. We define the following two sets:

$$\Lambda = \{G = \delta_{2^m}[p_1 \ \cdots \ p_{2^n}] : p_i \in P_i, i = 1, \cdots, 2^n\} \tag{2.43}$$

and

$$\Theta = \{K = \delta_{2^m}[v_1 \ \cdots \ v_{2^p}] : KH \in \Lambda\}. \tag{2.44}$$

For each integer $1 \le k \le 2^p$, denote by $O(k)$ the set of states whose output are $\delta_{2^p}^k$. Noting that $H \ltimes \delta_{2^n}^i = Col_i(H)$, we have:

$$O(k) = \{\delta_{2^n}^i : Col_i(H) = \delta_{2^p}^k\}. \tag{2.45}$$

Obviously, $O(k_1) \bigcap O(k_2) = \emptyset$, $\forall\, k_1 \neq k_2$, and $\bigcup_{k=1}^{2^p} O(k) = \Delta_{2^n}$.

For each $O(k)$, we construct a set, denoted by $I(k)$, as:

$$I(k) = \begin{cases} \bigcap_{\delta_{2^n}^i \in O(k)} P_i, & O(k) \neq \emptyset, \\ \{1, 2, \cdots, 2^m\}, & O(k) = \emptyset. \end{cases} \tag{2.46}$$

Then, we have the following result on the existence of output feedback stabilizers.

Theorem 2.10

System (2.34) is globally stabilizable to $x_e = \delta_{2^n}^{\alpha}$ by an output feedback control $u(t) = Ky(t), K \in \Theta$, if and only if,

$$I(k) \neq \emptyset, \ \forall\, k = 1, 2, \cdots, 2^p. \tag{2.47}$$

■

Proof. (Sufficiency) Assuming that (2.47) holds, we prove that system (2.34) is globally stabilizable to x_e by an output feedback control $u(t) = Ky(t), K \in \Theta$.

In fact, we construct the output feedback control $u(t) = Ky(t)$ with,

$$K = \delta_{2^m}[v_1 \ v_2 \ \cdots \ v_{2^p}], \ v_k \in I(k).$$

Then, for $H = \delta_{2^p}[w_1 \ w_2 \ \cdots \ w_{2^n}]$, one can obtain that:

$$KH = \delta_{2^m}[v_{w_1} \ v_{w_2} \ \cdots \ v_{w_{2^n}}]. \tag{2.48}$$

Since $v_{w_i} \in I(w_i) \subseteq P_i$, $\forall\ i = 1,2,\cdots,2^n$, we conclude that $K \in \Theta$. Hence, $u(t) = (KH)x(t)$ is a state feedback stabilizer for system (2.34). Therefore, $u(t) = Ky(t), K \in \Theta$ is an output feedback stabilizer of system (2.34), which implies that the sufficiency holds.

(Necessity) Suppose that system (2.34) is stabilizable to x_e by an output feedback control, say, $u(t) = Ky(t) = \delta_{2^m}[v_1\ v_2\ \cdots\ v_{2^p}]y(t), K \in \Theta$. We prove that (2.47) holds.

In fact, if (2.47) does not hold, then there exists an integer $1 \leq k \leq 2^p$ such that $I(k) = \emptyset$. In this case, one can see that $O(k) \neq \emptyset$ and $\bigcap_{\delta_{2^n}^i \in O(k)} P_i = \emptyset$. Denote by $O(k) = \{\delta_{2^n}^{i_1}, \cdots, \delta_{2^n}^{i_q}\}$.

On the other hand, since $u(t) = Ky(t), K \in \Theta$ is an output feedback stabilizer of system (2.34), we conclude that $u(t) = (KH)x(t), KH \in \Lambda$ is a state feedback stabilizer.

Set $H = \delta_{2^p}[w_1\ w_2\ \cdots\ w_{2^n}]$. We have:

$$v_{w_{i_1}} = \cdots = v_{w_{i_q}} = v_k,$$

which implies that $v_k \in \bigcap_{\delta_{2^n}^i \in O(k)} P_i = \emptyset$, a contradiction.

Thus, $I(k) \neq \emptyset$, $\forall\ k = 1,2,\cdots,2^p$.

From the proof of Theorem 2.10, we can design output feedback stabilizers for system (2.34).

Theorem 2.11

Suppose that (2.47) holds. Then, the output feedback gain matrices of system (2.34) can be designed in the form of:

$$K = \delta_{2^m}[v_1\ v_2\ \cdots\ v_{2^p}],\ v_k \in I(k), \tag{2.49}$$

where $I(k)$ is given in (2.46). ■

ALGORITHM 2.2

One can design output feedback stabilizers for system (2.34) according to the following steps:

1) Calculate P_i, $i = 1,2,\cdots,2^n$ by Theorem 2.9.
2) Calculate $O(k)$ and $I(k)$, $k = 1,2,\cdots,2^p$ from (2.45) and (2.46), respectively.
3) Judge whether or not $I(k) \neq \emptyset$, $\forall\ k = 1,2,\cdots,2^p$. If yes, go to 4); otherwise, system (2.34) has no output feedback gain matrix $K \in \Theta$, and end the calculation.

4) The output feedback gain matrices can be designed as:

$$K = \delta_{2^m}[v_1 \ v_2 \ \cdots \ v_{2^p}], \ v_k \in I(k).$$

Example 2.5:

Consider the following Boolean control network, which is a reduced model for the lac operon in the bacterium Escherichia coli:

$$\begin{cases} x_1(t+1) = \neg u_1(t) \wedge (x_2(t) \vee x_3(t)), \\ x_2(t+1) = \neg u_1(t) \wedge u_2(t) \wedge x_1(t), \\ x_3(t+1) = \neg u_1(t) \wedge (u_2(t) \vee (u_3(t) \wedge x_1(t))), \end{cases} \tag{2.50}$$

where x_1, x_2 and x_3 are state variables which denote the lac mRNA, the lactose in high concentrations, and the lactose in medium concentrations, respectively; u_1, u_2 and u_3 are control inputs which represent the extracellular glucose, the high extracellular lactose, and the medium extracellular lactose, respectively.

In this example, the measured outputs are assumed as:

$$\begin{cases} y_1(t) = x_1(t), \\ y_2(t) = x_2(t). \end{cases} \tag{2.51}$$

Our objective is to design output feedback controllers such that the system is stabilized to $X_e = (1,0,1)$.

Using the vector form of logical variables and setting $x(t) = \ltimes_{i=1}^{3} x_i(t)$, $u(t) = \ltimes_{i=1}^{3} u_i(t)$ and $y(t) = \ltimes_{i=1}^{2} y_i(t)$, by the semi-tensor product of matrices, we have the following algebraic form:

$$\begin{cases} x(t+1) = Lu(t)x(t), \\ y(t) = Hx(t), \end{cases} \tag{2.52}$$

where,

$$\begin{aligned} L = \ & \delta_8[8\ 8\ 8\ 8\ 8\ 8\ 8\ 8\ 8\ 8\ 8\ 8\ 8\ 8\ 8\ 8 \\ & 8\ 8\ 8\ 8\ 8\ 8\ 8\ 8\ 8\ 8\ 8\ 8\ 8\ 8\ 8\ 8 \\ & 1\ 1\ 1\ 5\ 3\ 3\ 3\ 7\ 1\ 1\ 1\ 5\ 3\ 3\ 3\ 7 \\ & 3\ 3\ 3\ 7\ 4\ 4\ 4\ 8\ 4\ 4\ 4\ 8\ 4\ 4\ 4\ 8], \end{aligned}$$

and $H = \delta_4[1\ 1\ 2\ 2\ 3\ 3\ 4\ 4]$. Moreover, $x_e = \delta_2^1 \ltimes \delta_2^2 \ltimes \delta_2^1 = \delta_8^3$.

One can obtain from Theorem 2.9 that $P_1 = \{7\}$, $P_2 = \{7\}$, $P_3 = \{7\}$, $P_4 = \{5,6,7\}$, $P_5 = \{5,6\}$, $P_6 = \{5,6\}$, $P_7 = \{5,6\}$ and $P_8 = \{5,6\}$.

A straightforward calculation shows that $O(1) = \{\delta_8^1, \delta_8^2\}$, $O(2) = \{\delta_8^3, \delta_8^4\}$, $O(3) = \{\delta_8^5, \delta_8^6\}$ and $O(4) = \{\delta_8^7, \delta_8^8\}$. Thus, $I(1) = P_1 \bigcap P_2 = \{7\}$, $I(2) = P_3 \bigcap P_4 = \{7\}$, $I(3) = P_5 \bigcap P_6 = \{5,6\}$ and $I(4) = P_7 \bigcap P_8 = \{5,6\}$.

Finally, according to Theorem 2.10, we can design the output feedback gain matrices as $K_1 = \delta_8[7\ 7\ 5\ 5]$, $K_2 = \delta_8[7\ 7\ 5\ 6]$, $K_3 = \delta_8[7\ 7\ 6\ 5]$ and $K_4 = \delta_8[7\ 7\ 6\ 6]$. The corresponding output feedback stabilizers for K_i, $i = 1,2,3,4$ are:

$$\begin{cases} u_1(t) = 0, \\ u_2(t) = \neg y_1(t), \\ u_3(t) = h_i(y_1(t), y_2(t)), \end{cases} \tag{2.53}$$

where $h_1 = 1$, $h_2 = y_1(t) \vee y_2(t)$, $h_3 = \neg y_1(t) \rightarrow \neg y_2(t)$ and $h_4 = 0$.

2.6 OUTPUT TRACKING OF BOOLEAN CONTROL NETWORKS

In this section, we study the output tracking control of BCNs, including constant reference signal case and time-varying reference signal case.

2.6.1 CONSTANT REFERENCE SIGNAL

Consider system (2.34). Given a constant reference signal $Y_r = (y_1^r, \cdots, y_p^r) \in \mathscr{D}^p$, we aim to design a state feedback control in the form of:

$$\begin{cases} u_1(t) = k_1(X(t)), \\ \vdots \\ u_m(t) = k_m(X(t)), \end{cases} \tag{2.54}$$

such that the output of the closed-loop system consisting of the system (2.34) and the control (2.54) tracks Y_r, that is, there exists an integer $\tau > 0$ such that $Y(t; X(0), U) = Y_r$ holds $\forall\ X(0) \in \mathscr{D}^n$ and $\forall\ t \geq \tau$, where $k_i : \mathscr{D}^n \mapsto \mathscr{D}$, $i = 1, \cdots, m$ are logical functions to be determined.

Using the vector form of logical variables and setting $x(t) = \ltimes_{i=1}^n x_i(t) \in \Delta_{2^n}$, $u(t) = \ltimes_{i=1}^m u_i(t) \in \Delta_{2^m}$ and $y(t) = \ltimes_{i=1}^p y_i(t) \in \Delta_{2^p}$, one can convert (2.54) into the following algebraic form:

$$u(t) = Kx(t), \tag{2.55}$$

where $K \in \mathscr{L}_{2^m \times 2^n}$. Moreover, the reference signal becomes $y_r = \ltimes_{i=1}^p y_i^r = \delta_{2^p}^{\alpha}$, where α is uniquely determined by $y_i^r, i = 1, \cdots, p$.

For the algebraic form (2.37) with $L = \delta_{2^n}[i_1\ i_2\ \cdots\ i_{2^{m+n}}]$, we define a set, denoted by $\mathscr{O}(\alpha)$, as:

$$\mathscr{O}(\alpha) = \{r \in \mathbb{N} : Col_r(H) = \delta_{2^p}^{\alpha}, 1 \leq r \leq 2^n\}. \tag{2.56}$$

Note that the set $\{\delta_{2^n}^r : r \in \mathscr{O}(\alpha)\}$ contains all the states of system (2.37) whose outputs form the vector y_r. We presuppose $\mathscr{O}(\alpha) \neq \emptyset$ in the following. Otherwise, if $\mathscr{O}(\alpha) = \emptyset$, the output tracking control problem is not solvable.

For $S \subseteq \Delta_{2^n}, S \neq \emptyset$ and $k = 1, 2, \cdots$, let $R_k(S)$ denote the set of all states controllable at S in k steps, that is,

$$R_k(S) = \Big\{x(0) \in \Delta_{2^n} : \text{ there exists } \{u(t) \in \Delta_{2^m} : \\ t = 0, \cdots, k-1\} \text{ such that } x(k; x(0), u) \in S\Big\}. \tag{2.57}$$

Then, we have the following result.

Theorem 2.12

The output of system (2.34) tracks the reference signal $y_r = \delta_{2^p}^{\alpha}$ by a state feedback control, if and only if there exist a nonempty set $S \subseteq \{\delta_{2^n}^r : r \in \mathscr{O}(\alpha)\}$ and an integer

$1 \leq \tau \leq 2^n$ such that:

$$\begin{cases} S \subseteq R_1(S), \\ R_\tau(S) = \Delta_{2^n}. \end{cases} \tag{2.58}$$

■

Proof. (Sufficiency) Assuming that (2.58) holds, we prove that the output of system (2.34) tracks y_r by a constructed state feedback control.

Set,

$$R_k^\circ(S) = R_k(S) \setminus R_{k-1}(S), k = 1, \cdots, \tau, \tag{2.59}$$

where $R_0(S) := \emptyset$. Then, it is easy to see that $R_{k_1}^\circ(S) \bigcap R_{k_2}^\circ(S) = \emptyset$, $\forall\ k_1, k_2 \in \{1, \cdots, \tau\}, k_1 \neq k_2$, and $\bigcup_{k=1}^{\tau} R_k^\circ(S) = \Delta_{2^n}$. Thus, for any integer $1 \leq j \leq 2^n$, there exists a unique integer $1 \leq k_j \leq \tau$ such that $\delta_{2^n}^j \in R_{k_j}^\circ(S)$.

For the integer $1 \leq j \leq 2^n$ with $k_j = 1$, there exists an integer $1 \leq p_j \leq 2^m$ such that the integer $l := (p_j - 1)2^n + j$ satisfies $1 \leq l \leq 2^{m+n}$ and $\delta_{2^n}^{i_l} \in S$. Similarly, for the integer $1 \leq j \leq 2^n$ with $2 \leq k_j \leq \tau$, there exists an integer $1 \leq p_j \leq 2^m$ such that the integer $l := (p_j - 1)2^n + j$ satisfies $1 \leq l \leq 2^{m+n}$ and $\delta_{2^n}^{i_l} \in R_{k_j - 1}(S)$.

Now, we set $K = \delta_{2^m}[p_1\ p_2\ \cdots\ p_{2^n}] \in \mathscr{L}_{2^m \times 2^n}$. Then, under the control $u(t) = Kx(t)$, along the trajectory of system (2.37) starting from any initial state $x(0) = \delta_{2^n}^j \in \Delta_{2^n}$, we have:

$$\begin{aligned} x(1; x(0), u) &= LKx(0)x(0) = \delta_{2^n}^{i_l} \\ &\in \begin{cases} S, & \text{if } k_j = 1, \\ R_{k_j - 1}(S), & \text{if } 2 \leq k_j \leq \tau. \end{cases} \end{aligned}$$

Thus, $x(k_j; x(0), u) \in S$, $\forall\ 1 \leq j \leq 2^n$. Since $S \subseteq R_1(S)$, one can see that:

$$x(t; x(0), u) \in S,\ \forall\, t \geq \tau,\ \forall\, x(0) \in \Delta_{2^n},$$

which implies that:

$$y(t; x(0), u) = Hx(t; x(0), u) = y_r$$

holds $\forall\, t \geq \tau$ and $\forall\, x(0) \in \Delta_{2^n}$.

Therefore, the output of system (2.34) tracks y_r by the state feedback control:

$$u(t) = \delta_{2^m}[p_1\ p_2\ \cdots\ p_{2^n}]x(t).$$

(Necessity) Suppose that the output of system (2.34) tracks $y_r = \delta_{2^p}^\alpha$ by a state feedback control, say, $u(t) = Kx(t)$, $K \in \mathscr{L}_{2^m \times 2^n}$. Then, the closed-loop system consisting of system (2.37) and the control $u(t) = Kx(t)$ becomes:

$$\begin{cases} x(t+1) = \widehat{L}x(t), \\ y(t) = Hx(t), \end{cases} \tag{2.60}$$

where $\widehat{L} = LK\Phi_n$, and $\Phi_n = \text{Diag}\{\delta_{2^n}^1, \delta_{2^n}^2, \cdots, \delta_{2^n}^{2^n}\}$ is the power-reducing matrix satisfying $x \ltimes x = \Phi_n \ltimes x$, $\forall\ x \in \Delta_{2^n}$. Denote the state trajectory of system (2.60) starting from an initial state $x(0) \in \Delta_{2^n}$ by $x(t;x(0))$, and the output trajectory of system (2.60) starting from $x(0) \in \Delta_{2^n}$ by $y(t;x(0))$.

For the Boolean network (2.60), denote the set of states in the limit set (all the fixed points and cycles) by S. In addition, let T_t be the transient period of system (2.60). Then, it is easy to see that (2.58) holds for S and $\tau = T_t \leq 2^n$.

Now, we prove that $S \subseteq \{\delta_{2^n}^r : r \in \mathcal{O}(\alpha)\}$.

In fact, if $S \nsubseteq \{\delta_{2^n}^r : r \in \mathcal{O}(\alpha)\}$, then there exists $\delta_{2^n}^i \in S$ with $i \notin \mathcal{O}(\alpha)$. Since $\delta_{2^n}^i$ is a state located in some fixed point or cycle of the system (2.60), there exists a positive integer T such that $\delta_{2^n}^i = x(nT;\delta_{2^n}^i)$ holds for all $n \in \mathbb{N}$. Thus, $y(nT;\delta_{2^n}^i) = Hx(nT;\delta_{2^n}^i) \neq y_r$, $\forall\ n \in \mathbb{N}$, which is a contradiction to the fact that the output of system (2.34) tracks $y_r = \delta_{2^p}^{\alpha}$ by $u(t) = Kx(t)$. Therefore, $S \subseteq \{\delta_{2^n}^r : r \in \mathcal{O}(\alpha)\}$. This completes the proof.

From the proof of Theorem 2.12, we can design a state feedback based output tracking controller for system (2.34) as follows.

ALGORITHM 2.3

One can design state feedback based output tracking controller for system (2.34) according to the following steps:

1) For each integer $1 \leq j \leq 2^n$, calculate the unique integer $1 \leq k_j \leq \tau$ such that $\delta_{2^n}^j \in R_{k_j}^{\circ}(S)$.
2) Calculate $1 \leq p_j \leq 2^m$ such that:

$$\begin{cases} \delta_{2^n}^{i_l} \in S, & \text{for } k_j = 1, \\ \delta_{2^n}^{i_l} \in R_{k_j - 1}(S), & \text{for } 2 \leq k_j \leq \tau, \end{cases} \tag{2.61}$$

where $l = (p_j - 1)2^n + j$.
3) The state feedback based output tracking control can be designed as $u(t) = Kx(t)$ with:

$$K = \delta_{2^m}[p_1\ p_2\ \cdots\ p_{2^n}]. \tag{2.62}$$

2.6.2 TIME-VARYING REFERENCE SIGNAL

Consider system (2.34). The time-varying reference signal is generated by the following reference Boolean network:

$$\begin{cases} \widehat{x}_1(t+1) = \widehat{f}_1(\widehat{X}(t)), \\ \widehat{x}_2(t+1) = \widehat{f}_2(\widehat{X}(t)), \\ \vdots \\ \widehat{x}_{n_1}(t+1) = \widehat{f}_{n_1}(\widehat{X}(t)); \\ \widehat{y}_j(t) = \widehat{h}_j(\widehat{X}(t)), \quad j = 1, \cdots, p, \end{cases} \tag{2.63}$$

where $\widehat{X}(t) = (\widehat{x}_1(t), \widehat{x}_2(t), \cdots, \widehat{x}_{n_1}(t)) \in \mathscr{D}^{n_1}$ and $\widehat{Y}(t) = (\widehat{y}_1(t), \cdots, \widehat{y}_p(t)) \in \mathscr{D}^p$ are the state and the output of system (2.63), respectively, and $\widehat{f}_i : \mathscr{D}^{n_1} \mapsto \mathscr{D}, i = 1, \cdots, n_1$ and $\widehat{h}_j : \mathscr{D}^{n_1} \mapsto \mathscr{D}, j = 1, \cdots, p$ are Boolean functions. Given an initial state $\widehat{X}(0) \in \mathscr{D}^{n_1}$, the state trajectory of system (2.63) is denoted by $\widehat{X}(t; \widehat{X}(0))$, and the output trajectory of system (2.63) is denoted by $\widehat{Y}(t; \widehat{X}(0))$.

The output regulation problem is to find a state feedback control in the form of:

$$\begin{cases} u_1(t) = g_1(X(t), \widehat{X}(t)), \\ \vdots \\ u_m(t) = g_m(X(t), \widehat{X}(t)), \end{cases} \tag{2.64}$$

where $g_i : \mathscr{D}^{n+n_1} \mapsto \mathscr{D}, i = 1, \cdots, m$ are Boolean functions, under which there exists an integer $\tau > 0$ such that:

$$Y(t; X_0, U) = \widehat{Y}(t; \widehat{X}_0)$$

holds $\forall\, t \geq \tau$, $\forall\, X_0 \in \mathscr{D}^n$ and $\forall\, \widehat{X}_0 \in \mathscr{D}^{n_1}$.

In the following section, we convert (2.63) and (2.64) into equivalent algebraic forms, respectively, by using the semi-tensor product of matrices.

Using the vector form of Boolean values and setting $\widehat{x}(t) = \ltimes_{i=1}^{n_1} \widehat{x}_i(t) \in \Delta_{2^{n_1}}$ and $\widehat{y}(t) = \ltimes_{i=1}^{p} \widehat{y}_i(t) \in \Delta_{2^p}$, system (2.63) and control (2.64) can be converted to:

$$\begin{cases} \widehat{x}(t+1) = \widehat{L}\widehat{x}(t), \\ \widehat{y}(t) = \widehat{H}\widehat{x}(t), \end{cases} \tag{2.65}$$

and

$$u(t) = Gx(t)\widehat{x}(t), \tag{2.66}$$

respectively, where $\widehat{L} \in \mathscr{L}_{2^{n_1} \times 2^{n_1}}$, $\widehat{H} \in \mathscr{L}_{2^p \times 2^{n_1}}$ and $G \in \mathscr{L}_{2^m \times 2^{n+n_1}}$.

The following two propositions are necessary for the further investigation.

PROPOSITION 2.6

Given a state feedback control $u(t) = Gx(t)\widehat{x}(t)$, then:

$$x(t)\widehat{x}(t) = R^t x(0)\widehat{x}(0) \tag{2.67}$$

holds for any $t \in \mathbb{Z}_+$, where:

$$R = LGW_{[2^n, 2^{n+n_1}]} M_{r,2^n} (I_{2^{n+n_1}} \otimes \widehat{L})(I_{2^n} \otimes M_{r,2^{n_1}}), \tag{2.68}$$

and

$$M_{r,2^n} = \text{Diag}\{\delta_{2^n}^1, \delta_{2^n}^2, \cdots, \delta_{2^n}^{2^n}\} \in \mathscr{L}_{2^{2n} \times 2^n} \tag{2.69}$$

is the power-reducing matrix satisfying $x \ltimes x = M_{r,2^n} \ltimes x$, $\forall\, x \in \Delta_{2^n}$.

Proof. We prove it by induction.

When $t = 1$, it is easy to see that,

$$\begin{aligned} & x(1)\widehat{x}(1) = Lu(0)x(0)\widehat{L}\widehat{x}(0) \\ = & LGx(0)\widehat{x}(0)x(0)\widehat{L}\widehat{x}(0) \\ = & LGW_{[2^n,2^{n+n_1}]}x(0)x(0)\widehat{x}(0)\widehat{L}\widehat{x}(0) \\ = & LGW_{[2^n,2^{n+n_1}]}M_{r,2^n}x(0)\widehat{x}(0)\widehat{L}\widehat{x}(0) \\ = & LGW_{[2^n,2^{n+n_1}]}M_{r,2^n}(I_{2^{n+n_1}} \otimes \widehat{L})x(0)\widehat{x}(0)\widehat{x}(0) \\ = & LGW_{[2^n,2^{n+n_1}]}M_{r,2^n}(I_{2^{n+n_1}} \otimes \widehat{L}) \ltimes \\ & (I_{2^n} \otimes M_{r,2^{n_1}})x(0)\widehat{x}(0) \\ = & Rx(0)\widehat{x}(0), \end{aligned}$$

which implies that (2.67) holds for $t = 1$.

Assume that (2.67) holds for $t = k$, that is,

$$x(k)\widehat{x}(k) = R^k x(0)\widehat{x}(0).$$

Then, for $t = k+1$, we have:

$$\begin{aligned} & x(k+1)\widehat{x}(k+1) = Lu(k)x(k)\widehat{L}\widehat{x}(k) \\ & = Rx(k)\widehat{x}(k) = R^{k+1}x(0)\widehat{x}(0). \end{aligned}$$

Thus, (2.67) holds for $t = k+1$.

By induction, (2.67) holds for any $t \in \mathbb{Z}_+$.

PROPOSITION 2.7

For the matrix R defined in (2.68), there exist two integers $0 \leq r_0 < 2^{n+n_1}$ and $T > 0$, such that:

$$R^{r_0} = R^{r_0+T}. \tag{2.70}$$

In the following section, we study the solvability of the output regulation problem based on Propositions 2.6 and 2.7.

For a given state feedback control in the form of (2.66) and an integer $t \geq 1$, it is easy to see that:

$$\begin{aligned} y(t) &= Hx(t) = HLu(t-1)x(t-1) \\ &= HLGx(t-1)\widehat{x}(t-1)x(t-1) \\ &= HLGW_{[2^n,2^{n+n_1}]}x(t-1)x(t-1)\widehat{x}(t-1) \\ &= HLGW_{[2^n,2^{n+n_1}]}M_{r,2^n}x(t-1)\widehat{x}(t-1), \end{aligned}$$

and

$$\begin{aligned}\widehat{y}(t) &= \widehat{H}\widehat{x}(t) = \widehat{H}\widehat{L}\widehat{x}(t-1) \\ &= \widehat{H}\widehat{L}E_d^n x(t-1)\widehat{x}(t-1).\end{aligned}$$

Thus,

$$y(t) - \widehat{y}(t) = Qx(t-1)\widehat{x}(t-1), \tag{2.71}$$

where,

$$Q = HLGW_{[2^n, 2^{n+n_1}]}M_{r,2^n} - \widehat{H}\widehat{L}E_d^n \in \mathbb{R}^{2^p \times 2^{n+n_1}}. \tag{2.72}$$

From Proposition 2.6, one can obtain:

$$y(t) - \widehat{y}(t) = Qx(t-1)\widehat{x}(t-1) = QR^{t-1}x(0)\widehat{x}(0). \tag{2.73}$$

Based on the above analysis, we have the following result on the solvability of the output regulation problem.

Theorem 2.13

The output regulation problem is solvable, if and only if there exists a logical matrix $G \in \mathscr{L}_{2^m \times 2^{n+n_1}}$ and an integer $1 \leq \tau \leq 2^{n+n_1}$ such that:

$$QR^{\tau-1} = \mathbf{0}_{2^p \times 2^{n+n_1}}, \tag{2.74}$$

where Q and R are given in (2.72) and (2.68), respectively, and $\mathbf{0}_{2^p \times 2^{n+n_1}}$ denotes the $2^p \times 2^{n+n_1}$ matrix with all entries being 0. ■

Proof. (Sufficiency) Suppose that there exists a logical matrix $G \in \mathscr{L}_{2^m \times 2^{n+n_1}}$ and an integer $1 \leq \tau \leq 2^{n+n_1}$ such that (2.74) holds. Then, for any integer $t \geq \tau$, we have:

$$QR^{t-1} = (QR^{\tau-1})R^{t-\tau} = \mathbf{0}_{2^p \times 2^{n+n_1}}R^{t-\tau} = \mathbf{0}_{2^p \times 2^{n+n_1}}.$$

Thus, under the state feedback control $u(t) = Gx(t)\widehat{x}(t)$, it is easy to see from (2.73) that:

$$y(t) - \widehat{y}(t) = QR^{t-1}x(0)\widehat{x}(0) = \mathbf{0}_{2^p \times 1}$$

holds $\forall\, t \geq \tau$, $\forall\, x(0) \in \Delta_{2^n}$ and $\forall\, \widehat{x}(0) \in \Delta_{2^{n_1}}$, which implies that the output regulation problem is solvable by the state feedback control $u(t) = Gx(t)\widehat{x}(t)$.

(Necessity) Assuming that the output regulation problem is solvable by a state feedback control, say, $u(t) = Gx(t)\widehat{x}(t)$, $G \in \mathscr{L}_{2^m \times 2^{n+n_1}}$. Then, there exists a positive integer $\widehat{\tau}$ such that:

$$y(t; x(0), u) = \widehat{y}(t; \widehat{x}(0)) \tag{2.75}$$

holds $\forall\, t \geq \widehat{\tau}$, $\forall\, x(0) \in \Delta_{2^n}$ and $\forall\, \widehat{x}(0) \in \Delta_{2^{n_1}}$.

Combining (2.73) and (2.75), one can obtain:

$$y(t;x(0),u) - \widehat{y}(t;\widehat{x}(0)) = QR^{t-1}x(0)\widehat{x}(0) = \mathbf{0}_{2^p \times 1}$$

holds $\forall\ t \geq \widehat{\tau}$, $\forall\ x(0) \in \Delta_{2^n}$ and $\forall\ \widehat{x}(0) \in \Delta_{2^{n_1}}$. From the arbitrariness of $x(0)$ and $\widehat{x}(0)$, it is easy to see that:

$$QR^{t-1} = \mathbf{0}_{2^p \times 2^{n+n_1}},\ \forall\ t \geq \widehat{\tau}.$$

Next, we find an integer $1 \leq \tau \leq 2^{n+n_1}$ such that (2.74) holds.

In fact, if $1 \leq \widehat{\tau} \leq 2^{n+n_1}$, we set $\tau = \widehat{\tau}$, and complete the proof. Otherwise, if $\widehat{\tau} > 2^{n+n_1}$, by Proposition 2.7, we have:

$$QR^{r_0} = QR^{r_0+T} = \cdots = QR^{r_0+kT},\ \forall\ k \in \mathbb{N}.$$

When k is large enough such that $r_0 + kT + 1 \geq \widehat{\tau}$, one can see that $QR^{r_0} = QR^{r_0+kT} = \mathbf{0}_{2^p \times 2^{n+n_1}}$. We set $\tau = r_0 + 1$, and complete the proof.

Finally, we apply the results obtained in this paper to the regulation of the lactose operon in Escherichia coli.

Example 2.6:

Consider the "minimal" Boolean model for the lactose operon in Escherichia coli (1.3) and the five-variable model of the lactose operon in Escherichia coli (1.2).

It is noted that in both models, the mRNA determines the behavior of the lactose operon regulation. The operon is "on" when the mRNA is produced ($x_1 = \widehat{x}_1 = 1$). On the other hand, when the mRNA is not made ($x_1 = \widehat{x}_1 = 0$), the operon is "off". Thus, in both models, we treat the mRNA as the output variable.

Our objective is to check whether or not a given sequence of control inputs containing the information of state variables of both models can make the mRNA take the same value for both models after some time instance, that is, to check whether or not the output regulation problem is solvable for (1.2) and (1.3) by a given state feedback control.

Using the vector form of Boolean values and setting $x(t) = \ltimes_{i=1}^{3} x_i(t)$, $u(t) = \ltimes_{i=1}^{2} u_i(t)$ and $\widehat{x}(t) = \ltimes_{i=1}^{5} \widehat{x}_i(t)$, one can convert (1.3) and (1.2) into:

$$\begin{cases} x(t+1) = Lu(t)x(t), \\ y(t) = Hx(t), \end{cases} \tag{2.76}$$

and

$$\begin{cases} \widehat{x}(t+1) = \widehat{L}\widehat{x}(t), \\ \widehat{y}(t) = \widehat{H}\widehat{x}(t), \end{cases} \tag{2.77}$$

respectively, where:

$$L = \delta_8[6\ 6\ 6\ 6\ 8\ 8\ 8\ 8\ 6\ 6\ 6\ 6\ 8\ 8\ 8\ 8\\ 1\ 1\ 1\ 2\ 3\ 3\ 3\ 4\ 2\ 6\ 1\ 6\ 4\ 8\ 3\ 8],$$

$$\widehat{L} = \delta_{32}[1\ 3\ 1\ 3\ 17\ 19\ 21\ 23\\ 1\ 1\ 1\ 3\ 21\ 21\ 21\ 23\\ 10\ 12\ 10\ 12\ 26\ 28\ 30\ 32\\ 10\ 10\ 10\ 12\ 30\ 30\ 30\ 32],$$

$H = \delta_2[1\ 1\ 1\ 1\ 2\ 2\ 2\ 2]$, and,

$$\widehat{H} = \delta_2[1\ 1\ 1\ 1\ 1\ 1\ 1\ 1\ 1\ 1\ 1\ 1\ 1\ 1\ 1\ 1\\ 2\ 2\ 2\ 2\ 2\ 2\ 2\ 2\ 2\ 2\ 2\ 2\ 2\ 2\ 2\ 2].$$

Set,

$$G = \delta_4[3\ 1\ 1\ 1\ 1\ 1\ 1\ 1\ 2\ 2\ 2\ 2\ 2\ 2\ 2\ 2\\ 1\ 1\ 1\ 2\ 3\ 3\ 3\ 4\ 2\ 2\ 1\ 2\ 4\ 2\ 3\ 2\\ 3\ 1\ 1\ 1\ 1\ 1\ 1\ 1\ 1\ 1\ 1\ 1\ 1\ 1\ 1\ 1\\ 2\ 2\ 2\ 2\ 2\ 2\ 2\ 2\ 2\ 2\ 2\ 2\ 2\ 1\ 1\ 1\\ 3\ 1\ 1\ 1\ 1\ 1\ 1\ 1\ 1\ 1\ 1\ 1\ 1\ 1\ 1\ 1\\ 1\ 1\ 1\ 1\ 1\ 1\ 1\ 1\ 1\ 1\ 1\ 1\ 1\ 2\ 2\ 2\\ 3\ 1\ 1\ 1\ 1\ 1\ 1\ 1\ 1\ 1\ 1\ 1\ 1\ 1\ 1\ 1\\ 1\ 1\ 1\ 1\ 1\ 1\ 1\ 1\ 1\ 1\ 1\ 1\ 1\ 2\ 2\ 2\\ 3\ 1\ 1\ 1\ 1\ 1\ 1\ 1\ 1\ 1\ 1\ 1\ 1\ 1\ 1\ 1\\ 1\ 1\ 1\ 1\ 1\ 1\ 1\ 1\ 1\ 1\ 1\ 1\ 1\ 2\ 2\ 2\\ 3\ 1\ 1\ 1\ 1\ 1\ 1\ 1\ 1\ 1\ 1\ 1\ 1\ 1\ 1\ 1\\ 1\ 1\ 1\ 1\ 1\ 1\ 1\ 1\ 1\ 1\ 1\ 1\ 1\ 2\ 2\ 2\\ 3\ 1\ 1\ 1\ 1\ 1\ 1\ 1\ 1\ 1\ 1\ 1\ 1\ 1\ 1\ 1\\ 1\ 1\ 1\ 1\ 1\ 1\ 1\ 1\ 1\ 1\ 1\ 1\ 1\ 2\ 2\ 2\\ 3\ 1\ 1\ 1\ 1\ 1\ 1\ 1\ 1\ 1\ 1\ 1\ 1\ 1\ 1\ 1\\ 1\ 1\ 1\ 1\ 1\ 1\ 1\ 1\ 1\ 1\ 1\ 1\ 1\ 2\ 2\ 2].$$

For this example, a simple calculation gives:

$$QR^2 = \mathbf{0}_{2\times 256}.$$

By Theorem 2.13, the output regulation problem is solvable by $u(t) = Gx(t)\widehat{x}(t)$.

REFERENCES

1. Kauffman, S. (1969). Metabolic stability and epigenesis in randomly constructed genetic nets. Journal of Theoretical Biology, 22(3): 437–467.
2. Akutsu, T., Hayashida, M., Ching, W. et al. (2007). Control of Boolean networks: Hardness results and algorithms for tree structured networks. Journal of Theoretical Biology, 244: 670–679.
3. Chaves, M. (2009). Methods for qualitative analysis of genetic networks. Proc. 10th European Control Conference, 671–676.
4. Cheng, D., Qi, H. and Li, Z. (2011). Analysis and Control of Boolean Networks: A Semi-Tensor Product Approach. London, Springer.
5. Hammer, P. L. and Rudeanu, S. (1968). Boolean Methods in Operations Research and Related Areas. Berlin, Springer.
6. Xiao, Y. and Dougherty, E. R. (2007). The impact of function perturbations in Boolean networks. Bioinformatics, 23(10): 1265–1273.

3 Mix-Valued Logical Networks

3.1 INTRODUCTION TO MIX-VALUED LOGICAL NETWORKS

Boolean networks, which were introduced by Kauffman for investigating cellular networks [1], are networks with nodes taking logical values from $\{1,0\}$ at the discrete-time sequence $\{0,1,\cdots\}$. However, the most general logical network in practice is the so-called mix-valued logical one. For example, in game theory every player can have an action set of different size from those of others. It is obvious that the mix-valued logical network is more complex and has wider applications than Boolean networks.

The semi-tensor product of matrices [3] have been successfully used in the study of the Boolean (control) network, and many fundamental and essential results have been presented. It is noted that the semi-tensor product method can be also applied to general logical (control) networks with k-valued nodes, which can be regarded as a generalization of Boolean (control) networks and have a structure similar to that of a Boolean network [6, 10, 11].

Section 3.2 studies the topological structure of mix-valued logic and logical networks. The mix-valued logical function is first introduced and its algebraic form is obtained. Then, mix-valued logical network is investigated. Based on the semi-tensor product, a logical (control) network is converted into an algebraic form.

Section 3.3 investigates the disturbance decoupling problem (DDP) of mix-valued logical networks [8, 9] by using the semi-tensor product. First, in order to solve the DDP, a new algorithm is proposed to construct a kind of Y-friendly subspaces of the state space for mix-valued logical networks. Secondly, based on the Y-friendly subspace, several necessary and sufficient conditions are obtained for the existence of decoupling controllers, and a new control design procedure is established to find all the controllers on the DDP. Finally, an illustrative example is studied by the method proposed in this paper, and two kinds of decoupling controllers are designed.

It is noted that the pseudo-Boolean functions are applied to solve some optimization problems existing in the decision-making process, thinking logic, networks theory, graph theory and operations research. The pseudo logical function whose variables take values from multi-valued (or mix-valued) logic can express payoff functions of a game. As a special case, the pseudo Boolean function has been studied in [4] and can be uniquely presented in the multi-linear polynomial form [5]. It is easy to see that the pseudo logical function is a natural generalization of the pseudo-

Boolean function. But it is quite difficult to present the pseudo logical function in the polynomial form since the logical operator "Conjunction $\wedge$" is not equivalent to the algebraic computation "Multiplication $\times$".

Section 3.4 investigates the pseudo logical function and presents its general expression [7]. First, we define the pseudo logical function and give its general expression. Moreover, using the semi-tensor product, the pseudo logical function can be converted into an algebraic form, based on which the structural matrix is obtained. Second, we generalize the pseudo logical function to the pseudo mix-valued logical function. The main difference is that the variables of pseudo mix-valued logical function take values from logical sets with different dimensions. Besides, its general expression and structural matrix are given. Third, the pseudo mix-valued logical equations are considered and the existence of solutions is studied. Finally, an illustrative example is studied to support the new results.

3.2 TOPOLOGICAL STRUCTURE OF MIX-VALUED LOGICAL NETWORKS

This section introduces some basic concepts of mix-valued logical networks and their algebraic forms by using the semi-tensor product. First, we state some definitions of the mix-valued logical function.

DEFINITION 3.1

A n-ary function $f(x_1,x_2\cdots,x_n)$ is called a mix-valued logical function if it maps from $\mathscr{D}_{k_1}\times\mathscr{D}_{k_2}\times\cdots\times\mathscr{D}_{k_n}$ to $\mathscr{D}_{k_0}$, and $k_i, i=0,1,2\cdots,n$ are not identically equal, where $x_i\in\mathscr{D}_{k_i}=\{0,1,\cdots,k_i-1\}$, $i=1,2\cdots,n$.

DEFINITION 3.2

The projection $\phi_{[q,p]}:\mathscr{D}_p\to\mathscr{D}_q$ is defined as follows: Let $x\in\mathscr{D}_p$, then $\phi_{[q,p]}(x)=\xi$, where $\xi\in\mathscr{D}_q$, satisfying:

$$|\xi-x|=\min_{y\in\mathscr{D}_q}|x-y|.$$

If there are two solutions $\xi_1\geq x$ and $\xi_2\leq x$, then $\phi_{[q,p]}(x)=\xi_1$ is called the up-round projection and $\phi_{[q,p]}(x)=\xi_2$ is called the down-round projection. In the sequel, we assume that the default projection is the up-round projection unless otherwise noted. The structural matrix of projection $\phi_{[q,p]}$ is denoted by $\Phi_{[q,p]}\in\mathscr{L}_{p\times q}$.

Lemma 3.1

Let $y = f(P_1, P_2, \cdots, P_n)$ be a mix-valued logical function with logical variables $P_i \in \mathcal{D}_{k_i}$, $i = 1, 2, \cdots, n$ and $y \in \mathcal{D}_{k_0}$. Then there exists a unique structural matrix $M_f \in \mathcal{L}_{k_0 \times (k_1 k_2 \cdots k_n)}$ such that:

$$y = f(P_1, \cdots, P_n) = M_f \ltimes P_1 \ltimes \cdots \ltimes P_n, \tag{3.1}$$

where $y \in \Delta_{k_0}$, $P_i \in \Delta_{k_i}$, $i = 1, 2, \cdots, n$.

Now, we give some important algebraic expressions of k-valued logical operators. These operators take values from the set $\mathcal{D}_k = \{0, 1, \cdots, k-1\}$.

Negative operator ($\neg$): $\neg x = k - x$, which has a structural matrix as $M_{n,k}$;

Conjunction operator ($\wedge$): $x \wedge y = \min\{x, y\}$, which has a structural matrix as $M_{c,k}$;

Disjunction operator ($\vee$): $x \vee y = \max\{x, y\}$, which has a structural matrix as $M_{d,k}$;

Conditional operator ($\rightarrow$): $x \rightarrow y = \neg x \vee y$, which has a structural matrix as $M_{i,k}$;

Biconditional operator ($\leftrightarrow$): $x \leftrightarrow y = (x \rightarrow y) \wedge (y \rightarrow x)$, which has a structural matrix as $M_{e,k}$.

For example, when $k = 3$, $\Delta_3 = \{\delta_3^1, \delta_3^2, \delta_3^3\}$

$$M_{n,3} = \delta_3[3\ 2\ 1],$$

$$M_{c,3} = \delta_3[1\ 2\ 3\ 2\ 2\ 3\ 3\ 3\ 3],$$
$$M_{d,3} = \delta_3[1\ 1\ 1\ 1\ 2\ 2\ 1\ 2\ 3],$$
$$M_{i,3} = \delta_3[1\ 2\ 3\ 1\ 2\ 2\ 1\ 1\ 1],$$
$$M_{e,3} = \delta_3[1\ 2\ 3\ 2\ 2\ 2\ 3\ 2\ 1].$$

DEFINITION 3.3

Let σ be a binary operator on $\mathcal{D}_k$, $x \in \mathcal{D}_p$ and $y \in \mathcal{D}_q$. Then, define,

$$x\sigma y := (\phi_{[k,p]}(x))\sigma(\phi_{[k,q]}(y)) \in \mathcal{D}_k.$$

Next, we give a simple example to explain this.

Example 3.1:

1. Set $\mathcal{D} = \{0, 1\}$, and $\mathcal{D}_3 = \{0, 1, 2\}$, then,

$$\phi_{[2,3]}(0) = 0,\ \phi_{[2,3]}(1) = 1,\ \phi_{[2,3]}(2) = 1,$$

and

$$\phi_{[3,2]}(0) = 0,\ \phi_{[3,2]}(1) = 1.$$

Thus, $\Phi_{[2,3]} = \delta_2[1\ 1\ 2]$ and $\Phi_{[3,2]} = \delta_3[2\ 3]$

2. Consider the following mix-valued logical function:

$$y = f(x_1,x_2,x_3) = x_1 \wedge (x_2 \leftrightarrow x_3),$$

where, $x_1,x_3 \in \mathscr{D}$, $x_2,y \in \mathscr{D}_3$.

In fact, the function y should be converted into:

$$y = \phi_{[3,2]}(x_1) \wedge [x_2 \leftrightarrow \phi_{[3,2]}(x_3)]. \tag{3.2}$$

According to Lemma 3.1, we can obtain the algebraic form of y as:

$$\begin{aligned} y &= \phi_{[3,2]}(x_1) \wedge [x_2 \leftrightarrow \phi_{[3,2]}(x_3)] \\ &= M_{c,3} \ltimes \phi_{[3,2]}(x_1) \ltimes [M_{e,3} \ltimes x_2 \ltimes \phi_{[3,2]}(x_3)] \\ &= M_{c,3}\Phi_{[3,2]}x_1 M_{e,3}x_2\Phi_{[3,2]}x_3 \\ &= M_{c,3}\Phi_{[3,2]}(I_2 \otimes M_{e,3})x_1x_2\Phi_{[3,2]}x_3 \\ &= M_{c,3}\Phi_{[3,2]}(I_2 \otimes M_{e,3})\left(I_6 \otimes \Phi_{[3,2]}\right)x_1x_2x_3 \\ &= M_f x_1x_2x_3, \end{aligned} \tag{3.3}$$

where $\Phi_{[3,2]} = \delta_3[2\ 3]$ is the structural matrix of the logical operator $\phi_{[3,2]}$, and

$$\begin{aligned} M_f &= M_{c,3}\Phi_{[3,2]}(I_2 \otimes M_{e,3})\left(I_6 \otimes \Phi_{[3,2]}\right) \\ &= \delta_3[1\ 2\ 3\ 2\ 2\ 3\ 3\ 3\ 3] \ltimes \delta_3[3\ 2](I_2 \otimes \delta_3[1\ 2\ 3\ 2\ 2\ 2\ 3\ 2\ 1])(I_6 \otimes \delta_3[2\ 3]) \\ &= \delta_3[3\ 3\ 3\ 3\ 2\ 2]\delta_6[1\ 2\ 3\ 2\ 2\ 2\ 3\ 2\ 1\ 4\ 5\ 6\ 5\ 5\ 5\ 6\ 5\ 4](I_6 \otimes \delta_3[2\ 3]) \\ &= \delta_3[3\ 3\ 3\ 3\ 3\ 3\ 3\ 3\ 3\ 3\ 2\ 2\ 2\ 2\ 2\ 2\ 2\ 2\ 3]\delta_{18}[2\ 3\ 5\ 6\ 8\ 9\ 11\ 12\ 14\ 15\ 17\ 18] \\ &= \delta_3[3\ 3\ 3\ 3\ 3\ 3\ 2\ 2\ 2\ 2\ 2\ 3]. \end{aligned} \tag{3.4}$$

In fact, when $x_1 = 0 \in \mathscr{D}$, $x_2 = 2 \in \mathscr{D}_3$, and $x_3 = 1 \in \mathscr{D}$, that is $x_1 = \delta_2^2, x_2 = \delta_3^1, x_3 = \delta_2^1$, we obtain:

$$\begin{aligned} y &= M_f x_1x_2x_3 = \delta_3[3\ 3\ 3\ 3\ 3\ 3\ 2\ 2\ 2\ 2\ 2\ 3]\delta_2^2\delta_3^1\delta_2^1 \\ &= \delta_3[2\ 2\ 2\ 2\ 2\ 3]\delta_3^1\delta_2^1 = \delta_3[2\ 2]\delta_2^1 \\ &= \delta_3^2. \end{aligned} \tag{3.5}$$

That is, $y = 1 \in \mathscr{D}_3$.

Now, we state some definitions and algebraic forms of mix-valued logical network and mix-valued logical control network.

DEFINITION 3.4

1. Consider the following logical system:

$$\begin{cases} x_1(t+1) = & f_1(x_1(t),x_2(t),\cdots,x_n(t)), \\ x_2(t+1) = & f_2(x_1(t),x_2(t),\cdots,x_n(t)), \\ & \vdots \\ x_n(t+1) = & f_n(x_1(t),x_2(t),\cdots,x_n(t)). \end{cases} \tag{3.6}$$

If $f_i : \Pi_{j=1}^{n} \mathscr{D}_{k_j} \to \mathscr{D}_{k_i}$, $i = 1, 2, \cdots, n$ are mix-valued logical functions, system (3.6) is called a mix-valued logical network.

2. Consider the following logical control system:

$$\begin{cases} x_1(t+1) = & f_1(x_1(t), x_2(t), \cdots, x_n(t), u_1(t), \cdots, u_m(t)) \\ x_2(t+1) = & f_2(x_1(t), x_2(t), \cdots, x_n(t), u_1(t), \cdots, u_m(t)), \\ & \vdots \\ x_n(t+1) = & f_n(x_1(t), x_2(t), \cdots, x_n(t), u_1(t), \cdots, u_m(t)), \end{cases} \tag{3.7}$$

where $u_i \in \mathscr{D}_{s_i}$, $i = 1, \cdots, m$. If $f_i : \Pi_{j=1}^{n} \mathscr{D}_{k_j} \times \Pi_{j=1}^{m} \mathscr{D}_{s_j} \to \mathscr{D}_{k_i}$, $i = 1, 2, \cdots, n$ are mix-valued logical functions, system (3.7) is called a mix-valued logical control network.

Next, we can convert systems (3.6) and (3.7) to their algebraic forms. Using the semi-tensor product and Lemma 3.1, each equation of (3.6) can be converted into $x_i(t+1) = L_i x_1(t) x_2(t) \cdots x_n(t)$, $i = 1, 2, \cdots, n$. Then, we have:

$$x(t+1) = Lx(t), \tag{3.8}$$

where $x(t) = x_1(t)x_2(t) \cdots x_n(t) \in \Delta_k$, $k = k_1 k_2 \cdots k_n$, $L \in \mathscr{L}_{k \times k}$ is called the transition matrix of system (3.6), which is given by:

$$Col_i(L) = Col_i(L_1) \ltimes Col_i(L_2) \ltimes \cdots \ltimes Col_i(L_n), i = 1, 2, \cdots, k. \tag{3.9}$$

and $L_i \in \mathscr{L}_{k_i \times k}$ is the structural matrix of function f_i, $i = 1, 2, \cdots, n$. Similarly, the algebraic form of system (3.7) is:

$$x(t+1) = Lu(t)x(t), \tag{3.10}$$

where $x(t) = x_1(t)x_2(t) \cdots x_n(t) \in \Delta_k$, $u(t) = u_1(t)u_2(t) \cdots u_m(t) \in \Delta_s$, $k = k_1 k_2 \cdots k_n$, $s = s_1 s_2 \cdots s_m$, $L \in \mathscr{L}_{k \times (ks)}$, and $L_i \in \mathscr{L}_{k_i \times (ks)}$, $i = 1, 2, \cdots, n$.

3.3 DISTURBANCE DECOUPLING OF MIX-VALUED LOGICAL CONTROL NETWORKS

This section addresses the disturbance decoupling problem (DDP) of mix-valued logical networks by using both the semi-tensor product of matrices and matrix expression of logical functions. By the results in this section, one can obtain a general state feedback control law, which is different from the constant controller design method presented in [8]. Moreover, the results can be also applied to the DDP of Boolean networks [2] for designing state feedback disturbance decoupling control laws.

3.3.1 PROBLEM FORMULATION

Consider the following disturbed mix-valued logical network:

$$\begin{cases} x_1(t+1) &= f_1(x(t),u(t),\xi(t)), \\ x_2(t+1) &= f_2(x(t),u(t),\xi(t)), \\ &\vdots \\ x_n(t+1) &= f_n(x(t),u(t),\xi(t)), \\ y_j(t) &= h_j(x(t)), \qquad j=1,2,\cdots,p, \end{cases} \tag{3.11}$$

where $x(t)=(x_1(t),x_2(t),\cdots,x_n(t))$, $x_i(t)\in\mathcal{D}_{k_i}$, $i=1,2,\cdots,n$, are states, $u(t)=(u_1(t),u_2(t),\ \cdots,u_m(t))$, $u_i(t)\in\mathcal{D}_{s_i}$, $i=1,2,\cdots,m$, are control inputs, $\xi(t)=(\xi_1(t),\xi_2(t),\cdots,\xi_q(t))$, $\xi_i(t)\in\mathcal{D}_{w_i}$, $i=1,2,\cdots,q$, are external disturbances, and $y_i(t)\in\mathcal{D}_{v_i}$, $i=1,2,\cdots,p$, are system's outputs.

Disturbance Decoupling: The objective of this section is to find a control law $u(t)$ and a coordinate transformation $z=T(x)$, such that under the coordinate frame z, the system (3.11) can be expressed in the form of:

$$\begin{cases} z^1(t+1)=F^1(z(t),u(t),\xi(t)), \\ z^2(t+1)=F^2(z^2(t)), \\ y(t)=G(z^2(t)), \end{cases} \tag{3.12}$$

where $z=(z_1,z_2,\cdots,z_n)^T:=[(z^1)^T,(z^2)^T]^T$ and $y=(y_1,y_2,\cdots,y_p)^T$.

This section focuses on solving the DDP of mix-valued logical networks. First, we give some new concepts on the state space and the Y-friendly subspace, and establish a novel algorithm to construct the Y-friendly subspace. Then, we present several necessary and sufficient conditions for the existence of decoupling controllers, and propose a new technique to design all decoupling controllers.

3.3.2 Y-FRIENDLY SUBSPACE

DEFINITION 3.5

Consider mix-valued logical network (3.6).

1. The state space, $\mathcal{X}$, of the system (3.6) is defined as the set of all the mix-valued logical operators of $x_1,x_2,\cdots,x_n$, denoted by $\mathcal{X}=\mathcal{F}_l\{x_1,x_2,\cdots,x_n\}$.
2. Let $z_1,z_2,\cdots,z_r\in\mathcal{X}$. The subspace $\mathcal{Y}$ generated by $\{z_1,z_2,\cdots,z_r\}$ is the set of all the logical operators of $z_1,z_2,\cdots,z_r$, denoted by $\mathcal{Y}=\mathcal{F}_l\{z_1,z_2,\cdots,z_r\}$.

DEFINITION 3.6

Let $\{z_1, z_2, \cdots, z_n\} \subset \mathscr{X}$, where $z_j \in \Delta_{k_{i_j}}$, $j = 1, 2, \cdots, n$. The mapping $G : \mathscr{D}_{k_1} \times \cdots \times \mathscr{D}_{k_n} \to \mathscr{D}_{k_{i_1}} \times \cdots \times \mathscr{D}_{k_{i_n}}$, $\{i_1, i_2, \cdots, i_n\} = \{1, 2, \cdots, n\}$, that is, $G : \{x_1, x_2, \cdots, x_n\} \mapsto \{z_1, z_2, \cdots, z_n\}$, is called a coordinate change, if G is one-to-one and onto.

Let $z = \ltimes_{i=1}^{n} z_i = T \ltimes_{i=1}^{n} x_i$, where $T \in \mathscr{L}_{k \times k}$. It is easy to prove that T is nonsingular and orthogonal.

DEFINITION 3.7

A subspace $\mathscr{Y} = \mathscr{F}_l\{z_1, \cdots, z_r\}$ is called a regular subspace with dimension r if there exists $\{z_{r+1}, \cdots, z_n\}$ such that $\{z_1, \cdots, z_n\}$ is a coordinate frame, where $z_j \in \Delta_{k_{i_j}}$, $\{i_1, i_2, \cdots, i_n\} = \{1, 2, \cdots, n\}$. Moreover, $\{z_1, \cdots, z_r\}$ is called a regular basis of $\mathscr{Y}$.

DEFINITION 3.8

Let $\mathscr{X} = \mathscr{F}_l\{x_1, x_2, \cdots, x_n\}$ be the state space, where $x_i \in \Delta_{k_i}$, $i = 1, 2, \cdots, n$, and $Y = \{y_1, y_2, \cdots, y_p\} \subset \mathscr{X}$. $\mathscr{Y} = \mathscr{F}_l\{z_1, z_2, \cdots, z_r\} \subset \mathscr{X}$ is called a Y-friendly subspace, if $\mathscr{F}_l\{z_1, z_2, \cdots, z_r\}$ is a regular subspace and $y_i \in \mathscr{Y}$, $i = 1, 2, \cdots, p$. A Y-friendly subspace is also called an output-friendly subspace.

It is necessary that the limitation of the states' dimension, $\{i_1, i_2, \cdots, i_n\} = \{1, 2, \cdots, n\}$, is added in definitions of the state spaces and the Y-friendly subspaces of mix-valued logical networks, but it isn't needed in Boolean networks. Based on this, the study of the Y-friendly for mix-valued logical is different from that of Boolean networks and is necessary.

Following results can be obtained from the regular subspace of mix-valued logical networks.

Theorem 3.1

Assume that $z = T_0 x$, where $x = \ltimes_{i=1}^{n} x_i$, $z = \ltimes_{i=1}^{r} z_i$, $x_i \in \Delta_{k_i}$, $i = 1, 2, \cdots, n$, $z_j \in \Delta_{k_{i_j}}$, $j = 1, 2, \cdots, r$, $T_0 = (t_{ij}) \in \mathscr{L}_{l \times k}$ and $l = \prod_{j=1}^{r} k_{i_j}$. Then, $\mathscr{Y} = \mathscr{F}_l\{z_1, \cdots, z_r\}$ is a regular subspace iff the corresponding coefficient matrix T_0 satisfies:

$$\sum_{j=1}^{k} t_{ij} = k/l, \quad i = 1, 2, \cdots, l. \tag{3.13}$$

Proof. The proof is similar to Theorem 11 in [2], thus it is omitted. ∎

Since $y_i \in \Delta_{v_i}$, $i = 1, 2, \cdots, p$, $y = \ltimes_{i=1}^{p} y_i$ can be expressed in an algebraic form as:

$$y = \delta_v[i_1, i_2, \cdots, i_k]x := Hx, \quad v = v_1 v_2 \cdots v_p. \tag{3.14}$$

Set $n_j = |\{t \mid i_t = j,\ t = 1, 2, \cdots, k\}|$, $j = 1, 2, \cdots, v$. It is obvious that $n_1 + n_2 + \cdots + n_v = k$. Then, we have a result to find the Y-friendly subspaces.

Theorem 3.2

Consider the system (3.11). Assume that $y = \ltimes_{i=1}^{p} y_i$ has its algebraic form (3.14). Then, there is a Y-friendly subspace of dimension r iff $n_1, \cdots, n_v$ have a common factor:

$$k/l = \frac{k_1 k_2 \cdots k_n}{k_{i_1} k_{i_2} \cdots k_{i_r}}.$$

Proof. (Necessity) Let $\mathscr{Y} = \mathscr{F}_l\{z_1, z_2, \cdots, z_r\} \subset \mathscr{X}$ $(z_j \in \Delta_{k_{i_j}},\ j = 1, 2, \cdots, r)$ be a Y-friendly subspace. Set $z^2 = T_0 x = (t_{ij})x$, where $z^2 = \ltimes_{i=1}^{r} z_i$, $x = \ltimes_{i=1}^{n} x_i$, $T_0 \in \mathscr{L}_{l \times k}$. Since $\mathscr{Y}$ is a Y-friendly subspace, $y \in \mathscr{Y}$, we have $y = Gz^2 = GT_0 x$, where $G \in \mathscr{L}_{v \times l}$ can be expressed as $G = \delta_v[j_1, j_2, \cdots, j_l]$. Hence,

$$H = \delta_v[i_1, i_2, \cdots, i_k] = \delta_v[j_1, j_2, \cdots, j_l]T_0.$$

Denoted by $m_j = \{t \mid j_t = j,\ t = 1, 2, \cdots, l\}$, $j = 1, 2, \cdots, v$. Using Theorem 3.1, a straightforward computation shows that $n_j = m_j k/l$, $j = 1, 2, \cdots, v$, which implies that $n_1, n_2, \cdots, n_v$ have a common factor k/l.

(Sufficiency) Since $n_1, n_2, \cdots, n_v$ have a common factor k/l, let $n_j = m_j k/l$. Assume that there are n_j columns of H equal to δ_v^j, $j = 1, 2, \cdots, v$. Then, we can construct a logical matrix $T_0 \in \mathscr{L}_{l \times k}$ as follows. Set,

$$\begin{aligned}
J_1 &= \{t \mid H_t = \delta_v^1\} &&= \{t_1, \cdots, t_{n_1}\}, \\
J_2 &= \{t \mid H_t = \delta_v^2\} &&= \{t_{n_1+1}, \cdots, t_{n_1+n_2}\}, \\
&\vdots \\
J_v &= \{t \mid H_t = \delta_v^v\} &&= \{t_{n_1+\cdots+n_{v-1}+1}, \cdots, t_{n_1+\cdots+n_v}\},
\end{aligned}$$

where H_t denotes the t-th column of H. Split J_j into m_j equal blocks, denoted by $J_j = \{J_{j1}, J_{j2}, \cdots, J_{jm_j}\}$, $j = 1, 2, \cdots, v$. For simplicity, let $I_1 = \{1, \cdots, m_1\}$, $I_2 = \{m_1 + 1, \cdots, m_1 + m_2\}$, $\cdots$, $I_v = \{m_1 + \cdots + m_{v-1} + 1, \cdots, l\}$. Then, we can construct

$T_0 \in \mathcal{L}_{l\times k}$ as follows:

$$Col_j(T_0) = \begin{cases} \delta_l^1, & j \in J_{11}, \\ \delta_l^2, & j \in J_{12}, \\ & \vdots \\ \delta_l^{m_1}, & j \in J_{1m_1}, \\ \delta_l^{m_1+1}, & j \in J_{21}, \\ & \vdots \\ \delta_l^{m_1+m_2}, & j \in J_{2m_2}, \\ & \vdots \\ \delta_l^l, & j \in J_{\nu m_\nu}. \end{cases} \tag{3.15}$$

It is obvious that T_0, constructed in this way, satisfies (3.13). According to Theorem 3.1, $z = T_0 x$ forms a regular basis. Choose $G \in \mathcal{L}_{\nu\times l}$ as:

$$Col_j(G) = \begin{cases} \delta_\nu^1, & j = i_1, \cdots, i_{m_1}, \\ \delta_\nu^2, & j = i_{m_1+1}, \cdots, i_{m_1+m_2}, \\ & \vdots \\ \delta_\nu^\nu, & j = i_{m_1+\cdots+m_{\nu-1}+1}, \cdots, i_l. \end{cases}$$

A straightforward computation shows that $GT_0 = H$, which means $GT_0 x = Hx = y$. ∎

It is noted that the proof of Theorem 3.2 itself provides an algorithm to construct a Y-friendly subspace for mix-valued logical networks.

In Theorem 3.2, the method of constructing T_0 is not unique. With the Y-friendly subspace constructed above, we can choose a coordinate change $z(t) = T(x(t))$, such that under the coordinate frame z the system (3.11) can be expressed as:

$$\begin{cases} z^1(t+1) = F^1(z(t), u(t), \xi(t)), \\ z^2(t+1) = F^2(z(t), u(t), \xi(t)), \\ y(t) = G(z^2(t)). \end{cases} \tag{3.16}$$

In the next subsection, we need the following result.

Theorem 3.3

Let $f(x_1, x_2, \cdots, x_n)$ be a mix-valued logical function, and its structural matrix $M_f \in \mathcal{L}_{k_0\times k}$ can be split into k_1 equal blocks as:

$$M_f = [B_1\ B_2\ \cdots\ B_{k_1}].$$

Then, $f(x_1, x_2, \cdots, x_n)$ can be expressed as:

$$[\nabla_1(x_1) \wedge f_1(x_2, x_3, \cdots, x_n)] \vee \cdots \vee [\nabla_{k_1}(x_1) \wedge f_{k_1}(x_2, x_3, \cdots, x_n)], \tag{3.17}$$

where $B_i \in \mathscr{L}_{k_0 \times \frac{k}{k_1}}$ is the structural matrix of $f_i(x_2, x_3, \cdots, x_n)$, $i = 1, 2, \cdots, k_1$, and $\nabla_i(z)$, $i = 1, 2, \cdots, m$ are m-valued logical operators defined as:

$$\nabla_i(z) = \begin{cases} m-1, & z = m-i, \\ 0, & z = m-j, \ j \neq i. \end{cases}$$

Proof. The proof is similar to that of k-valued logical networks [6], and thus it is omitted. ∎

3.3.3 CONTROL DESIGN

Assume that there exists a Y-friendly subspace and a coordinate change $z(t) = T(x(t))$ such that the mix-valued logical network (3.11) can be changed into (3.16). Then, it is easy to see that solving the DDP is reduced to finding $u(t)$ such that:

$$z^2(t+1) = F^2(z(t), u(t), \xi(t)) = \widetilde{F}^2(z^2(t)), \tag{3.18}$$

where $z^2 = (z_1, z_2, \cdots, z_r)^T$, $z_j \in \mathscr{D}_{k_{i_j}}$, $j = 1, 2, \cdots, r$, and $F^2 = \{F_1^2, F_2^2, \cdots, F_r^2\}$. Define a set of functions as:

$$\begin{aligned}
e_1(z^2) &= \nabla_1(z_1) \wedge \cdots \wedge \nabla_1(z_{r-1}) \wedge \nabla_1(z_r), \\
&\vdots \\
e_{k_{i_r}}(z^2) &= \nabla_1(z_1) \wedge \cdots \wedge \nabla_1(z_{r-1}) \wedge \nabla_{k_{i_r}}(z_r), \\
e_{k_{i_r}+1}(z^2) &= \nabla_1(z_1) \wedge \cdots \wedge \nabla_2(z_{r-1}) \wedge \nabla_1(z_r), \\
&\vdots \\
e_{k_{i_r} k_{i_{r-1}}}(z^2) &= \nabla_1(z_1) \wedge \cdots \wedge \nabla_{k_{i_{r-1}}}(z_{r-1}) \wedge \nabla_{k_{i_r}}(z_r), \\
&\vdots \\
e_l(z^2) &= \nabla_{k_{i_1}}(z_1) \wedge \nabla_{k_{i_2}}(z_2) \wedge \cdots \wedge \nabla_{k_{i_r}}(z_r).
\end{aligned}$$

Using Theorem 3.3, each F_j^2 can be expressed as:

$$F_j^2 = \vee_{i=1}^{l} \left[e_i(z^2(t)) \wedge P_j^i\left(z^1(t), u(t), \xi(t)\right) \right]. \tag{3.19}$$

PROPOSITION 3.1

$F^2(z(t), u(t), \xi(t)) = \widetilde{F}^2(z^2(t))$ iff, in the expression (3.2), there exists a control $u(t)$ such that:

$$P_j^i(z^1(t), u(t), \xi(t)) = \widetilde{P}_j^{\,i}(z^2(t)), \tag{3.20}$$

where $j = 1, 2, \cdots, r$; $i = 1, 2, \cdots, l$.

Based on Proposition 3.1 and the above analysis, we have the following result.

Theorem 3.4

Consider the system (3.11). There exists a control $u(t)$ such that the DDP is solvable iff,

(i) there exists a Y-friendly subspace such that the system can be expressed as (3.16), and
(ii) there is a control $u(t)$ such that (3.20) holds.

Next, we propose a new technique to design the control,

$$u(t) = W(z(t)) = Wz^1(t)z^2(t)$$

such that (3.20) holds. Assume that the structural matrix of F_j^2 is M_j, that is,

$$F_j^2(z(t), u(t), \xi(t)) = M_j z^2(t)\xi(t)z^1(t)u(t),$$

where $z^2(t) = \ltimes_{i=1}^r z_i(t)$, $z^1(t) = \ltimes_{i=r+1}^n z_i(t)$, $u(t) = \ltimes_{i=1}^m u_i(t)$ and $\xi(t) = \ltimes_{i=1}^q \xi_i(t)$. Let $l_1 = \prod_{j=r+1}^n k_{i_j}$, $s = \prod_{i=1}^m s_i$, $w = \prod_{i=1}^q w_i$ and $l_2 = l_1 sw$. Split M_j into l equal blocks:

$$M_j = [M_j^1 \; M_j^2 \cdots M_j^l], j = 1, 2, \cdots, r.$$

From (3.19) and Theorem 3.3, it is easy to know that $P_j^i(z^1(t), u(t), \xi(t))$ has a structural matrix $M_j^i \in \mathscr{L}_{k_{i_j} \times l_2}$ such that:

$$P_j^i(z^1(t), u(t), \xi(t)) = M_j^i \xi(t) z^1(t) u(t).$$

Set:

$$Q(z^1(t), u(t), \xi(t)) = \ltimes_{j=1}^r \ltimes_{i=1}^l P_j^i(z^1(t), u(t), \xi(t)).$$

Then, we obtain:

$$Q(z^1(t), u(t), \xi(t)) = Q\xi(t)z^1(t)u(t),$$

where $Col_a(Q) = \ltimes_{j=1}^r \ltimes_{i=1}^l Col_a(M_j^i)$, $a = 1, 2, \cdots, l_2$.

Split Q into w equal blocks $Q = [Q^1 \; Q^2 \cdots Q^w]$ first, and then split Q^j into l_1 equal blocks:

$$Q^j = [Q_1^j \; Q_2^j \cdots Q_{l_1}^j],$$

where $Q^j \in \mathscr{L}_{l^l \times (l_1 s)}$ and $Q_i^j \in \mathscr{L}_{l^l \times s}$, $j = 1, 2, \cdots, w$, $i = 1, 2, \cdots, l_1$. Then, we have the following result.

PROPOSITION 3.2

There is a control law $u(t) = W(z(t))$ such that for any $z^1(t)$ and $\xi(t)$,

$$Q(z^1(t), u(t), \xi(t)) = \widetilde{Q}(z^2(t)) \tag{3.21}$$

iff there are l_1 integers j_i, $1 \leq j_i \leq s$, $i = 1, \cdots, l_1$ and at least one vector δ_{ll}^{v} such that:

$$Col_{j_i}\{Q_i^1\} = \cdots = Col_{j_i}\{Q_i^w\} \quad = \quad \delta_{ll}^{v}, \ i = 1, \cdots, l_1. \tag{3.22}$$

Proof. (Necessity) Assume that there exists,

$$u(t) = W(z(t)) = Wz^1z^2$$

such that (3.21) holds for any $z^1(t)$ and $\xi(t)$. Split W into l_1 equal blocks:

$$W = [W_1 \ \ W_2 \cdots W_{l_1}].$$

Then, for any $\xi(t) = \delta_w^j$ and $z^1(t) = \delta_{l_1}^i$, $Q\xi z^1 = Q^j z^1 = Q_i^j$ and $u(t) = Wz^1z^2 = W_i z^2$. Hence,

$$Q(z^1(t), u(t), \xi(t)) = Q_i^j W_i z^2(t).$$

Since $Q_i^j W_i$, $j = 1, 2, \cdots, w$ are equal for each i, by the property of the semi-tensor product, there is an integer $1 \leq j_i \leq s$ such that:

$$Col_{j_i}\{Q_i^1\} = Col_{j_i}\{Q_i^2\} = \cdots = Col_{j_i}\{Q_i^w\}.$$

On the other hand, the matrices $Q_i^j W_i$, $j = 1, 2, \cdots, w$, $i = 1, 2, \cdots, l_1$, are equal. Thus, (3.22) holds and the necessity is completed.

(Sufficiency) We give a constructive proof for the sufficiency. Suppose that (3.22) holds. Denote by S the set of all δ_{ll}^{v} satisfying (3.22). It is easy to see that all the columns of $Q_i^j W_i$ are in S, for any $j = 1, 2, \cdots, w$, $i = 1, 2, \cdots, l_1$.

Then, we can choose u as follows: Let $Col_a(Q_i^j W_i) \in S$, $1 \leq a \leq l$, then there are j_i, $i = 1, 2, \cdots, l_1$ such that (3.22) holds. Thus, we can choose $Col_a(W_i) = \delta_s^{j_i}$. Therefore, a control law $u(t) = Wz^1(t)z^2(t)$ can be determined and the sufficiency is completed. ∎

It is noted that if (3.22) holds, we can set $Q_i^j W_i = \widetilde{Q}$, where $\widetilde{Q}$ is the structural matrix of $\widetilde{Q}(z^2(t))$. Then, the set of all the columns of $\widetilde{Q}$, $Col(\widetilde{Q})$, is a subset of S.

According to Proposition 3.2, we can prove the following result.

Theorem 3.5

Consider the system (3.11). There exists a control $u(t) = W(z(t))$ such that the DDP is solvable iff,

(i) there exists a Y-friendly subspace such that the system can be expressed as (3.16), and
(ii) there are l_1 integers j_i, $1 \leq j_i \leq s$, $i = 1, \cdots, l_1$, and at least one vector $\delta_{l^l}^{y}$ such that (3.22) holds.

3.3.4 AN ILLUSTRATIVE EXAMPLE

In this subsection, we present an illustrative example to show how to use the method to solve the DDP of mix-valued logical networks.

Example 3.2:

Consider the following system:

$$\begin{cases} x_1(t+1) &= [((x_1(t) \wedge_2 x_2(t)) \vee_2 \xi(t)) \rightarrow_2 u(t)] \\ & \quad \leftrightarrow_2 [x_1(t) \wedge_2 \neg_2 (x_2(t) \leftrightarrow_3 x_3(t))], \\ x_2(t+1) &= [(x_1(t) \leftrightarrow_2 (x_2(t) \leftrightarrow_3 x_3(t))) \vee_2 \xi(t)] \\ & \quad \rightarrow_2 u(t), \\ x_3(t+1) &= \{[(x_1(t) \leftrightarrow_2 (x_2(t) \leftrightarrow_3 x_3(t))) \vee_2 \xi(t)] \\ & \quad \rightarrow_2 u(t)\} \leftrightarrow_3 \{[(x_2(t) \rightarrow_3 \xi(t)) \vee_3 u(t)] \\ & \quad \rightarrow_3 [x_1(t) \leftrightarrow_2 (x_2(t) \leftrightarrow_3 x_3(t))]\}, \\ y(t) &= x_1(t) \wedge_3 [x_2(t) \leftrightarrow_3 x_3(t)], \end{cases} \tag{3.23}$$

where x_1, $x_2 \in \mathscr{D}$, $x_3 \in \mathscr{D}_3$ are the states, $\xi \in \mathscr{D}$ is the disturbance, $u \in \mathscr{D}_3$ is the control input and $y \in \mathscr{D}_3$ is the system's output, "$\wedge_2$" denotes 2-valued conjunction operator, etc. (For details, please refer to [3], Chap 14.7).

Setting $x(t) = x_1(t) \ltimes x_2(t) \ltimes x_3(t)$, the mix-valued logical system (3.23) can be expressed as:

$$\begin{cases} x(t+1) &= Lx(t)\xi(t)u(t), \\ y(t) &= Hx(t), \end{cases} \tag{3.24}$$

where,

$$\begin{aligned} L = \delta_{12}[\; & 7\;7\;4\;7\;7\;4\;8\;8\;5\;8\;8\;5\;3\;3\;12\;3\;3\;9 \\ & 3\;3\;12\;3\;3\;3\;2\;2\;11\;2\;2\;2\;7\;7\;4\;7\;7\;10 \\ & 9\;9\;6\;9\;9\;9\;8\;8\;5\;8\;8\;8\;7\;7\;4\;7\;7\;10 \\ & 7\;7\;4\;7\;7\;10\;8\;8\;5\;8\;8\;11\;9\;9\;6\;9\;9\;9], \end{aligned}$$

and,

$$H = \delta_3[1\;2\;3\;3\;2\;1\;3\;3\;3\;3\;3\;3].$$

First, we construct a Y-friendly subspace for the system. Observing H, we have $n_1 = n_2 = 2$ and $n_3 = 8$. Obviously, their least common factor is 2, and according to Theorem 3.2, $l = k_{i_1} \cdots k_{i_r} = k_1 \cdots k_n / 2 = \frac{2\times 2\times 3}{2} = 6$ and $m_1 = m_2 = 1, m_3 = 4$. Thus, we can only decompose l into 2×3 or 3×2. For simplicity, we choose $l = 2 \times 3$. Using Theorem 3.2, we choose:

$$T_0 = \delta_6[1\ 2\ 3\ 3\ 2\ 1\ 4\ 5\ 6\ 6\ 5\ 4]$$

and,

$$G = \delta_3[1\ 2\ 3\ 3\ 3\ 3].$$

From T_0, we can find the output-friendly basis, denoted by $\{z_2, z_3\}$, $z_2 \in \mathscr{D}$ and $z_3 \in \mathscr{D}_3$ with $z_2 z_3 = T_0 x_1 x_2 x_3$. Let:

$$z_2(t) = T_2 x_1(t) x_2(t) x_3(t)$$

and

$$z_3(t) = T_3 x_1(t) x_2(t) x_3(t),$$

where,

$$\begin{aligned} T_2 &= \delta_2[1\ 1\ 1\ 1\ 1\ 1\ 2\ 2\ 2\ 2\ 2\ 2], \\ T_3 &= \delta_3[1\ 2\ 3\ 3\ 2\ 1\ 1\ 2\ 3\ 3\ 2\ 1]. \end{aligned}$$

Let $z_1 = T_1 x_1(t) x_2(t) x_3(t)$ and $z(t) = \ltimes_{i=1}^{3} z_i(t) = Tx(t)$. Since T is a nonsingular matrix, we can choose:

$$T_1 = \delta_2[1\ 1\ 1\ 2\ 2\ 2\ 1\ 1\ 1\ 2\ 2\ 2].$$

Hence,

$$T = \delta_{12}[1\ 2\ 3\ 9\ 8\ 7\ 4\ 5\ 6\ 12\ 11\ 10].$$

From T_i, $i = 1,2,3$, it is easy to obtain that the desired coordinate change is given as:

$$\begin{cases} z_1(t) &= x_2(t), \\ z_2(t) &= x_1(t), \\ z_3(t) &= x_2(t) \leftrightarrow_3 x_3(t). \end{cases} \tag{3.25}$$

Conversely, $x(t) = T^T z(t)$. Thus, under the coordinate frame (3.25), (3.24) becomes:

$$\begin{aligned} z(t+1) &= Tx(t+1) = TLx(t)\xi(t)u(t) \\ &= TLT^T z(t)\xi(t)u(t) := \widetilde{L}z(t)\xi(t)u(t), \\ y(t) &= Hx(t) = HT^T z(t) := \widetilde{H}z(t) \\ &= \delta_3[1\ 2\ 3\ 3\ 3\ 3]z_2 z_3. \end{aligned} \tag{3.26}$$

It is easy to see that the logical form of (3.26) can be expressed as:

$$\begin{cases} z_1(t+1) &= [(z_2(t) \leftrightarrow_2 z_3(t)) \vee_2 \xi(t)] \rightarrow_2 u(t), \\ z_2(t+1) &= [((z_1(t) \wedge_2 z_2(t)) \vee_2 \xi(t)) \rightarrow_2 u(t)] \leftrightarrow_2 [z_2(t) \wedge_2 \neg_2 z_3(t)], \\ z_3(t+1) &= [(z_1(t) \rightarrow_2 \xi(t)) \vee_2 u(t)] \rightarrow_2 [z_2(t) \leftrightarrow_2 z_3(t)], \\ y(t) &= z_2(t) \wedge_3 z_3(t). \end{cases}$$

Set:

$$z_2(t+1) = M_2 z_2(t) z_3(t) z_1(t) \xi(t) u(t)$$

and

$$z_3(t+1) = M_3 z_2(t) z_3(t) z_1(t) \xi(t) u(t)$$

with:

$$\begin{aligned} M_2 = \delta_2[\;& 2\,2\,1\,2\,2\,1\,2\,2\,1\,2\,2\,2\,2\,2\,1\,2\,2\,1\,2\,2\,1\,2\,2\,2 \\ & 1\,1\,2\,1\,1\,2\,1\,1\,2\,1\,1\,1\,2\,2\,1\,2\,2\,2\,2\,2\,1\,2\,2\,2 \\ & 2\,2\,1\,2\,2\,2\,2\,2\,1\,2\,2\,2\,2\,2\,1\,2\,2\,2\,2\,2\,1\,2\,2\,2], \\ M_3 = \delta_3[\;& 1\,1\,1\,1\,1\,1\,1\,1\,1\,1\,1\,1\,2\,2\,2\,2\,2\,1\,2\,2\,2\,2\,2\,2 \\ & 3\,3\,3\,3\,2\,1\,3\,3\,3\,3\,3\,3\,3\,3\,3\,3\,2\,1\,3\,3\,3\,3\,3\,3 \\ & 2\,2\,2\,2\,2\,1\,2\,2\,2\,2\,2\,2\,1\,1\,1\,1\,1\,1\,1\,1\,1\,1\,1\,1]. \end{aligned}$$

Split M_i into 6 equal blocks, $M_i = [M_{i1}\ M_{i2}\ M_{i3}\ M_{i4}\ M_{i5}\ M_{i6}]$, $i = 2,3$ and let,

$$\begin{aligned} e_1(z_2(t), z_3(t)) &= z_2(t) \wedge \nabla_1(z_3(t)), \\ e_2(z_2(t), z_3(t)) &= z_2(t) \wedge \nabla_2(z_3(t)), \\ e_3(z_2(t), z_3(t)) &= z_2(t) \wedge \nabla_3(z_3(t)), \\ e_4(z_2(t), z_3(t)) &= \neg z_2(t) \wedge \nabla_1(z_3(t)), \\ e_5(z_2(t), z_3(t)) &= \neg z_2(t) \wedge \nabla_2(z_3(t)), \\ e_6(z_2(t), z_3(t)) &= \neg z_2(t) \wedge \nabla_3(z_3(t)), \end{aligned}$$

then from Theorem 3.3 we have:

$$\begin{aligned} z_2(t+1) &= \vee_{j=1}^{6}[e_j(z_2(t), z_3(t)) \wedge Q_j^2(z_1(t), \xi(t), u(t))], \\ z_3(t+1) &= \vee_{j=1}^{6}[e_j(z_2(t), z_3(t)) \wedge Q_j^3(z_1(t), \xi(t), u(t))], \end{aligned}$$

where Q_j^i is a logical function with structural matrix M_{ij}, $i = 2,3$, $j = 1, \cdots, 6$. Thus,

$$\begin{aligned} Q(z_1(t), \xi(t), u(t)) &= \ltimes_{i=2}^{3} \ltimes_{j=1}^{6} Q_j^i(z_1(t), \xi(t), u(t)) \\ &= \ltimes_{i=2}^{3} \ltimes_{j=1}^{6} M_{ij} z_1(t) \xi(t) u(t) \\ &= P u(t) z_1(t) \xi(t), \end{aligned}$$

where:

$$\begin{aligned} P = \delta_{46656}[\;& 40252 \quad 40252 \quad 5989 \quad 40252 \quad 40252 \quad 5989 \\ & 40252 \quad 40216 \quad 10938 \quad 40252 \quad 40252 \quad 40252]. \end{aligned}$$

According to Proposition 3.2 and the matrix P, we can design two kinds of controllers to the DDP of system (3.23) as follows:

First, we propose a constant controller. Split P into two equal blocks P^1 and P^2, and split P^j into two equal blocks P_1^j and P_2^j. Hence, $P = [P_1^1\ P_2^1\ P_1^2\ P_2^2]$, where:

$$\begin{aligned} P_1^1 &= \delta_{46656}[40252\ 40252\ 5989], \\ P_2^1 &= \delta_{46656}[40252\ 40252\ 5989], \\ P_1^2 &= \delta_{46656}[40252\ 40216\ 10938], \\ P_2^2 &= \delta_{46656}[40252\ 40252\ 40252]. \end{aligned}$$

Noticing that the first columns of P_i^j, $i,j=1,2$ are equal, according to Proposition 3.2, we choose $W_1 = W_2 = \delta_3[1\ 1\ 1\ 1\ 1\ 1]$, where W_1 and W_2 are two equal blocks of $W = [W_1\ W_2]$. Then,

$$u(t) = \delta_3[1\ 1\ 1\ 1\ 1\ 1\ 1\ 1\ 1\ 1\ 1\ 1]z_1(t)z_2(t)z_3(t) = \delta_3^1$$

is a constant controller such that $Q(z_1(t),\xi(t),u(t)) = \delta_{46656}^{40252}$ for any $\xi(t)$ and $z_1(t)$. Under the control $u(t) = \delta_3^1$, the DDP of the system (3.23) is solved.

Second, we design a more general controller. According to Proposition 3.2 and the form of P, we can choose $W_1 = \delta_3[1\ 1\ 1\ 1\ 1\ 1]$ and $W_2 = \delta_3[1\ 1\ 1\ 2\ 2\ 2]$. Hence,

$$\begin{aligned} u(t) &= \delta_3[1\ 1\ 1\ 1\ 1\ 1\ 1\ 1\ 1\ 2\ 2\ 2]z_1(t)z_2(t)z_3(t) \\ &= \delta_3[1\ 1\ 1\ 2\ 1\ 1\ 1\ 2\ 1\ 1\ 1\ 2]z_3(t)z_1(t)z_2(t) \\ &= \delta_3[1\ 1\ 1\ 2]z_1(t)z_2(t). \end{aligned}$$

Under the control $u(t) = \delta_3[1\ 1\ 1\ 2]z_1(t)z_2(t)$, the term:

$$Q(z_1(t),\xi(t),u(t)) = \delta_{46656}^{40252}$$

for any $\xi(t)$ and $z_1(t)$. Therefore, according to Theorem 3.4, the output y is not affected by the disturbance ξ, that is, the DDP is solved by this control. ∎

It is noted that since only the first columns of P_1^1 and P_1^2 are equal to δ_{46656}^{40252}, W_1 must be $\delta_3[1\ 1\ 1\ 1\ 1\ 1]$. But the first and second columns of both P_2^1 and P_2^2 are δ_{46656}^{40252}. Hence, the column of W_2 can be arbitrarily chosen from δ_3^1 and δ_3^2. Considering the dimension of W_2, we can obtain 2^6 different controllers such that $Q(z_1(t),\xi(t),u(t)) = \delta_{46656}^{40252}$. Other controllers can be designed by the same technique as the above.

3.4 GENERAL EXPRESSION OF PSEUDO LOGICAL FUNCTION

In this section, we study the pseudo logical function and present its general expression. First, the definition of the pseudo logical function is given and its general expression is obtained. Since, the pseudo logical function can be converted into an algebraic form by using the semi-tensor product, based on which the structural matrix is obtained. Secondly, we generalize the pseudo logical function to the pseudo mix-valued logical function. Besides, the general expression and structural matrix of the pseudo mix-valued function are investigated. Thirdly, the pseudo mix-valued logical equations are considered and the existence of solutions is studied. Finally, an illustrative example is studied to support our new results.

3.4.1 GENERAL EXPRESSION OF PSEUDO LOGICAL FUNCTION

Consider an infinitely repeated game [11]. Both player 1 and player 2 have three actions, $\{L, M, R\}$. The payoff bi-matrix assumed is presented in the Table 3.1.

Table 3.1: The payoff bi-matrix

1 \ 2	L	M	F
L	3,3	0,4	9,2
M	4,0	4,4	5,3
R	2,9	3,5	6,6

Denote $L \sim 2$, $M \sim 1$, $R \sim 0$. It is easy to see that the game is a 3-valued logical network. But from the payoff bi-matrix, we conclude that each player's payoff is a real number, i.e, each player's payoff function will be a pseudo logical function which will be defined in the following section.

DEFINITION 3.9

A n-ary pseudo logical function $f(x_1, x_2, \cdots, x_n)$ is a mapping from $\mathscr{D}_k^n$ to $\mathbb{R}$, where $\underbrace{\mathscr{D}_k \times \cdots \mathscr{D}_k}_{n}$ and $x_i \in \mathscr{D}_k = \{0, 1, \cdots, k-1\}$, $i = 1, 2, \cdots, n$.

Now, set $x = (x_1, x_2, \cdots, x_n) \in \mathscr{D}_k^n$, and define the following logical operators.

$$\begin{aligned}
e_1(x) &= \nabla_1(x_1) \wedge \cdots \wedge \nabla_1(x_{n-1}) \wedge \nabla_1(x_n), \\
&\vdots \\
e_k(x) &= \nabla_1(x_1) \wedge \cdots \wedge \nabla_1(x_{n-1}) \wedge \nabla_k(x_n), \\
e_{k+1}(x) &= \nabla_1(x_1) \wedge \cdots \wedge \nabla_2(x_{n-1}) \wedge \nabla_1(x_n), \\
&\vdots \\
e_{2k}(x) &= \nabla_1(x_1) \wedge \cdots \wedge \nabla_2(x_{n-1}) \wedge \nabla_k(x_n), \\
&\vdots \\
e_{k^2}(x) &= \nabla_1(x_1) \wedge \cdots \wedge \nabla_k(x_{n-1}) \wedge \nabla_k(x_n), \\
&\vdots \\
e_{k^n}(x) &= \nabla_k(x_1) \wedge \cdots \wedge \nabla_k(x_{n-1}) \wedge \nabla_k(x_n).
\end{aligned} \tag{3.27}$$

Then, we obtain:

Theorem 3.6

Every pseudo logical function $f(x_1,x_2,\cdots, x_n)$ can be uniquely represented in the following form:

$$f(x_1,x_2,\cdots,x_n) = \sum_{t=1}^{k^n} b_t e_t(x) \tag{3.28}$$

$$= \sum_{t=1}^{k^n} b_t \nabla_{t_1}(x_1) \wedge \cdots \wedge \nabla_{t_{n-1}}(x_{n-1}) \wedge \nabla_{t_n}(x_n),$$

where $t_i \in \{1,2,\cdots,k\}$, $i = 1,2,\cdots,n$, and $t = t_n + k(t_{n-1}-1) + \cdots + k^{n-1}(t_1-1)$. Moreover,

$$b_t = \frac{1}{k-1} f(k-t_1,\cdots,k-t_{n-1},k-t_n)$$

is a real number, $t = 1,2\cdots,k^n$.

Proof. Apply mathematical induction to the number of the states. For $n = 1$, suppose that:

$$f(x_1) = b_1\nabla_1(x_1) + \cdots + b_{k-1}\nabla_{k-1}(x_1) + b_k\nabla_k(x_1).$$

For all $x_1 = k-j \in \mathscr{D}_k$, $\nabla_j(x_1) = k-1$ and $\nabla_i(x_1) = 0$, $i \neq j$. Thus, $b_j = \frac{1}{k-1} f(k-j)$. Then, (3.28) is correct for $n = 1$.

Now, assume that (3.28) is correct for n. In the following section, we prove that it is still correct for $n+1$. For $n+1$,

$$f(x_1,x_2,\cdots,x_n,x_{n+1}) \tag{3.29}$$

$$= \sum_{j=1}^{k} f(x_1,x_2,\cdots,x_n,k-j)\cdot \nabla_j(x_{n+1}).$$

Then, let $g_j(x_1,x_2,\cdots,x_n) = f(x_1,x_2,\cdots,x_n,k-j)$, $j = 1,2,\cdots,k$. By the assumption for n, each g_j can be converted in the form of (3.28). Hence, from (3.29) we conclude that for $n+1$, (3.28) is still correct. As for the real coefficient b_t, we can set,

$$x = (k-t_1,\cdots,k-t_{n-1},k-t_n).$$

By $\nabla_{t_j}(x_j) = k-1$ for all $j = 1,2,\cdots,n$, we obtain,

$$\nabla_{t_1}(x_1) \wedge \cdots \wedge \nabla_{t_{n-1}}(x_{n-1}) \wedge \nabla_{t_n}(x_n) = k-1,$$

and others is 0.

Therefore, $f(k-t_1,\cdots,k-t_{n-1},k-t_n) = b_t \times (k-1)$, that is, $b_t = \frac{1}{k-1} f(k-t_1,\cdots,k-t_{n-1},k-t_n)$. Thus, the proof is completed. ∎

If we use the vector form to express the logical variables, that is, $\Delta_k^n \sim \mathscr{D}_k^n$, then the pseudo logical function can be described as $f: \Delta_k^n \to \mathbb{R}$. Based on this, we have the following result.

Theorem 3.7

Assume that $f(x_1,x_2,\cdots,x_n):\Delta_k^n\to\mathbb{R}$ is a n-ary pseudo logical function. Then, there exists a unique matrix $M_f\in\mathbb{R}^{k\times k^n}$, such that:

$$f(x_1,x_2,\cdots,x_n)=J_kM_fx_1\ltimes x_2\ltimes\cdots\ltimes x_n, \tag{3.30}$$

where $J_k=[k-1\ k-2\ \cdots\ 1\ 0]$, $x_i\in\Delta_k$ and $i=1,2,\cdots,\ n$. Moreover, $J_kM_f=[b_1\ b_2\ \cdots\ b_{k^n}]$ is called the structural matrix of the pseudo logical function.

Proof. From the definition of $e_t(x)=\nabla_{t_1}(x_1)\wedge\cdots\wedge\nabla_{t_{n-1}}(x_{n-1})\wedge\nabla_{t_n}(x_n)$ and $t=t_n+k(t_{n-1}-1)+\cdots+k^{n-1}(t_1-1)$, one can see that $e_t(x)=\delta_k^1$ is satisfied only when $x_i=\delta_{k_i}^{t_i}$, $i=1,2\cdots,n$, i.e, $x=x_1\ltimes x_2\ltimes\cdots\ltimes x_n=\delta_{k^n}^t$. Else, $e_t(x)=\delta_k^k$ for any other $x\neq\delta_{k^n}^t$. Then, the structural matrix of $e_t(x)$ is $M_{e(t)}=\delta_k[k\ \cdots\ k\ 1\ k\cdots\ k]$ for $x\in\Delta_k$, where 1 is at the t-th position.

From Theorem 3.6, we obtain:

$$\begin{aligned}
& f(x_1,x_2,\cdots,x_n)\\
= & \sum_{t=1}^{k^n} b_tJ_kM_{e_t}x_1\ltimes x_2\ltimes\cdots\ltimes x_n\\
= & J_k\begin{bmatrix} b_1 & b_2 & \cdots & b_{k^n}\\ 0 & 0 & \cdots & 0\\ \vdots & \vdots & \cdots & \vdots\\ 0 & 0 & \cdots & 0\\ b-b_1 & b-b_2 & \cdots & b-b_{k^n}\end{bmatrix}x\\
= & J_kM_fx=[b_1\ b_2\ \cdots\ b_{k^n}]x,
\end{aligned} \tag{3.31}$$

where $b=b_1+b_2+\cdots+b_{k^n}$. Thus, the proof is completed. ∎

3.4.2 GENERAL EXPRESSION OF PSEUDO MIX-VALUED LOGICAL FUNCTION

Similar to the pseudo logical function, we can obtain the result about the pseudo mix-valued logical function. First, we state its definition.

DEFINITION 3.10

A n-ary pseudo mix-valued logical function $f(x_1,x_2\cdots,x_n)$ is a mapping from $D_{k_1}\times D_{k_2}\times\cdots\times D_{k_n}$ to $\mathbb{R}$, where $x_i\in D_{k_i}$, $i=1,2\cdots,n$.

Then, we give some logical functions about the logical state $x = (x_1, x_2, \cdots, x_n)$.

$$\begin{aligned}
e_1(x) &= \nabla_{1,k_1}(x_1) \wedge \cdots \wedge \nabla_{1,k_{n-1}}(x_{n-1}) \wedge \nabla_{1,k_n}(x_n), \\
&\vdots \\
e_{k_n}(x) &= \nabla_{1,k_1}(x_1) \wedge \cdots \wedge \nabla_{1,k_{n-1}}(x_{n-1}) \wedge \nabla_{k_n,k_n}(x_n), \\
e_{k_n+1}(x) &= \nabla_{1,k_1}(x_1) \wedge \cdots \wedge \nabla_{2,k_{n-1}}(x_{n-1}) \wedge \nabla_{1,k_n}(x_n), \\
&\vdots \\
e_{k_n \cdot k_{n-1}}(x) &= \nabla_{1,k_1}(x_1) \wedge \cdots \wedge \nabla_{k_{n-1},k_{n-1}}(x_{n-1}) \wedge \nabla_{k_n,k_n}(x_n), \\
&\vdots \\
e_l(x) &= \nabla_{k_1,k_1}(x_1) \wedge \cdots \wedge \nabla_{k_{n-1},k_{n-1}}(x_{n-1}) \wedge \nabla_{k_n,k_n}(x_n),
\end{aligned}$$

where $l = k_1 k_2 \cdots k_n$.

Hence, the general form of the pseudo mix-valued logical function can be given by the following result.

Theorem 3.8

Every pseudo mix-valued logical function $f(x_1, x_2, \cdots, x_n)$ can be uniquely represented in the following form:

$$\begin{aligned}
f(x_1, x_2, \cdots, x_n) &= \sum_{t=1}^{l} b_t e_t(x) \qquad (3.32) \\
&= \sum_{t=1}^{l} b_t \nabla_{t_1,k_1}(x_1) \wedge \cdots \wedge \nabla_{t_n,k_n}(x_n),
\end{aligned}$$

where $t_i \in \{1, 2, \cdots, k_i\}$, $i = 1, 2, \cdots, n$, and $t = t_n + k_n(t_{n-1} - 1) + \cdots + k_n \cdots k_2(t_1 - 1)$. Moreover, $b_t = f(k_1 - t_1, \cdots, k_{n-1} - t_{n-1}, k_n - t_n)$ is a real number, $t = 1, 2 \cdots, l$.

Proof. The Proof is similar to Theorem 3.6, thus we omit it. ∎

Fix the operator "$\wedge$" between two mix-valued logics in any operator $e_t(x)$ to one binary operator on $\mathscr{D} = \{1, 0\}$. Therefore, $e_t(x) \in \mathscr{D}$. If we use the vector to present the logical variables, we obtain that the structural matrix of the logical operator $e_t(x)$ is $\delta_2[\underbrace{2 \cdots 2}_{t-1}\ 1\ \underbrace{2 \cdots 2}_{l-t}]$. Then, we have the following result.

Theorem 3.9

Suppose that $f(x_1,x_2,\cdots,x_n)$ is a n-ary pseudo mix-valued logical function, $x_i \in \Delta_{k_i}$ $i = 1,2,\cdots,n$. Then, there exists one unique matrix $M_f \in \mathbb{R}^{2\times l}$, such that:

$$f(x_1,x_2,\cdots,x_n) = J_2 M_f x_1 \ltimes x_2 \ltimes \cdots \ltimes x_n, \tag{3.33}$$

where $J_2 = [1\ 0]$, $l = k_1 k_2 \cdots k_n$, $b = b_1 + b_2 + \cdots + b_l$ and

$$M_f = \begin{bmatrix} b_1 & b_2 & \cdots & b_l \\ b-b_1 & b-b_2 & \cdots & b-b_l \end{bmatrix}.$$

Obviously, from Theorem 3.9 we obtain $J_2 M_f = [b_1\ b_2\ \cdots\ b_l]$ which is called the structural matrix of the pseudo mix-valued logical function of f. Then,

$$f(x_1,x_2,\cdots,x_n) = [b_1\ b_2\ \cdots\ b_l] x_1 \ltimes x_2 \ltimes \cdots \ltimes x_n. \tag{3.34}$$

Next, we study the existence of solutions for the pseudo mix-valued logical equations. Consider the following pseudo mix-valued logical equation:

$$f(x_1,x_2,\cdots,x_n) = a, \tag{3.35}$$

where $x_i \in \mathscr{D}_{k_i}$, $i = 1,2,\cdots,n$, and $a \in \mathbb{R}$. It is easy to obtain:

Theorem 3.10

The pseudo mix-valued logical equation (3.35) has at least one solution $x = \delta_l^i$ if and only if there at least exists $1 \le i \le l$, such that $b_i = a$, where $l = k_1 k_2 \cdots k_n$, and $[b_1\ b_2\ \cdots\ b_l]$ is the structural matrix of f.
Proof. From (3.34), the proof follows directly the property of the structural matrix. ∎

COROLLARY 3.1

The pseudo mix-valued logical equation (3.35) has no solution if and only if $b_i \neq a$ for any $i = 1,2,\cdots,n$.

Consider the following pseudo mix-valued logical equations:

$$\begin{cases} f_1(x_1,x_2,\cdots,x_n) = a_1, \\ f_2(x_1,x_2,\cdots,x_n) = a_2, \\ \vdots \\ f_m(x_1,x_2,\cdots,x_n) = a_m. \end{cases} \tag{3.36}$$

From Theorem 3.10, we can assume that:

$$f_i(x_1, x_2, \cdots, x_n) = [b_{i1}\ b_{i2}\ \cdots\ b_{il}]x,\ i = 1, 2, \cdots, m,$$

where $x = x_1 \ltimes x_2 \ltimes \cdots \ltimes x_n$. Thus, we have:

Theorem 3.11

The pseudo mix-valued logical equations have at least one solution $x = \delta_l^j$, if and only if there at least exists $1 \leq j \leq l$ such that:

$$b_{ij} = a_i,\ i = 1, 2, \cdots, n.$$

From Theorems 3.10 and 3.11, it is easy to obtain the solutions of the pseudo mix-valued logical equation(s) by using the semi-tensor product. Furthermore, we can solve the the pseudo mix-valued logical inequalities by the same methods.

3.4.3 AN ILLUSTRATIVE EXAMPLE

In this subsection, we present an example to illustrate the effectiveness of the results obtained in this section.

Example 3.3:

Consider the following pseudo mix-valued logical equations:

$$\begin{cases} 2x_1 \vee x_2 + 2x_1 \wedge x_3 = 2, \\ 3x_1 \leftrightarrow x_2 + 4x_2 \vee x_3 = 4, \end{cases} \tag{3.37}$$

where $x_1, x_3 \in \mathscr{D} = \{1, 0\}$ and $x_2 \in \mathscr{D}_3 = \{2, 1, 0\}$.

Our objective is to study whether there exist solutions for system (3.37).

Suppose that,

$$f_1(x_1, x_2, x_3) = 2x_1 \vee x_2 + 2x_1 \wedge x_3$$

and,

$$f_2(x_1, x_2, x_3) = 3x_1 \leftrightarrow 1 + 4x_2 \vee x_3.$$

Then, we can obtain the structural matrices:

$$\begin{aligned} & f_1(x_1, x_2, x_3) = 2J_2 M_d x_1 \Phi_{[2,3]} x_2 + 2J_2 M_c x_1 x_3 \\ = \ & 2J_2[M_d(I_2 \otimes \Phi_{[2,3]})E_{d,2,2}W_{[6,2]} + M_c E_{d,3,2} W_{[2,3]}]x \\ = \ & [4\ 2\ 4\ 2\ 4\ 2\ \underline{2}\ \underline{2}\ 0\ 0\ 0\ 0]x \end{aligned} \tag{3.38}$$

and,

$$\begin{aligned} & f_2(x_1,x_2,x_3) \\ = \; & 3J_2M_ex_1\Phi_{[2,3]}x_2+4J_2M_d\Phi_{[2,3]}x_2x_3 \\ = \; & J_2[3M_e(I_2\otimes\Phi_{[2,3]})E_{d,2,2}W_{[6,2]}+4M_d\Phi_{[2,3]}E_{d,2,3}]x \\ = \; & [7\;7\;4\;0\;4\;0\;\underline{4}\;\underline{4}\;7\;3\;7\;3]x, \end{aligned} \tag{3.39}$$

where $E_{d,2,2}=\delta_2[1\;2\;1\;2]$, $E_{d,3,2}=\delta_2[1\;2\;1\;2\;1\;2]$, and $E_{d,2,3}=\delta_3[1\;2\;3\;1\;2\;3]$.

From (3.38), (3.39) and Theorem 3.11, we obtain that $f_1(x)=2$ and $f_2(x)=4$ for both $x=\delta_{12}^7$ and $x=\delta_{12}^8$. Then, $(x_1,x_2,x_3)=(0,1,1)$ and $(x_1,x_2,x_3)=(0,1,0)$ are solutions of the system (3.37).

REFERENCES

1. Kauffman, S. A. (1969). Metabolic stability and epigenesis in randomly constructed genetic nets. J. Theoretical Biology, 22(3): 437–467.
2. Cheng, D., Qi, H. and Li, Z. (2011). Disturbance decoupling of Boolean control networks. IEEE Trans. Aut. Contr., 56(1): 2–10.
3. Cheng, D., Qi, H. and Li, Z. (2011). Analysis and Control of Boolean Networks: A Semi-Tensor Product Approach. London, Springer.
4. Hammer, P. L. and Rudeanu, S. (1968). Boolean Methods in Operations Research and Related Areas. Berlin, Springer.
5. Hammer, P. L. and Holzman, R. (1992). Approximations of pseudo-Boolean functions: applications to game theory. Z. Oper. Res., 36(1): 3C21.
6. Li, Z. and Cheng, D. (2010). Algebraic approach to dynamics of multivalued networks. Int. J. Bifurcat. Chaos, 20(3): 561–582.
7. Liu, Z. and Wang, Y. (2012). General logical expression of k-valued and mix-valued pseudological functions. Proc. the 31th Chinese Control Conference, 66–71.
8. Liu, Z., Wang, Y. and Li, H. (2011). Disturbance decoupling of multi-valued logical networks. Proc. the 30th Chinese Control Conference, 93–96.
9. Liu, Z. and Wang, Y. (2012). Disturbance decoupling of mix-valued logical networks via the semi-tensor product method. Automatica, 48(8): 1839–1844.
10. Zhao, Y. and Cheng, D. (2010). Optimal control of mix-valued logical control networks. Proc. the 29th Chinese Control Conference, 1618–1623.
11. Zhao, Y., Li, Z. and Cheng, D. (2011). Optimal control of logical control networks. IEEE Trans. Aut. Contr., 56(8): 1766–1776.

4 Delayed Logical Networks

4.1 INTRODUCTION TO DELAYED LOGICAL NETWORKS

This chapter introduces some basic concepts of delayed logical networks.

DEFINITION 4.1

A logical network is called a μ-th order delayed logical network, if the current states depend on μ length histories. Precisely, its dynamics can be described as:

$$\begin{cases} x_1(t+1) = f_1(x_1(t), \cdots, x_n(t), \cdots, x_1(t-\mu+1), \cdots, x_n(t-\mu+1)), \\ x_2(t+1) = f_2(x_1(t), \cdots, x_n(t), \cdots, x_1(t-\mu+1), \cdots, x_n(t-\mu+1)), \\ \quad \vdots \\ x_n(t+1) = f_n(x_1(t), \cdots, x_n(t), \cdots, x_1(t-\mu+1), \cdots, x_n(t-\mu+1)), \\ \quad y_j(t) = h_j(x_1(t), x_2(t), \cdots, x_n(t)), \quad j = 1, 2, \cdots, q. \end{cases} \tag{4.1}$$

where $x_i(t), x_i(t-1), \cdots, x_i(t-\mu+1) \in \Delta_k$, $f_i : \Delta_k^{\mu(m+n)} \to \Delta_k$, $h_j : \Delta_k^n \to \Delta_k$, initial states $x_i(d) = x_{i,d} \in \Delta_k$, $i = 1, 2, \cdots, n$, $j = 1, 2, \cdots, q$, and $d = -\mu+1, -\mu+2, \cdots, 0$.

We give an example to illustrate this kind of systems. It is a biochemical network of coupled oscillations in the cell cycle [1].

Example 4.1: Coupled oscillations

Consider the following Boolean network:

$$\begin{cases} A(t+3) = \neg(A(t-2) \wedge B(t-1)); \\ B(t+3) = \neg(A(t-1) \wedge B(t-2)), \quad t \geq 2. \end{cases} \tag{4.2}$$

It can be easily converted into the canonical form (4.1) as:

$$\begin{cases} A(t+1) = \neg(A(t-2) \wedge B(t-1)); \\ B(t+1) = \neg(A(t-1) \wedge B(t-2)), \quad t \geq 2. \end{cases} \tag{4.3}$$

This is a 3rd order delayed Boolean network.

The second example comes from [2], which proposed a model, where the infinitely repeated game between a human and a machine based on the standard prisoners' dilemma model is considered. The following example describes it.

Table 4.1: Payoff bi-matrix of the standard prisoners' dilemma model in the given infinitely repeated game

Human \ Machine	M	F
M	(3, 3)	(0, 5)
F	(5, 0)	(1, 1)

Example 4.2: Infinitely repeated prisoners' dilemma game

We consider the model of infinitely repeated prisoners' dilemma game. The player m is a machine and player h is a person. The payoff bi-matrix is shown in Table 4.1. Assume the machine's updating law, which depends on the μ-memory, is fixed. It is defined as:

$$\begin{aligned} m(t+1) &= f_m(m(t-\mu+1), m(t-\mu+2), \cdots, m(t), \\ &\quad h(t-\mu+1), h(t-\mu+2), \cdots, h(t)), \end{aligned} \tag{4.4}$$

where $m(t)$ is the machine's strategy at time t and f_m is a fixed logical function. The human strategy is $h(t)$. It was proved in [2] that the human's best policy, f_h, can be obtained by also using μ-memory. That is

$$\begin{aligned} h(t+1) &= f_h(m(t-\mu+1), m(t-\mu+2), \cdots, m(t), \\ &\quad h(t-\mu+1), h(t-\mu+2), \cdots, h(t)). \end{aligned}$$

Putting them together, we have a delayed logical network as:

$$\begin{cases} m(t+1) = f_m(m(t-\mu+1), m(t-\mu+2), \cdots, m(t), \\ \qquad h(t-\mu+1), h(t-\mu+2), \cdots, h(t)), \\ h(t+1) = f_h(m(t-\mu+1), m(t-\mu+2), \cdots, m(t), \\ \qquad h(t-\mu+1), h(t-\mu+2), \cdots, h(t)). \end{cases}$$

As for standard Boolean networks, the following chapter would explore the topological structure of delayed logical networks. These two examples will be referred to later.

4.2 TOPOLOGICAL STRUCTURE OF DELAYED LOGICAL NETWORKS

Lemma 4.1: see [3]

Assume $x = \ltimes_{i=1}^{l} x_i$, where $x_i \in \Delta_k$ and $i = 1, 2, \cdots, l$. Define:

$$\Phi_{l,k} = \prod_{i=1}^{l} \left(I_{k^{i-1}} \otimes \left[(I_k \otimes W_{[k,k^{l-i}]}) M_{r,k} \right] \right),$$

then $x^2 = \Psi_{l,k}x$ holds, where:

$$M_{r,k} = \begin{bmatrix} \delta_k^1 & 0_k & \cdots & 0_k \\ 0_k & \delta_k^2 & \cdots & 0_k \\ \vdots & \vdots & & \vdots \\ 0_k & 0_k & \cdots & \delta_k^k \end{bmatrix}$$

is the base-k power-reducing matrix satisfying $z^2 = M_{r,k}z$, $z \in \Delta_k$, and $0_k \in \mathbb{R}^{k\times 1}$ is a zero vector. ■

Lemma 4.2: see [3]

Assume $X \in \Delta_p$ and $Y \in \Delta_q$. Define two dummy matrices, named by "front-maintaining operator" (FMO) and "rear-maintaining operator" (RMO) respectively, as: $D_f^{p,q} = \delta_p[\underbrace{1\cdots 1}_{q}\ \underbrace{2\cdots 2}_{q}\cdots\underbrace{p\cdots p}_{q}]$, $D_r^{p,q} = \delta_q[\underbrace{\underbrace{1\ 2\cdots q}\ \underbrace{1\ 2\cdots q}\cdots\underbrace{1\ 2\cdots q}}_{p}]$. Then,

$D_f^{p,q}XY = X$, $D_r^{p,q}XY = Y$. ■

Lemma 4.3: see [3]

1. The number of cycles of length s for the dynamics of the NEG, denoted by N_s, is inductively determined by:

$$\begin{cases} N_1 = tr(L), \\ N_s = \frac{tr(L^s) - \sum_{k\in\mathscr{P}(s)} kN_k}{s}, \quad 2 \le s \le k^n, \end{cases}$$

   where $\mathscr{P}(s)$ denotes the set of proper factors of s, the proper factor of s is a positive integer $k < s$ satisfying $s/k \in \mathbb{Z}_+$, and $\mathbb{Z}_+$ is the set of positive integers.
2. The set of elements on cycles of length s, denoted by $\mathscr{C}_s$, is:

$$\mathscr{C}_s = \mathscr{D}_a(L^s) \setminus \bigcup_{t\in\mathscr{P}(s)} \mathscr{D}_a(L^t),$$

   where $\mathscr{D}_a(L)$ is the set of diagonal nonzero columns of L. ■

As for standard logical networks, the investigation of the topological structure for delayed logical networks is very important and meaningful. This chapter explores the topological structure of delayed logical networks.

At first, there is a rigorous definition for cycles and fixed points of delayed logical networks in the following.

DEFINITION 4.2

Consider system (4.1). Denote the state space by:

$$\mathscr{X} = \{X \mid X = (x_1, \cdots, x_n) \in \Delta^n\}.$$

1 Let $X^i = (x_1^i, \cdots, x_n^i)$, $X^j = (x_1^j, \cdots, x_n^j) \in \mathscr{X}$. (X^i, X^j) is said to be a directed edge, if there exist X^{j_α}, $\alpha = 1, \cdots, \mu - 1$ such that X^i, X^j, $\{X^{j_\alpha}\}$ satisfy (4.1). Precisely,

$$x_k^j = f_k(X^{j_1}, X^{j_2}, \cdots, X^{j_{\mu-1}}, X^i), \quad k = 1, \cdots, n.$$

The set of edges is denoted by $\mathscr{E} \subset \mathscr{X} \times \mathscr{X}$.

2 $(X^1, X^2, \cdots, X^\ell)$ is called a path, if $(X^i, X^{i+1}) \in \mathscr{X}$, $i = 1, 2, \cdots, \ell - 1$.

3 A path $(X^1, X^2, \cdots)$ is called a cycle if $X^{i+\ell} = X^i$ for all i, the smallest ℓ is called the length of the cycle. Particularly, the cycle of length 1 is called a fixed point.

Standard logical network can be expressed formally as a delayed logical network with order $\mu = 1$. Hence, Definition 4.2 is also applicable for standard logical networks when $\mu = 1$ holds.

To explore the topological structure of a delayed logical network, we first attempt to convert it into its algebraic form as follows:

Using a vector form, we define:

$$\begin{cases} x(t) = \ltimes_{i=1}^n x_i(t) \in \Delta_{k^n}; \\ z(t) = \ltimes_{i=t}^{t+\mu-1} x(i) \Delta_{k^{\mu n}}, \quad t = 0, 1, \cdots. \end{cases}$$

Assume that the structure matrix of f_i is $M_i \in \mathscr{L}_{k \times k^{\mu n}}$. Then, we can express (4.1) in its component-wise algebraic form as:

$$x_i(t+1) = M_i z(t - \mu + 1), \quad i = 1, \cdots, n; \; t = \mu - 1, \mu, \mu + 1, \cdots. \tag{4.5}$$

Multiplying the equations in (4.5) together yields:

$$x(t+1) = L_0 z(t - \mu + 1), \quad t \geq \mu, \tag{4.6}$$

where:

$$L_0 = M_1 \ltimes_{j=2}^n \left[(I_{k^{\mu n}} \otimes M_j) \Phi_{\mu n, k} \right].$$

Note that the L_0 here can be calculated by a standard procedure explained before and $\Phi_{l,k}$ is defined in Lemma 4.1.

Using some properties of the semi-tensor product of matrix, we have:

$$\begin{aligned} z(t+1) &= \ltimes_{i=t+1}^{t+\mu} x(i) \\ &= (D_r^{k,k})^n \ltimes_{i=t}^{t+\mu-1} x(i) \left(L_0 \ltimes_{i=t}^{t+\mu-1} x(i) \right) \\ &= (D_r^{k,k})^n (I_{k^{\mu n}} \otimes L_0) \Phi_{\mu n, k} \ltimes_{i=t}^{t+\mu-1} x(i) \\ &:= L z(t), \end{aligned} \tag{4.7}$$

where $D_r^{k,k}$ is defined in Lemma 4.2.

We give an example to illustrate the above results.

Example 4.3:

Consider the following logical network:

$$\begin{cases} A(t+1) = C(t-1) \vee (A(t) \wedge B(t)), \\ B(t+1) = \neg (C(t-1) \wedge A(t)), \\ C(t+1) = B(t-1) \wedge B(t). \end{cases} \tag{4.8}$$

Using vector form, we rewrite (4.8) as:

$$\begin{cases} A(t+1) = M_d C(t-1) M_c A(t) B(t), \\ B(t+1) = M_n M_c C(t-1) A(t), \\ C(t+1) = M_c B(t-1) B(t). \end{cases} \tag{4.9}$$

Let $x(t) = A(t)B(t)C(t)$. Then (4.9) can be converted into its component-wise algebraic form as:

$$\begin{cases} A(t+1) = M_1 x(t-1)x(t), \\ B(t+1) = M_2 x(t-1)x(t), \\ C(t+1) = M_3 x(t-1)x(t), \end{cases} \tag{4.10}$$

where:

$$\begin{aligned} M_1 = \delta_4[\quad & 1\,1\,1\,1\,1\,1\,1\,1\,1\,1\,2\,2\,2\,2\,2\,2\,1\,1\,1\,1\,1\,1\,1\,1\,1\,1\,2\,2\,2\,2\,2\,2 \\ & 1\,1\,1\,1\,1\,1\,1\,1\,1\,1\,2\,2\,2\,2\,2\,2\,1\,1\,1\,1\,1\,1\,1\,1\,1\,1\,2\,2\,2\,2\,2\,2]; \\ M_2 = \delta_4[\quad & 2\,2\,2\,2\,1\,1\,1\,1\,1\,1\,1\,1\,1\,1\,1\,1\,2\,2\,2\,2\,1\,1\,1\,1\,1\,1\,1\,1\,1\,1\,1\,1 \\ & 2\,2\,2\,2\,1\,1\,1\,1\,1\,1\,1\,1\,1\,1\,1\,1\,2\,2\,2\,2\,1\,1\,1\,1\,1\,1\,1\,1\,1\,1\,1\,1]; \\ M_3 = \delta_4[\quad & 1\,1\,2\,2\,1\,1\,2\,2\,1\,1\,2\,2\,1\,1\,2\,2\,2\,2\,2\,2\,2\,2\,2\,2\,2\,2\,2\,2\,2\,2\,2\,2 \\ & 1\,1\,2\,2\,1\,1\,2\,2\,1\,1\,2\,2\,1\,1\,2\,2\,2\,2\,2\,2\,2\,2\,2\,2\,2\,2\,2\,2\,2\,2\,2\,2]. \end{aligned}$$

Multiplying three equations in (4.10) together yields:

$$x(t+1) = L_0 x(t-1)x(t), \tag{4.11}$$

where:

$$\begin{aligned} L_0 = \delta_8[\quad & 3\,3\,4\,4\,1\,1\,2\,2\,1\,1\,6\,6\,5\,5\,6\,6\,4\,4\,4\,4\,2\,2\,2\,2\,2\,2\,6\,6\,6\,6\,6\,6 \\ & 3\,3\,4\,4\,1\,1\,2\,2\,1\,1\,6\,6\,5\,5\,6\,6\,4\,4\,4\,4\,2\,2\,2\,2\,2\,2\,6\,6\,6\,6\,6\,6]. \end{aligned}$$

Setting $z(t) = x(t)x(t+1)$, $t \geq 1$, we finally have:

$$\begin{aligned} z(t+1) &= x(t+1)x(t+2) \\ &= (D_r^{2,2})^3 x(t)x(t+1)x(t+2) \\ &= (D_r^{2,2})^3 x(t)x(t+1)L_0 x(t)x(t+1) \\ &= (D_r^{2,2})^3 (I_{2^6} \otimes L_0)\Phi_{6,2} x(t)x(t+1) \\ &:= Lz(t) \end{aligned} \tag{4.12}$$

where:

$$L = \delta_{64}[\begin{array}{l} 3\ 11\ 20\ 28\ 33\ 41\ 50\ 58\ 1\ 9\ 22\ 30\ 37\ 45\ 54\ 62 \\ 4\ 12\ 20\ 28\ 34\ 42\ 50\ 58\ 2\ 10\ 22\ 30\ 38\ 46\ 54\ 62 \\ 3\ 11\ 20\ 28\ 33\ 41\ 50\ 58\ 1\ 9\ 22\ 30\ 37\ 45\ 54\ 62 \\ 4\ 12\ 20\ 28\ 34\ 42\ 50\ 58\ 2\ 10\ 22\ 30\ 38\ 46\ 54\ 62]. \end{array}$$

The following result shows the equivalence between the topological structures of (4.1) and (4.7).

Lemma 4.4: see [4]

1. Each trajectory ξ^x of (4.6) can be obtained from a trajectory ξ^z of (4.7).
2. Each cycle C_x of (4.6) can be obtained from a cycle of (4.7).
3. The transient period of network (4.6) equals to the transient period of network (4.7). ■

Lemma 4.4 shows that to find the cycles of (4.6) it is enough to find the cycles of (4.7). Hence the method developed in the previous sections of this chapter can be used for system (4.7).

We consider the following example.

Example 4.4:

Recall Example 4.1. Set $x(t) = A(t)B(t)$. Using vector form, (4.3) can be expressed as:

$$x(t+1) = L_0 x(t-2)x(t-1)x(t), \tag{4.13}$$

where:

$$L_0 = \delta_4[\begin{array}{l} 4\ 4\ 4\ 4\ 2\ 2\ 2\ 2\ 3\ 3\ 3\ 3\ 1\ 1\ 1\ 1 \\ 3\ 3\ 3\ 3\ 1\ 1\ 1\ 1\ 3\ 3\ 3\ 3\ 1\ 1\ 1\ 1 \\ 2\ 2\ 2\ 2\ 2\ 2\ 2\ 2\ 1\ 1\ 1\ 1\ 1\ 1\ 1\ 1 \\ 1\ 1\ 1\ 1\ 1\ 1\ 1\ 1\ 1\ 1\ 1\ 1\ 1\ 1\ 1\ 1]. \end{array} \tag{4.14}$$

Set $z(t) = x(t)x(t+1)x(t+2)$. Then:

$$\begin{array}{rcl} z(t+1) & = & x(t+1)x(t+2)x(t+3) \\ & = & (D_r^{2,2})^2 x(t)x(t+1)x(t+2)x(t+3) \\ & = & (D_r^{2,2})^2 x(t)x(t+1)x(t+2)L_0 x(t)x(t+1)x(t+2) \\ & = & (D_r^{2,2})^2 (I_{26} \otimes L_0)\Phi_{6,2} x(t)x(t+1)x(t+2) \\ & := & Lz(t), \end{array} \tag{4.15}$$

where:

$$L = \delta_{26}[\begin{array}{l} 4\ 8\ 12\ 16\ 18\ 22\ 26\ 30\ 35\ 39\ 43\ 47\ 49\ 53\ 57\ 61 \\ 3\ 7\ 11\ 15\ 17\ 21\ 25\ 29\ 35\ 39\ 43\ 47\ 49\ 53\ 57\ 61 \\ 2\ 6\ 10\ 14\ 18\ 22\ 26\ 30\ 33\ 37\ 41\ 45\ 49\ 53\ 57\ 61 \\ 1\ 5\ 9\ 13\ 17\ 21\ 25\ 29\ 33\ 37\ 41\ 45\ 49\ 53\ 57\ 61].\end{array} \tag{4.16}$$

To find the cycles of (4.13), it is enough to find all the cycles in system (4.15). We can check $tr(L^k)$, $k = 1, 2, \cdots, 64$ and look for nontrivial power s. They can be easily calculated as:

$$tr(L^2) = 2,\ tr(L^5) = 5,\ tr(L^{10}) = 17.$$

Using Lemma 4.3, we conclude that the system does not have fixed point, but it has one cycle of length 2, one cycle of length 5 and one cycle of length 10.

Now, we find out the cycles of (4.15). First we consider L^2. It is easy to figure out that the 26-th column, $Col_{26}(L^2)$ is a diagonal nonzero column. Then we can use it to generate the cycle of length 2. Since $L\delta_{64}^{26} = \delta_{64}^{29}$, and $L\delta_{64}^{29} = \delta_{64}^{26}$, we have a cycle of length 2. Define $\pi(z) = \Gamma z$, where:

$$\Gamma = I_4 \otimes \mathbf{1}_{16}^T.$$

Using Lemma 4.3, the cycle of system (4.3) with length 2 is:

$$\pi(\delta_{64}^{26}) \to \pi(\delta_{64}^{39}) \to \pi(\delta_{64}^{26}).$$

Equivalently, $\delta_4^2 \to \delta_4^2 \to \delta_4^2$. Using the scalar form, it is show in Figure 4.1.

Similarly, since $Col_1(L^5) = d_{64}^1$ is a diagonal nonzero column of L^5, then δ_{64}^1, $L\delta_{64}^1 = \delta_{64}^4$, $L^2\delta_{64}^1 = \delta_{64}^{16}$, $L^3\delta_{64}^1 = \delta_{64}^{61}$, $L^4\delta_{64}^1 = \delta_{64}^{49}$ form a cycle of length 5. Using Lemma 4.3, the cycle of system (4.3) with length 5 is:

$$\pi(\delta_{64}^1) \to \pi(\delta_{64}^4) \to \pi(\delta_{64}^{16}) \to \pi(\delta_{64}^{61}) \to \pi(\delta_{64}^{49}) \to \pi(\delta_{64}^1).$$

Equivalently, it is:

$$\delta_4^1 \to \delta_4^1 \to \delta_4^1 \to \delta_4^4 \to \delta_4^4 \to \delta_4^1.$$

Using the scalar form, it is the cycle depicted in Figure 4.2.

Since $Col_2(L^{10}) = \delta_{64}^2$ is a diagonal nonzero column of L^{10}, then δ_{64}^2, $L\delta_{64}^2 = \delta_{64}^8$, $L^2\delta_{64}^2 = \delta_{64}^{30}$, $L^3\delta_{64}^2 = \delta_{64}^{53}$, $L^4\delta_{64}^2 = \delta_{64}^{17}$, $L^5\delta_{64}^2 = \delta_{64}^3$, $L^6\delta_{64}^2 = \delta_{64}^{12}$, $L^7\delta_{64}^2 = \delta_{64}^{47}$, $L^8\delta_{64}^2 = \delta_{64}^5$, $L^9\delta_{64}^2 = \delta_{64}^{33}$ form a cycle with length 10.

Using Lemma 4.3, the cycle of system (4.3) with length 10 is:

$$\begin{array}{ll} & \pi(\delta_{64}^2) \to \pi(\delta_{64}^8) \to \pi(\delta_{64}^{30}) \to \pi(\delta_{64}^{53}) \to \pi(\delta_{64}^{17}) \to \pi(\delta_{64}^3) \to \pi(\delta_{64}^{12}) \\ \to & \pi(\delta_{64}^{47}) \to \pi(\delta_{64}^5) \to \pi(\delta_{64}^{33}) \to \pi(\delta_{64}^2). \end{array}$$

Equivalently,

$$\delta_4^1 \to \delta_4^1 \to \delta_4^2 \to \delta_4^4 \to \delta_4^2 \to \delta_4^1 \to \delta_4^1 \to \delta_4^3 \to \delta_4^4 \to \delta_4^3 \to \delta_4^1.$$

Using scalar form, it is the cycle depicted in Figure 4.3.

It is easy to calculate the transient period of (4.15), which is 4. From Lemma 4.4, we know that the transient time of network (4.3) is 4. It is said that, for any initial state $(A(t_0), B(t_0))$, the state of the network will enter into certain cycle. The result coincides with the one in [6].

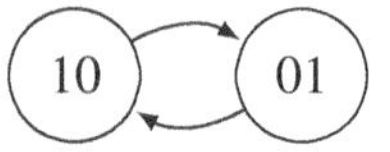

Figure 4.1 The Cycle of (4.3) with Length 2

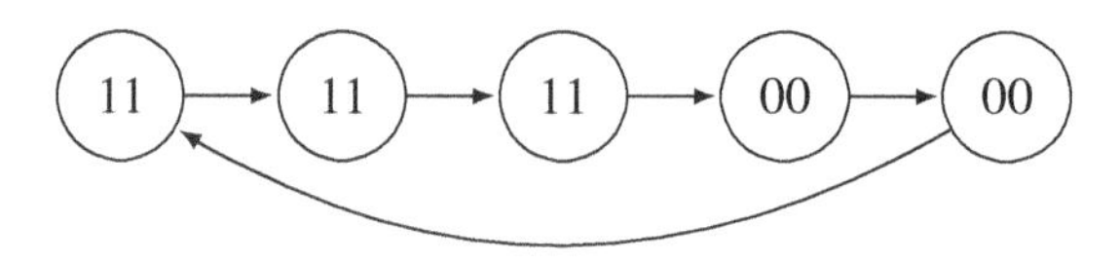

Figure 4.2 The Cycle of (4.3) with Length 5

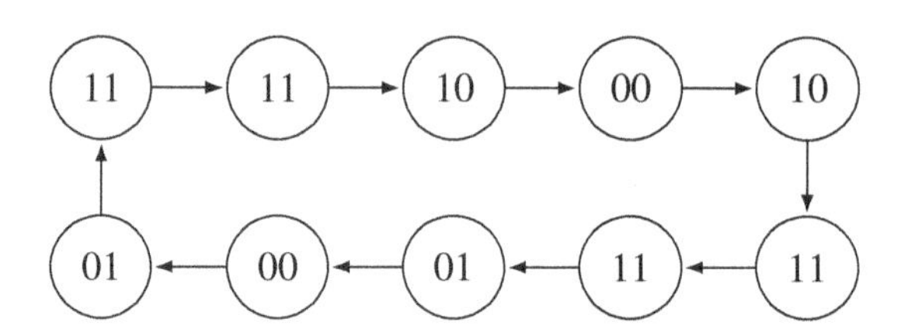

Figure 4.3 The Cycle of (4.3) with Length 10

4.3 TRAJECTORY CONTROLLABILITY OF DELAYED LOGICAL CONTROL NETWORKS

Lemma 4.5

Let $f(x_1, x_2, \cdots, x_s) : \mathscr{D}^s \to \mathscr{D}$ be a logical function. Then, there exists a unique matrix $M_f \in \mathscr{L}_{2\times 2^s}$, called the structural matrix of f, such that:

$$f(x_1, x_2, \cdots, x_s) = M_f \ltimes_{i=1}^{s} x_i, \quad x_i \in \Delta. \tag{4.17}$$

■

Lemma 4.6

(1) Let $X \in \mathbb{R}_m$ and $Y \in \mathbb{R}_n$ be two column vectors. Then, $W_{[m,n]}XY = YX$, where $W_{[m,n]}$ is called the swap matrix. Especially $W_{[n,n]} := W_{[n]}$.
(2) (pseudo-commutative property) Let $X \in \mathbb{R}_t$ and $A \in \mathbb{R}_{m\times n}$. Then, $XA = (I_t \otimes A)X$ holds. ■

After the introduction of delayed logical networks, this chapter concentrates on the research of delayed logical networks. When we talk about Example 4.2, if we consider $h(t)$ as the input strategy sequence $u(t)$, (4.4) can be rewritten as:

$$\begin{aligned} m(t+1) &= f_m(m(t-\mu+1), m(t-\mu+2), \cdots, m(t), \\ &\quad u(t-\mu+1), u(t-\mu+2), \cdots, u(t)), \end{aligned}$$

which is a simple delayed logical control network.

By introducing control in our model, we study the delayed logical control networks as follows:

DEFINITION 4.3

In general, the μ-th delayed logical control networks with n state nodes, m input nodes and q output nodes can be described as:

$$\begin{cases} x_1(t+1) = f_1(u_1(t), \cdots, u_m(t), \cdots, u_1(t-\mu+1), \cdots, u_m(t-\mu+1), \\ \qquad\qquad x_1(t), \cdots, x_n(t), \cdots, x_1(t-\mu+1), \cdots, x_n(t-\mu+1)), \\ x_2(t+1) = f_2(u_1(t), \cdots, u_m(t), \cdots, u_1(t-\mu+1), \cdots, u_m(t-\mu+1), \\ \qquad\qquad x_1(t), \cdots, x_n(t), \cdots, x_1(t-\mu+1), \cdots, x_n(t-\mu+1)), \\ \qquad \vdots \\ x_n(t+1) = f_n(u_1(t), \cdots, u_m(t), \cdots, u_1(t-\mu+1), \cdots, u_m(t-\mu+1), \\ \qquad\qquad x_1(t), \cdots, x_n(t), \cdots, x_1(t-\mu+1), \cdots, x_n(t-\mu+1)), \\ \quad y_j(t) = h_j(x_1(t), x_2(t), \cdots, x_n(t)), \quad j = 1, 2, \cdots, q. \end{cases} \tag{4.18}$$

where $x_i(t), x_i(t-1), \cdots, x_i(t-\mu+1) \in \Delta_k$, $u_v(t), u_v(t-1), \cdots, u_v(t-\mu+1) \in \Delta_k$, $f_i : \Delta_k^{\mu(m+n)} \to \Delta_k$, $h_j : \Delta_k^n \to \Delta_k$, initial states $x_i(d) = x_{i,d} \in \Delta_k$, $u_v(s) = u_{v,s} \in \Delta_k$, $i = 1, 2, \cdots, n$, $j = 1, 2, \cdots, q$, $v = 1, 2, \cdots, m$, $d = -\mu+1, -\mu+2, \cdots, 0$, and $s = -\mu+1, -\mu+2, \cdots, -1$.

Just like in (4.10), (4.11) and (4.12), using STP method, we convert system (4.18) into the following form:

$$\begin{cases} x(t+1) = Lu(t)\cdots u(t-\mu+1)x(t)\cdots x(t-\mu+1), \\ y(t) = Hx(t), \end{cases} \tag{4.19}$$

where $x(t), x(t-1), \cdots, x(t-\mu+1) \in \Delta_{k^n}$, $u(t), u(t-1), \cdots, u(t-\mu+1) \in \Delta_{k^m}$, $L \in \mathscr{L}_{k^n \times k^{\mu(n+m)}}$, $H \in \mathscr{L}_{k^q \times k^n}$, initial states $x(d) = \ltimes_{i=1}^n x_{i,d} = x_d \in \Delta_{k^n}$, $u(s) = u_s = \ltimes_{j=1}^m u_{j,s} \in \Delta_{k^m}$, $d = -\mu+1, -\mu+2, \cdots, 0$, and $s = -\mu+1, -\mu+2, \cdots, -1$.

This chapter briefly introduces the necessary basic knowledge on symbolic dynamics, which can be found in [7,8]. We mainly use the compactness of the following symbolic space $S^{\mathbb{N}}$ to prove some of the results of this chapter.

Let S be a finite nonempty set, for any given $p \in \mathbb{Z}_+$, we use S^p to denote the set of all sequences of length p over S, i.e., $S^p = \{u_1 u_2 \cdots u_p \mid u_i \in S, i = 1, 2, \cdots, p\}$. For all $1 \leq i \leq p$, and all $u \in S^p$, $u(i)$ or $u[i]$ denotes the i-th element of u. Furthermore, for all $1 \leq i \leq j \leq n$, we use $u[i, j]$ to denote the subsequence $u(i)u(i+1) \cdots u(j) \in S^{j-i+1}$. Given $s \in S^p$ for some $p \in \mathbb{Z}_+$, s^∞ denotes concatenation of infinite copies of s. We use $S^{\mathbb{N}}$ to denote set of all infinite sequences (also called configurations) over S, i.e., $\{u_0 u_1 \cdots \mid u_i \in S, i \in \mathbb{N}\}$. Note that $S^{\mathbb{N}}$ is an infinite set. Given $u \in S^{\mathbb{N}}$, and $0 \leq i \leq j$, we use $u(i)$ and $u[i, j]$ to denote the i-th element of u and subsequence $u(i)u(i+1) \cdots u(j) \in S^{j-i+1}$.

The configuration space $S^{\mathbb{N}}$ forms a compact metric space, if $S^{\mathbb{N}}$ is endowed with the Cantor metric:

For any configuration $c, e \in S^{\mathbb{N}}$,

$$d(e,c) = \begin{cases} 0, & \text{if } c = e, \\ 2^{-min\{|v| \mid v \in \mathbb{N}, c(v) \neq e(v)\}}, & \text{if } c \neq e. \end{cases}$$

A configuration sequence $c_1, c_2, \cdots \in S^{\mathbb{N}}$ is convergent to a configuration $c \in S^{\mathbb{N}}$, if $\forall \varepsilon > 0$, there exists $P > 0$ such that $d(c_p, c) < \varepsilon$ for all $p > P$. Note that a configuration sequence $c_1, c_2, \cdots \in S^{\mathbb{N}}$ converges to a configuration $c \in S^{\mathbb{N}}$, iff for all nodes $i \in \mathbb{N}$, there is a positive integer j such that $c_k(i) = c(i)$ for all $k > j$. The compactness of $S^{\mathbb{N}}$ is defined as that any configuration sequence $c_1, c_2, \cdots \in S^{\mathbb{N}}$ has a converging subsequence.

Before we investigate the trajectory controllability of delayed logical networks [5, 9], we consider the delayed logical networks as the mappings from the space of input trajectories to the space of output trajectories, based on which the continuity, injectivity and surjectivity of higher order k-valued logical control networks are analyzed via the theory of symbolic dynamics. After that, the concept for trajectory controllability of delayed logical control networks is defined.

Consider Δ_M, Δ_N, and Δ_Q as alphabets, where $M = k^m$, $N = k^n$ and $Q = k^q$. Define two classes of mappings generated by the μ-th order logical control network (4.19) from the initial states $\bar{x}_0$ and $\bar{u}_0$, where we define $\bar{x}_0 = x_{-\mu+1} x_{-\mu+2} \cdots x_0$ and $\bar{u}_0 = u_{-\mu+1} u_{-\mu+2} \cdots u_{-1}$:

For $\forall \bar{x}_0 \in \Delta_N^\mu$, $\forall \bar{u}_0 \in \Delta_M^{\mu-1}$, and all $p \in \mathbb{Z}_+$,

$$\begin{aligned} L^p_{\bar{x}_0, \bar{u}_0} &: \Delta_M^p \to \Delta_N^p, \quad u_0 u_1 \cdots u_{p-1} \mapsto x_1 x_2 \cdots x_p; \\ H^p_{\bar{x}_0, \bar{u}_0} &: \Delta_N^p \to \Delta_Q^p, \quad x_1 x_2 \cdots x_p \mapsto y_1 y_2 \cdots y_p; \\ HL^p_{\bar{x}_0, \bar{u}_0} &= H^p_{\bar{x}_0, \bar{u}_0} \circ L^p_{\bar{x}_0, \bar{u}_0}, \quad u_0 u_1 \cdots u_{p-1} \mapsto y_1 y_2 \cdots y_p. \end{aligned} \tag{4.20}$$

For $\forall \bar{x}_0 \in \Delta_N^{\mu}$ and $\forall \bar{u}_0 \in \Delta_M^{\mu-1}$,

$$
\begin{aligned}
&L_{\bar{x}_0,\bar{u}_0}^{\mathbb{N}} : \Delta_M^{\mathbb{N}} \to \Delta_N^{\mathbb{N}}, \quad u_0 u_1 \cdots u_{p-1} \cdots \mapsto x_1 x_2 \cdots x_p \cdots ;\\
&H_{\bar{x}_0,\bar{u}_0}^{\mathbb{N}} : \Delta_N^{\mathbb{N}} \to \Delta_Q^{\mathbb{N}}, \quad x_1 x_2 \cdots x_p \cdots \mapsto y_1 y_2 \cdots y_p \cdots ;\\
&HL_{\bar{x}_0,\bar{u}_0}^{\mathbb{N}} = H_{\bar{x}_0,\bar{u}_0}^{\mathbb{N}} \circ L_{\bar{x}_0,\bar{u}_0}^{\mathbb{N}}, \quad u_0 u_1 \cdots u_{p-1} \cdots \mapsto y_1 y_2 \cdots y_p \cdots .
\end{aligned} \tag{4.21}
$$

Where, $\Delta_M^{\mathbb{N}}$, $L_{\bar{x}_0,\bar{u}_0}^{\mathbb{N}}(\Delta_M^{\mathbb{N}})$, and $HL_{\bar{x}_0,\bar{u}_0}^{\mathbb{N}}(\Delta_M^{\mathbb{N}})$ denote the spaces of input trajectories, state trajectories, and output trajectories respectively. Thus, the following proposition proves that the mappings (4.21) are continuous, which is very essential for rest of this paper.

PROPOSITION 4.1

For $\forall \bar{x}_0 \in \Delta_N^{\mu}$ and $\forall \bar{u}_0 \in \Delta_M^{\mu-1}$, if the mappings $L_{\bar{x}_0,\bar{u}_0}^{\mathbb{N}}$, $H_{\bar{x}_0,\bar{u}_0}^{\mathbb{N}}$, and $HL_{\bar{x}_0,\bar{u}_0}^{\mathbb{N}}$ are generated by system (4.18), then $L_{\bar{x}_0,\bar{u}_0}^{\mathbb{N}}$, $H_{\bar{x}_0,\bar{u}_0}^{\mathbb{N}}$, and $HL_{\bar{x}_0,\bar{u}_0}^{\mathbb{N}}$ are continuous.

Proof. First, we prove that, for $\forall \bar{x}_0 \in \Delta_N^{\mu}$ and $\forall \bar{u}_0 \in \Delta_M^{\mu-1}$, $L_{\bar{x}_0,\bar{u}_0}^{\mathbb{N}}$ is continuous.

Choose any converging configuration sequence $U_0, U_1, \cdots \in \Delta_M^{\mathbb{N}}$ and let $U = \lim_{t\to\infty} U_t$, where $U \in \Delta_M^{\mathbb{N}}$. Namely, we need to prove that $\lim_{t\to\infty} L_{\bar{x}_0,\bar{u}_0}^{\mathbb{N}}(U_t) = L_{\bar{x}_0,\bar{u}_0}^{\mathbb{N}}(U)$ holds. By the definition of convergence, for $\forall \varepsilon > 0$, $\exists s \in \mathbb{N}$ such that $d(U_j, U) < \varepsilon$ for any $j > s$. Namely, there exists s' in $\mathbb{N}$ such that $U_j[0,s'] = U[0,s']$ and $2^{-s'} < \varepsilon$, for any $j > s$. Then, for $\forall \bar{x}_0 \in \Delta_N^{\mu}$ and $\forall \bar{u}_0 \in \Delta_M^{\mu-1}$, one obtains:

$$
L_{\bar{x}_0,\bar{u}_0}^{\mathbb{N}}(U_j)[0,s'] = L_{\bar{x}_0,\bar{u}_0}^{s'+1}(U_j)[0,s'] = L_{\bar{x}_0,\bar{u}_0}^{s'+1}(U)[0,s'] = L_{\bar{x}_0,\bar{u}_0}^{\mathbb{N}}(U)[0,s'],
$$

for any $j > s$, that is, $d(L_{\bar{x}_0,\bar{u}_0}^{\mathbb{N}}(U_j), L_{\bar{x}_0,\bar{u}_0}^{\mathbb{N}}(U)) = 2^{-s'} < \varepsilon$. Then, we have $\lim_{t\to\infty} L_{\bar{x}_0,\bar{u}_0}^{\mathbb{N}}(U_t) = L_{\bar{x}_0,\bar{u}_0}^{\mathbb{N}}(U)$. Hence, $L_{\bar{x}_0,\bar{u}_0}^{\mathbb{N}}$ is continuous. Similarly, we can prove that $H_{\bar{x}_0,\bar{u}_0}^{\mathbb{N}}$ and $HL_{\bar{x}_0,\bar{u}_0}^{\mathbb{N}}$ are continuous. ∎

The following two propositions reveal some properties about injectivity and surjectivity of the mappings (4.18) and (4.19).

PROPOSITION 4.2

The following four items satisfy the implications $(a) \Rightarrow (b) \Rightarrow (c) \Rightarrow (d)$.

(a) For $\forall \bar{x}_0 \in \Delta_N^{\mu}$ and $\forall \bar{u}_0 \in \Delta_M^{\mu-1}$, the mapping $HL_{\bar{x}_0,\bar{u}_0}^{1}$ is injective.

(b) For $\forall \bar{x}_0 \in \Delta_N^{\mu}$, $\forall \bar{u}_0 \in \Delta_M^{\mu-1}$, and $\forall p \in \mathbb{Z}_+$, the mapping $HL_{\bar{x}_0,\bar{u}_0}^{p}$ is injective.

(c) For $\forall \bar{x}_0 \in \Delta_N^{\mu}$ and $\forall \bar{u}_0 \in \Delta_M^{\mu-1}$, there is an integer $p \in \mathbb{Z}_+$ such that the mapping $HL^{p}_{\bar{x}_0,\bar{u}_0}$ is injective.

(d) For $\forall \bar{x}_0 \in \Delta_N^{\mu}$ and $\forall \bar{u}_0 \in \Delta_M^{\mu-1}$, the mapping $HL^{\mathbb{N}}_{\bar{x}_0,\bar{u}_0}$ is injective.

Proof. We omit the obvious proof of $(b) \Rightarrow (c)$.

$(a) \Rightarrow (b)$: We prove it by induction. When $p = 1$, it is obvious that (b) holds. Hence, (b) is true for $p \le s$. For $\forall \bar{x}_0 \in \Delta_N^{\mu}$ and $\forall \bar{u}_0 \in \Delta_M^{\mu-1}$, choose arbitrarily $U_1, U_2 \in \Delta_N^{s+1}$, where $U_1 \neq U_2$. If $U_1[0,s-1] \neq U_2[0,s-1]$, by induction assumption, we have $HL^{s}_{\bar{x}_0,\bar{u}_0}(U_1[0,s-1]) \neq HL^{s}_{\bar{x}_0,\bar{u}_0}(U_2[0,s-1])$; else if $U_1[s] \neq U_2[s]$, by induction assumption, we have $HL^{1}_{\bar{x}'_0,\bar{u}'_0}(U_1[s]) \neq HL^{1}_{\bar{x}'_0,\bar{u}'_0}(U_2[s])$, where:

$$\bar{x}'_0 = \begin{cases} L^{\mathbb{N}}_{\bar{x}_0,\bar{u}_0}(U_1)[s-\mu+1,s], & s > \mu, \\ x_{s-\mu+1}x_{s-\mu+2}\cdots x_0 L^{\mathbb{N}}_{\bar{x}_0,\bar{u}_0}(U_1)[1,s], & s \le \mu, \end{cases}$$

and

$$\bar{u}'_0 = \begin{cases} U_1[s-\mu+1,s-1], & s > \mu, \\ u_{s-\mu+1}u_{s-\mu+2}\cdots u_0 U_1[1,s-1], & s \le \mu. \end{cases}$$

Thus, the conclusion follows.

$(c) \Rightarrow (d)$: For $\forall \bar{x}_0 \in \Delta_N^{\mu}$ and $\forall \bar{u}_0 \in \Delta_M^{\mu-1}$, choose arbitrarily $U_1, U_2 \in \Delta_M^{\mathbb{N}}$, where $U_1 \neq U_2$. Define $s = \min\{j \in \mathbb{N} \mid U_1(j) \neq U_2(j)\} \ge 0$. Then $L^{s}_{\bar{x}_0,\bar{u}_0}(U_1[0,s-1]) = L^{s}_{\bar{x}_0,\bar{u}_0}(U_2[0,s-1])$. Define $\bar{x}'_0$ and $\bar{u}'_0$ as in (4.20) and (4.21). There is $p \in \mathbb{Z}_+$ such that $HL^{p}_{\bar{x}'_0,\bar{u}'_0}$ is injective. Then $HL^{p}_{\bar{x}'_0,\bar{u}'_0}(U_1[s,s+p-1]) \neq HL^{p}_{\bar{x}'_0,\bar{u}'_0}(U_2[s,s+p-1])$. Hence $HL^{\mathbb{N}}_{\bar{x}_0,\bar{u}_0}(U_1) \neq HL^{\mathbb{N}}_{\bar{x}_0,\bar{u}_0}(U_2)$ holds. Namely, $HL^{\mathbb{N}}_{\bar{x}_0,\bar{u}_0}$ is injective. ∎

PROPOSITION 4.3

The following three items are equivalent:

(a) For $\forall \bar{x}_0 \in \Delta_N^{\mu}$ and $\forall \bar{u}_0 \in \Delta_M^{\mu-1}$, the mapping $HL^{\mathbb{N}}_{\bar{x}_0,\bar{u}_0}$ is surjective.

(b) For $\forall \bar{x}_0 \in \Delta_N^{\mu}$, $\forall \bar{u}_0 \in \Delta_M^{\mu-1}$, and $\forall p \in \mathbb{Z}_+$, the mapping $HL^{p}_{\bar{x}_0,\bar{u}_0}$ is surjective.

(c) For $\forall \bar{x}_0 \in \Delta_N^{\mu}$ and $\forall \bar{u}_0 \in \Delta_M^{\mu-1}$, the mapping $HL^{1}_{\bar{x}_0,\bar{u}_0}$ is surjective.

Proof. $(a) \Rightarrow (b)$: Assume that, for some $\bar{x}_0 \in \Delta_N^{\mu}$, $\bar{u}_0 \in \Delta_M^{\mu-1}$, and $p \in \mathbb{Z}_+$, $HL^{p}_{\bar{x}_0,\bar{u}_0}$ is not surjective. Then, there exists $e \in \Delta_Q^{p}$ that has no preimages under $HL^{p}_{\bar{x}_0,\bar{u}_0}$. And then for any $c \in \Delta_Q^{\mathbb{N}}$, $ec \in \Delta_Q^{\mathbb{N}}$ has no preimages under $HL^{\mathbb{N}}_{\bar{x}_0,\bar{u}_0}$. It contradicts (a).

$(b) \Rightarrow (c)$: It holds obviously.

$(c) \Rightarrow (b)$: Similar to the proof of $(a) \Rightarrow (b)$ of Proposition 4.2, it can be proved by induction. We omit the similar proof.

$(b) \Rightarrow (a)$: This proof is based on the compactness of the space of input trajectories $\Delta_M^{\mathbb{N}}$.

For $\forall c \in \Delta_Q^{\mathbb{N}}$. Because of (b), for $\forall \bar{x}_0 \in \Delta_N^{\mu}$, $\forall \bar{u}_0 \in \Delta_M^{\mu-1}$, and $\forall i \in \mathbb{N}$, $c[0,i]$ has a preimage under $HL_{\bar{x}_0,\bar{u}_0}^{i+1}$. Denote a preimage of $c[0,i]$ under $HL_{\bar{x}_0,\bar{u}_0}^{i+1}$ by f_i', that is $HL_{\bar{x}_0,\bar{u}_0}^{i+1}(f_i') = c[0,i]$.

Construct configuration sequence $f_0, f_1, \cdots$ such that for all $i \in \mathbb{N}$, $f_i = f_i'(\delta_M^1)^\infty$. Then $HL_{\bar{x}_0,\bar{u}_0}^{\mathbb{N}}(f_i)[0,i] = c[0,i]$. The compactness of the configuration space $\Delta_M^{\mathbb{N}}$ shows that there exists a converging subsequence $f_{l(i)}$ of $\{f_i\}_{0 \leq i < +\infty}$, where $l : \mathbb{N} \to \mathbb{N}$ is a strictly monotonically increasing function. Define $f = \lim_{i \to \infty} f_{h(i)}$. Then $c = \lim_{i \to \infty} HL_{\bar{x}_0,\bar{u}_0}^{\mathbb{N}}(f_i) = \lim_{i \to \infty} HL_{\bar{x}_0,\bar{u}_0}^{\mathbb{N}}(f_{l(i)}) = HL_{\bar{x}_0,\bar{u}_0}^{\mathbb{N}}(\lim_{i \to \infty} f_{l(i)}) = HL_{\bar{x}_0,\bar{u}_0}^{\mathbb{N}}(f)$ by Proposition 4.1. That is, c has a preimage f. The conclusion follows. ∎

The following proposition studies the bijectivity of the mappings (4.20) and (4.21).

PROPOSITION 4.4

The following three items are equivalent:

(a) For $\forall \bar{x}_0 \in \Delta_N^{\mu}$ and $\forall \bar{u}_0 \in \Delta_M^{\mu-1}$, $HL_{\bar{x}_0,\bar{u}_0}^{\mathbb{N}}$ is bijective.

(b) For $\forall \bar{x}_0 \in \Delta_N^{\mu}$, $\forall \bar{u}_0 \in \Delta_M^{\mu-1}$, and $\forall p \in \mathbb{Z}_+$, $HL_{\bar{x}_0,\bar{u}_0}^{p}$ is bijective.

(c) For $\forall \bar{x}_0 \in \Delta_N^{\mu}$ and $\forall \bar{u}_0 \in \Delta_M^{\mu-1}$, $HL_{\bar{x}_0,\bar{u}_0}^{1}$ is bijective.

Proof. $(a) \Rightarrow (b)$: Because of Proposition 4.3, one has that, for $\forall \bar{x}_0 \in \Delta_N^{\mu}$, $\forall \bar{u}_0 \in \Delta_M^{\mu-1}$, and $\forall p \in \mathbb{Z}_+$, $HL_{\bar{x}_0,\bar{u}_0}^{p}$ is surjective. In the following, we prove that, for $\forall \bar{x}_0 \in \Delta_N^{\mu}$, $\forall \bar{u}_0 \in \Delta_M^{\mu-1}$, and $\forall p \in \mathbb{Z}_+$, $HL_{\bar{x}_0,\bar{u}_0}^{p}$ is injective by reduction to absurdity. Fix $\bar{x}_0 \in \Delta_N^{\mu}$ and $\bar{u}_0 \in \Delta_M^{\mu-1}$, we construct two input sequences $\{f_i\}_{0 \leq i \leq +\infty}$ and $\{g_i\}_{0 \leq i \leq +\infty} \subset \Delta_M^{\mathbb{N}}$ as follows:

$f_0 = u_0(\delta_M^1)^\infty$, $g_0 = u_0'(\delta_M^1)^\infty$ such that $u_0 \neq u_0'$ and $HL_{\bar{x}_0,\bar{u}_0}^{1}(u_0) = HL_{\bar{x}_0,\bar{u}_0}^{1}(u_0')$. Define $x_1 = L_{\bar{x}_0,\bar{u}_0}^{1}(u_0)$ and $x_1' = L_{\bar{x}_0,\bar{u}_0}^{1}(u_0')$.

Arbitrarily choose $y_2 \in \Delta_Q$. Then there exist $u_1, u_1' \in \Delta_M$ such that $HL_{\bar{x}_1,\bar{u}_1}^{1}(u_1) = HL_{\bar{x}_1',\bar{u}_1'}^{1}(u_1') = y_2$, where:

$$\begin{aligned} \bar{x}_1 &= x_{-\mu+2}x_{-\mu+3}\cdots x_0x_1, \quad & \bar{u}_1 &= u_{-\mu+2}u_{-\mu+3}\cdots u_{-1}u_0, \\ \bar{x}_1' &= x_{-\mu+2}x_{-\mu+3}\cdots x_0x_1', \quad & \bar{u}_1' &= u_{-\mu+2}u_{-\mu+3}\cdots u_{-1}u_0'. \end{aligned} \tag{4.22}$$

Define $x_2 = L_{\bar{x}_1,\bar{u}_1}^{1}(u_1)$ and $x_2' = L_{\bar{x}_1',\bar{u}_1'}^{1}(u_1')$. Construct $f_1 = u_0u_1(\delta_M^1)^\infty$ and $g_1 = \bar{u}_0'u_1'(\delta_M^1)^\infty$.

...

Arbitrarily choose $y_{i+1} \in \Delta_Q$. Then there exist $u_i, u_i' \in \Delta_M$ such that $HL_{\bar{x}_i,\bar{u}_i}^{1}(u_i) = HL_{\bar{x}_i',\bar{u}_i'}^{1}(u_i') = y_{i+1}$ where:

$$\begin{aligned} \bar{x}_i &= x_{i-\mu+1}x_{i-\mu+2}\cdots x_i, \quad & \bar{u}_i &= u_{i-\mu+1}u_{i-\mu+2}\cdots u_{i-1}, \\ \bar{x}_i' &= x_{i-\mu+1}'x_{i-\mu+2}'\cdots x_i', \quad & \bar{u}_i' &= u_{i-\mu+1}'u_{i-\mu+2}'\cdots u_{i-1}'. \end{aligned} \tag{4.23}$$

Define $x_{i+1} = L^1_{\bar{x}_i,\bar{u}_i}(u_i)$ and $x'_{i+1} = L^1_{\bar{x}'_i,\bar{u}'_i}(u'_i)$. Construct $f_i = u_0 u_1 \cdots u_i (\delta^1_M)^\infty$ and $g_i = u'_0 u'_1 \cdots u'_i (\delta^1_M)^\infty$.
...

Now take the input sequences $\{f_i\}_{0\leq i<+\infty}$ and $\{g_i\}_{0\leq i<+\infty}$ into consideration. By the compactness of the space of input trajectories $\Delta^{\mathbb{N}}_M$, there exist two converging subsequences $\{f_{l(i)}\}_{0\leq i<\infty}$ and $\{g_{l(i)}\}_{0\leq i<\infty}$, where $l : \mathbb{N} \to \mathbb{N}$ is a strictly monotonically increasing function.

Define $f = \lim_{i\to\infty} f_{l(i)}$ and $g = \lim_{i\to\infty} g_{l(i)}$. By the process of constructing $\{f_i\}_{0\leq i\leq+\infty}$, $\{g_i\}_{0\leq i\leq+\infty}$, one has $f(0) \neq g(0)$, namely, $f \neq g$, and $HL^{\mathbb{N}}_{x_0}(f_{l(i)})[0, l(i)] = HL^{\mathbb{N}}_{x_0}(g_{l(i)})[0, l(i)]$, where $i \in \mathbb{N}$. Thus, by virtue of Proposition 4.1, one gets:

$$\begin{aligned}
&\lim_{t\to\infty} HL^{\mathbb{N}}_{\bar{x}_0,\bar{u}_0}(f_{l(i)}) = HL^{\mathbb{N}}_{\bar{x}_0,\bar{u}_0}(\lim_{t\to\infty} f_{l(i)}) = HL^{\mathbb{N}}_{\bar{x}_0,\bar{u}_0}(c) \\
= &\lim_{t\to\infty} HL^{\mathbb{N}}_{\bar{x}_0,\bar{u}_0}(g_{l(i)}) = HL^{\mathbb{N}}_{\bar{x}_0,\bar{u}_0}(\lim_{t\to\infty} g_{l(i)}) = HL^{\mathbb{N}}_{\bar{x}_0,\bar{u}_0}(g).
\end{aligned}$$

Hence $HL^{\mathbb{N}}_{\bar{x}_0,\bar{u}_0}$ is not injective, which is a contradiction. Then, for $\forall \bar{x}_0 \in \Delta^{\mu}_N$, $\forall \bar{u}_0 \in \Delta^{\mu-1}_M$, and $\forall p \in \mathbb{Z}_+$, $HL^{p}_{\bar{x}_0,\bar{u}_0}$ is injective. Thus, (b) is proved.

$(b) \Rightarrow (a)$: This implication is obtained directly by using Proposition 4.2 and Proposition 4.3.

$(b) \Rightarrow (c)$: This implication holds obviously.

$(c) \Rightarrow (b)$: This implication is also obtained directly by using Proposition 4.2 and Proposition 4.3. ∎

In the following paragraph, we give the definition of trajectory controllable.

DEFINITION 4.4: Trajectory Controllable

The system (4.19) is said to be trajectory controllable if for $\forall Z \in (\Delta_Q)^{\mathbb{N}}$, $\forall \bar{x}_0 \in \Delta^{\mu}_N$, and $\forall \bar{u}_0 \in \Delta^{\mu-1}_M$, there exists a control (or input) sequence $U \in (\Delta_M)^{\mathbb{N}}$ such that the corresponding output sequence $Y = HL^{\mathbb{N}}_{\bar{x}_0,\bar{u}_0}(U) \in (\Delta_Q)^{\mathbb{N}}$ of (4.19) satisfies $Y = Z$.

Theorem 4.1

A delayed logical control network (4.19) is trajectory controllable, iff the delayed logical control network (4.19) is surjective. ∎

Proof. By Definition 4.4, it is obvious that the conclusion follows. ∎

The following result gives an equivalent criterion for the trajectory controllability of delayed logical control networks.

Theorem 4.2

A delayed logical control network (4.19) is trajectory controllable, iff $m \geq q$, $n \geq m$, and $rank(HLW_{[k^{m(\mu-1)+n\mu},k^m]}\delta^i_{k^{m(\mu-1)+n\mu}}) = k^q$ holds for $\forall i = 1,2,\cdots,k^{m(\mu-1)+n\mu}$. ■

Proof. By system (4.19), when $t = 0$, one has:

$$\begin{aligned} y(1) &= Hx(1) = HLu(0)u(-1)\cdots u(-\mu+1)x(0)x(-1) \\ &\quad \cdots x(-\mu+1) \\ &= HLW_{[k^{m(\mu-1)+n\mu},k^m]}u(-1)u(-2)\cdots u(-\mu+1) \\ &\quad \cdot x(0)x(-1)\cdots x(-\mu+1)u(0) \\ &= HL^1_{\bar{x}_0,\bar{u}_0}(u(0)). \end{aligned} \tag{4.24}$$

By the virtue of (4.24), it is easy to see that, for $\forall \bar{x}_0 \in \Delta^{\mu}_N$ and $\forall \bar{u}_0 \in \Delta^{\mu-1}_M$, $HL^1_{\bar{x}_0,\bar{u}_0}$ is surjective, iff:

$$rank(HLW_{[k^{m(\mu-1)+n\mu},k^m]}\delta^i_{k^{m(\mu-1)+n\mu}}) = k^q$$

and $n \geq m \geq q$ hold, $\forall i = 1,2,\cdots,k^{m(\mu-1)+n\mu}$. Thus, with Proposition 4.3 and Theorem 4.1, this conclusion follows. ■

Theorem 4.3

Assume that the given delayed logical control network (4.19) is trajectory controllable. For any output sequence $\{y(t)\}_{1\leq t<\infty}$, we can find a proper input sequence $\{u(t)\}_{0\leq t<\infty}$ of the given delayed logical control network (4.19) when we take $\{y(t)\}_{1\leq t<\infty}$ as input sequence of the following system:

$$\begin{cases} z(t+1) &= Qu(t-1)\cdots u(t-\mu+1) \\ &\quad x(t)\cdots x(t-\mu+1)\bar{u}(t), \\ x(t+1) &= P\bar{u}(t)z(t)\cdots z(t-\mu+2) \\ &\quad x(t)\cdots x(t-\mu+1), \\ u(t) &= z(t). \end{cases}$$

where $Q = [Q_1\ Q_2\ \cdots\ Q_{k^{m(\mu-1)+n\mu}}]$, $Q_i = \bar{Q}_i W_{[k^{m-q},k^q]}\delta^1_{k^{m-q}}$, $W = W_{[k^{m(\mu-1)+n\mu},k^m]}$, and

$$\begin{aligned} P &= LQW^T\left(I_{k^{m+m(\mu-1)}} \otimes W\right)\left(I_{k^m} \otimes \Psi_{\mu-1,k^m}\right) \\ &\quad \cdot \left(I_{k^{m+m(\mu-1)}} \otimes \Psi_{\mu,k^n}\right) \end{aligned} \tag{4.25}$$

with $\bar{Q}_i$ satisfying:

$$HLW\bar{Q}_i = [I \quad * \quad], \quad i = 1,2,\cdots,k^{m(\mu-1)+n\mu}.$$

■

Proof. First of all, by the definition of W, we consider a trajectory controllable delayed logical control network (4.19) with an output sequence $\{y(t)\}_{1\leq t<\infty}$ and $\{u(t)\}_{0\leq t<\infty}$, which is one of its corresponding input sequences, satisfying:

$$\begin{aligned} y(t+1) &= HLu(t)\cdots u(t-\mu+1)x(t)\cdots x(t-\mu+1) \\ &= HLWu(t-1)\cdots u(t-\mu+1) \\ &\quad \cdot x(t)\cdots x(t-\mu+1)u(t). \end{aligned}$$

By virtue of the property of logical matrices, fix $t \geq 0$, without loss of generality, let $u(t-1)\cdots u(t-\mu+1)x(t)\cdots x(t-\mu+1) = \delta^i_{k^{m(\mu-1)+n\mu}}$, then one has:

$$y(t+1) = HLW\delta^i_{k^{m(\mu-1)+n\mu}}u(t).$$

By Theorem 4.2, there exists a logical matrix $\bar{Q}_i$ such that:

$$y(t+1) = HLW\bar{Q}_i\bar{Q}_i^{-1}u(t) = [I \quad * \quad]\bar{Q}_i^{-1}u(t), \tag{4.26}$$

then one gets that:

$$\begin{aligned} u(t) &= \bar{Q}_i y(t+1) \ltimes \delta^1_{k^{m-q}} = \bar{Q}_i W_{[k^{m-q},k^q]}\delta^1_{k^{m-q}}y(t+1) \\ &:= Q_i y(t+1) \end{aligned}$$

makes (4.26) hold.

Thus, we have:

$$\begin{aligned} u(t) &= Qu(t-1)\cdots u(t-\mu+1) \\ &\quad \cdot x(t)\cdots x(t-\mu+1)y(t+1), \end{aligned} \tag{4.27}$$

where $Q = [Q_1 \; Q_2 \; \cdots \; Q_{k^{m(\mu-1)+n\mu}}]$.

With (4.27), one has:

$$\begin{aligned} x(t+1) &= Lu(t)\cdots u(t-\mu+1)x(t)\cdots x(t-\mu+1) \\ &= LQW^T y(t+1)u(t-1)\cdots u(t-\mu+1) \\ &\quad \cdot x(t)\cdots x(t-\mu+1)u(t-1)\cdots u(t-\mu+1) \\ &\quad \cdot x(t)\cdots x(t-\mu+1) \\ &= LQW^T \left(I_{k^{m+m(\mu-1)}} \otimes W_{[k^{(\mu-1)m},k^{n\mu}]}\right) \\ &\quad \cdot y(t+1)\left(u(t-1)\cdots u(t-\mu+1)\right)^2 \\ &\quad \cdot \left(x(t)\cdots x(t-\mu+1)\right)^2 \\ &= LQW^T \left(I_{k^{m+m(\mu-1)}} \otimes W_{[k^{(\mu-1)m},k^{n\mu}]}\right) \\ &\quad \cdot \left(I_{k^m} \otimes \Psi_{\mu-1,k^m}\right)\left(I_{k^{m+m(\mu-1)}} \otimes \Psi_{\mu,k^n}\right) \\ &\quad y(t+1)u(t-1)\cdots u(t-\mu+1) \\ &\quad \cdot x(t)\cdots x(t-\mu+1) \\ &:= Py(t+1)u(t-1)\cdots u(t-\mu+1) \\ &\quad \cdot x(t)\cdots x(t-\mu+1). \end{aligned} \tag{4.28}$$

Here after, we define $z(t+1) = u(t)$, $\bar{u}(t) = y(t+1)$, and $\bar{y}(t) = z(t)$ as the new state, the new input, and the new output, respectively, for the psedo inverse system. Thus, by (4.27) and (4.28), one obtains:

$$\begin{cases} z(t+1) &= Qu(t-1)\cdots u(t-\mu+1) \\ &\quad x(t)\cdots x(t-\mu+1)\bar{u}(t), \\ x(t+1) &= P\bar{u}(t)z(t)\cdots z(t-\mu+2) \\ &\quad x(t)\cdots x(t-\mu+1), \\ u(t) &= z(t). \end{cases}$$

The conclusion follows. ∎

The following example gives an illustrative example to show how to use our results to output a given strategy sequence.

Example 4.5:

Consider a finitely repeated game between a human and a machine [2]. The machine's strategy is assumed to be fixed with 2-step memory as follows:

$$m(t+1) = Lh(t)h(t-1)m(t)m(t-1), \tag{4.29}$$

where $h(t), h(t-1) \in \Delta_2$, $m(t), m(t-1) \in \Delta_2$, and:

$$L = \delta_2[1\ 1\ 2\ 2\ 2\ 2\ 1\ 1\ 2\ 1\ 1\ 2\ 1\ 1\ 2\ 2].$$

The rest of this section studies how to design the strategy sequence to track the output sequence $\{m(t)\}_{1\le t\le 18} = \{2\ 2\ 1\ 2\ 1\ 1\ 1\ 2\ 1\ 2\ 2\ 1\ 1\ 2\ 1\ 2\ 1\ 1\} \in \Delta_2^{18}$ under initial states $h(-1) = \delta_2^1$, $m(0) = \delta_2^1$, and $m(-1) = \delta_2^1$.

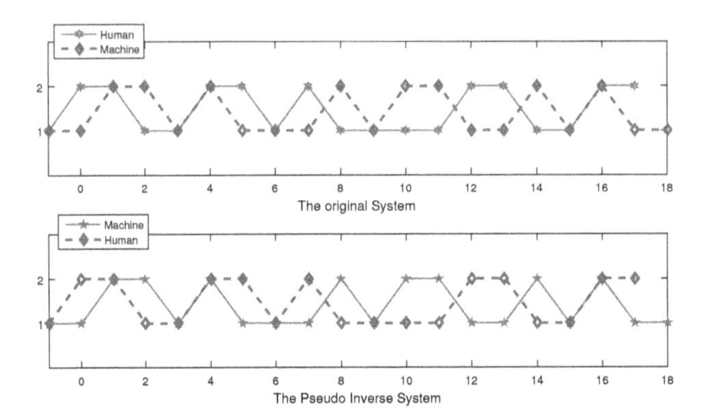

Figure 4.4 The inputs and outputs of the origin system (4.29) and the system (4.30)

It is obvious that the output of system (4.29) is $y(t) = m(t)$, and a simple calculation shows that $rank(LW_{[8,2]}\delta_8^i) = 2$ holds for $\forall i = 1, 2, \cdots, 8$. Hence, by Theorem 4.2, system 4.29 is trajectory controllable. Then, with the help of Theorem 4.3, one gets system as follows:

$$h(t) = Ph(t-1)m(t)m(t-1)m(t+1), \tag{4.30}$$

where:

$$P = \delta_2[1\ 2\ 2\ 1\ 2\ 1\ 1\ 2\ 1\ 2\ 2\ 1\ 1\ 2\ 1\ 2].$$

Considering the strategy sequence $\{m(t)\}_{1\leq t\leq 18} = \{2\,2\,1\,2\,1\,1\,1\,2\,1\,2\,2\,1\,1\,2\,1\,2\,1\,1\}$ $\in \Delta_2^{18}$ as input for system (4.30) under initial states $h(-1) = \delta_2^1$, $m(0) = \delta_2^1$, and $m(-1) = \delta_2^1$, one gets the output sequence $\{h(t)\}_{0\leq t\leq 17} = \{2\ 2\ 1\ 1\ 2\ 2\ 1\ 2\ 1\ 1\ 1\ 1\ 2\ 2\ 1\ 1\ 2\ 2\}$. Figure 4.4 demonstrates the effectiveness of the system (4.30).

4.4 CONTROL DESIGN FOR OUTPUT TRACKING OF DELAYED BOOLEAN CONTROL NETWORKS

This chapter investigates the output tracking [10–13] of delayed logical control networks. We, first convert the dynamics of delayed logical control networks into an equivalent algebraic form. Secondly, based on the algebraic form, we establish some necessary and sufficient conditions for the output tracking of delayed logical control networks without/with forbidden states, respectively.

We continue our research from Definition 4.3. Besides, given a constant reference signal $Y_r = (y_1^r, \cdots, y_p^r) \in \mathscr{D}_k^p$. *The output tracking control problem* is to design a state feedback control in the form of:

$$\begin{cases} u_1(t+1) = h_1(X(t-\tau+1), \cdots, X(t)), \\ \vdots \\ u_m(t+1) = h_m(X(t-\tau+1), \cdots, X(t)), \end{cases} \tag{4.31}$$

where $h_i : \mathscr{D}_k^{\tau n} \to \mathscr{D}_k$, $i = 1, \cdots, m$ are logical functions to be determined, under which there exists an integer $T > 0$ such that:

$$Y(t; Z(0), U(t)) = Y_r$$

holds for any initial trajectory $Z(0) \in \mathscr{D}_k^{\tau n}$ and any integer $t \geq T$.

In the following section, we convert the system (4.18) and the control (4.31) into equivalent algebraic forms, respectively.

Using the canonical vector form of logical variables and letting $x(t) = \ltimes_{i=1}^n x_i(t) \in \Delta_{k^n}$, $z(t) = \ltimes_{i=t-\tau+1}^t x(i) \in \Delta_{k^{\tau n}}$, $u(t) = \ltimes_{i=1}^m u_i(t) \in \Delta_{k^m}$, $y(t) = \ltimes_{i=1}^p y_i(t) \in \Delta_{k^p}$. By Lemma 4.5, the delayed logical control network (4.18) can be converted into the following component-wise algebraic form:

$$\begin{cases} x_i(t+1) = F_i u(t) z(t), i = 1, 2, \cdots, n; \\ y_j(t) = G_j z(t), j = 1, \cdots, p, \end{cases} \tag{4.32}$$

where $F_i \in \mathscr{L}_{k \times k^{\tau n+m}}$ and $G_j \in \mathscr{L}_{k \times k^{\tau n}}$ are the structural matrices of f_i, $i = 1, 2, \cdots, n$ and g_j, $j = 1, \cdots, p$, respectively. Multiplying the equations in (4.32) together, we have:

$$\begin{cases} x(t+1) = F u(t) z(t), \\ y(t) = G z(t), \end{cases}$$

where $F = F_1 * F_2 * \cdots * F_n \in \mathscr{L}_{k^n \times k^{\tau n+m}}$, and $G = G_1 * \cdots * G_p \in \mathscr{L}_{k^p \times k^{\tau n}}$.

By Lemma 4.5, we can obtain the algebraic form of delayed logical control network (4.18) as follows:

$$\begin{cases} z(t+1) = L u(t) z(t), \\ y(t) = G z(t), \end{cases} \tag{4.33}$$

where:

$$L = D_r^{[k^n, k^n]} (I_{k^{\tau n}} \otimes F) W_{[k^m, k^{\tau n}]} (I_{k^m} \otimes M_{r, k^{\tau n}}) u(t) z(t) \in \mathscr{L}_{k^n \times k^{\tau n+m}}.$$

In addition, the reference signal becomes $y_r = \ltimes_{i=1}^p y_i^r = \delta_{k^p}^\alpha$, where α is uniquely determined by y_i^r, $i = 1, \cdots, p$.

Using the same method, we can convert the control (4.31) into the following component-wise algebraic form:

$$\begin{cases} u_1(t) = H_1 z(t), \\ \vdots \\ u_m(t) = H_m z(t), \end{cases}$$

where $H_i \in \mathscr{L}_{k \times k^{\tau n}}$ is the structural matrix of h_i, $i = 1, \cdots, m$. Using the Khatri-Rao product, we can obtain the algebraic form of (4.31) as follows:

$$u(t) = H z(t),$$

where $H = H_1 * \cdots * H_m \in \mathscr{L}_{k^m \times k^{\tau n}}$ is called the state feedback gain matrix. Therefore, the output tracking control problem becomes one of designing the state feedback gain matrix $H \in \mathscr{L}_{k^m \times k^{\tau n}}$.

In the following section, we concentrate on the output tracking of delayed logical control networks.

Consider the system (4.33) with $L = \delta_{k^{\tau n}}[i_1\ i_2\ \cdots\ i_{k^{m+\tau n}}] \in \mathscr{L}_{k^{\tau n} \times k^{m+\tau n}}$ and $y_r = \delta_{k^p}^{\alpha}$. We define $\Gamma(\alpha)$ as:

$$\Gamma(\alpha) = \{a \in \mathbb{N} : Col_a(G) = \delta_{k^p}^{\alpha},\ 1 \leq a \leq k^{\tau n}\}.$$

It is easy to see that all the trajectories of the system (4.33) whose output form the vector y_r constitute the set $\{\delta_{k^{\tau n}}^{a} : a \in \Gamma(\alpha)\}$. If $\Gamma(\alpha) = \emptyset$, the output tracking control problem is not solvable. Therefore, we assume that $\Gamma(\alpha) \neq \emptyset$.

For a given non-empty set $M \in \Delta_{k^{\tau n}}$ and any integer $k \in \mathbb{Z}_+$, we denote by $J_k(M)$ the set of all the initial trajectories which can reach M in the k-th step, that is,

$$\begin{aligned} J_k(M) \quad = \quad & \{z(0) \in \Delta_{k^{\tau n}} : \text{ there exists a control sequence} \\ & \{u(t) : t = 0, \cdots, k-1\} \text{ such that } z(k; z(0), u) \in M\}. \end{aligned} \tag{4.34}$$

Split L into k^m equal blocks as $[l_1\ l_2\ \cdots\ l_{k^m}]$. Set:

$$A = \sum_{i=1}^{k^m} l_i \in \mathbb{R}^{k^{\tau n} \times k^{\tau n}}.$$

One can see from [9] that $z(k) = \delta_{k^{\tau n}}^{p}$ is reachable from $z(0) = \delta_{k^{\tau n}}^{q}$ at the k-th step, if and only if $(A^k)_{p,q} > 0$.

Given a non-empty set $M = \{\delta_{k^{\tau n}}^{j_1}, \cdots, \delta_{k^{\tau n}}^{j_v}\}$ with $1 \leq j_1 < \cdots < j_v \leq k^{\tau n}$ and any integer $\mu \in \mathbb{Z}_+$. We define:

$$A_M := \begin{bmatrix} A_{j_1,j_1} & \cdots & A_{j_1,j_v} \\ \vdots & \vdots & \vdots \\ A_{j_v,j_1} & \cdots & A_{j_v,j_v} \end{bmatrix}$$

and

$$A_M^{\mu} := \begin{bmatrix} Row_{j_1}(A^{\mu}) \\ \vdots \\ Row_{j_v}(A^{\mu}) \end{bmatrix}.$$

Denote:

$$B_M = \sum_{i=1}^{v} Row_i(A_M),\ B_M^{\mu} = \sum_{i=1}^{v} Row_i(A_M^{\mu}).$$

A simple calculation gives the following lemma.

Lemma 4.7

Given a non-empty set $M = \{\delta_{k^{\tau n}}^{j_1}, \cdots, \delta_{k^{\tau n}}^{j_v}\}$ with $1 \leq j_1 < \cdots < j_v \leq k^{\tau n}$ and an integer $\mu \in \mathbb{Z}_+$. Then,

i) $B_M > 0$ if and only if all the columns of A_M are nonzero.
ii) $B_M^{\mu} > 0$ if and only if all the columns of A_M^{μ} are nonzero. ■

Based on Lemma 4.7 and the reachability of delayed logical control networks, we have the following lemma.

Lemma 4.8

Given a non-empty set $M = \{\delta_{k^{\tau n}}^{j_1}, \cdots, \delta_{k^{\tau n}}^{j_v}\}$ with $1 \leq j_1 < \cdots < j_v \leq k^{\tau n}$ and an integer $\mu \in \mathbb{Z}_+$. Then,

i) $B_M > 0$ if and only if $M \subseteq J_1(M)$.
ii) $B_M^{\mu} > 0$ if and only if $J_\mu(M) = \Delta_{k^{\tau n}}$. ■

Proof. We, first prove Conclusion i).

(Necessity) Assume that $B_M > 0$, that is, all the columns of A_M are nonzero. For any $k = 1, \cdots, v$, since $Col_k(A_M) \neq \mathbf{0}_v$, there exists an integer $1 \leq l \leq v$ such that $A_{j_l, j_k} > 0$. That means $z(1) = \delta_{k^{\tau n}}^{j_l} \in M$ is reachable from $z(0) = \delta_{k^{\tau n}}^{j_k} \in M$ in one step. From the arbitrariness of k, we can obtain that $M \subseteq J_1(M)$.

(Sufficiency) Assume that $M \subseteq J_1(M)$. If there exists an integer $1 \leq k \leq v$ such that $Col_k(A_M) = \mathbf{0}_v$, then, any $z(1) = \delta_{k^{\tau n}}^{j_l} \in M$, $l = 1, \cdots, v$ is not reachable from $z(0) = \delta_{k^{\tau n}}^{j_k} \in M$ in one step, which is a contradiction to $z(0) \in J_1(M)$. Thus, all the columns of A_M are nonzero, which shows that $B_M > 0$.

Next, we prove Conclusion ii).

(Necessity) Suppose that $B_M^{\mu} > 0$. By Lemma 4.7, we know that all the columns of A_M^{μ} are nonzero. Since $Col_k(A_M^{\mu}) \neq \mathbf{0}_v$, $k = 1, \cdots, k^{\tau n}$, there exists an integer $1 \leq l \leq v$ such that $(A_M^{\mu})_{j_l, k} > 0$. Thus, $z(\mu) = \delta_{k^{\tau n}}^{j_l} \in M$ is reachable from $z(0) = \delta_{k^{\tau n}}^{k} \in \Delta_{k^{\tau n}}$ at the μ-th step. From the arbitrariness of k, we know that $\Delta_{k^{\tau n}} \subseteq J_\mu(M)$, which together with $J_\mu(M) \subseteq \Delta_{k^{\tau n}}$ implies that $J_\mu(M) = \Delta_{k^{\tau n}}$.

(Sufficiency) Assume that $J_\mu(M) = \Delta_{k^{\tau n}}$. If there exists an integer $1 \leq k \leq k^{\tau n}$ such that $Col_k(A_M^{\mu}) = \mathbf{0}_v$, then, any $z(\mu) = \delta_{k^{\tau n}}^{j_l} \in M$, $l = 1, \cdots, v$ is not reachable from $z(0) = \delta_{k^{\tau n}}^{k} \in \Delta_{k^{\tau n}}$ at the μ-th step, which is a contradiction to $\delta_{k^{\tau n}}^{k} \in J_\mu(M)$. Thus, all the columns of A_M^{μ} are nonzero, that is, $B_M^{\mu} > 0$. ■

Based on Lemma 4.7 and Lemma 4.8, we have the following result.

Theorem 4.4

The output of the system (4.18) tracks the reference signal $y_r = \delta_{k^p}^{\alpha}$ by a state feedback control, if and only if there exist an integer $1 \le \mu \le k^{\tau n}$ and integers $j_i \in \Gamma(\alpha), i = 1, \cdots, v$ satisfying $j_1 < \cdots < j_v$ such that $B_M > 0$ and $B_M^{\mu} > 0$. ■

Proof. (Sufficiency) Assume that there exist an integer $1 \le \mu \le k^{\tau n}$ and integers $j_i \in \Gamma(\alpha), i = 1, \cdots, v$ satisfying $j_1 < \cdots < j_v$ such that $B_M > 0$ and $B_M^{\mu} > 0$. By Lemma 4.8, we have:

$$\begin{cases} M \subseteq J_1(M), \\ J_{\mu}(M) = \Delta_{k^{\tau n}}. \end{cases}$$

Set:

$$J_k^{\circ}(M) = J_k(M) \setminus J_{k-1}(M), k = 1, \cdots, \mu,$$

where $J_0(M) := \emptyset$. Obviously, $J_i^{\circ}(M) \bigcap J_j^{\circ}(M) = \emptyset, \ \forall \ i \neq j \in \{1, \cdots, \mu\}$, and $\bigcup_{k=1}^{\mu} J_k^{\circ}(M) = \Delta_{k^{\tau n}}$. Therefore, for any integer $1 \le j \le k^{\tau n}$, there exists a unique integer $1 \le k_j \le \mu$ such that $\delta_{k^{\tau n}}^{j} \in J_{k_j}^{\circ}(M)$. We have the following two cases:

- Case 1: If $k_j = 1$, there exists an integer $1 \le p_j \le k^m$ such that $\delta_{k^{\tau n}}^{i_l} \in M$, where $l = (p_j - 1)k^{\tau n} + j$.
- Case 2: If $2 \le k_j \le \mu$, there exists an integer $1 \le p_j \le k^m$ such that $\delta_{k^{\tau n}}^{i_l} \in J_{k_j - 1}(M)$, where $l = (p_j - 1)k^{\tau n} + j$.

Let $H = \delta_{k^m}[p_1 \ p_2 \ \cdots \ p_{k^{\tau n}}] \in \mathcal{L}_{k^m \times k^{\tau n}}$. For any initial trajectory $z(0) = \delta_{k^{\tau n}}^{j} \in \Delta_{k^{\tau n}}$, under the control $u(t) = Hz(t)$, we have:

$$z(1; z(0), u) = LHz(0)z(0) = \delta_{k^{\tau n}}^{i_l} \in \begin{cases} M, \text{ if } k_j = 1, \\ J_{k_j - 1}(M), \text{ if } 2 \le k_j \le k^{\tau n}, \end{cases}$$

which implies that $z(k_j; z(0), u) \in M, \forall \ 1 \le j \le k^{\tau n}$. According to $M \subseteq J_1(M)$, one can see that $z(t; z(0), u) \in M$ holds for $\forall \ z(0) \in \Delta_{k^{\tau n}}$ and $\forall \ t \ge \mu$. Thus, $y(t; z(0), u) = Gz(t; z(0), u) = y_r$ holds for any $z(0) \in \Delta_{k^{\tau n}}$ and any $t \ge \mu$. Therefore, the output of the system (4.18) tracks the reference signal y_r by $u(t) = \delta_{k^m}[p_1 \ p_2 \cdots p_{k^{\tau n}}]z(t)$.

(Necessity) Suppose that the output of the system (4.18) tracks the reference signal $y_r = \delta_{k^p}^{\alpha}$ by a state feedback control, say, $H = \delta_{k^m}[p_1 \ p_2 \ \cdots \ p_{k^{\tau n}}]$. Then, the system (4.18) and the control $u(t) = Hz(t)$ form the following closed-loop system:

$$\begin{cases} z(t+1) = \hat{L}z(t), \\ y(t) = Gz(t), \end{cases} \tag{4.35}$$

where $\hat{L} = LHM_{r,k^{\tau n}}$, and $M_{r,k^{\tau n}}$ is the so-called power-reducing matrix satisfying $x \ltimes x = M_{r,k^{\tau n}} \ltimes x, \forall \ x \in \Delta_{k^{\tau n}}$.

Denote the set M by the limit set of the system (4.35) (all the fixed points and cycles). Let T_t be the transient period of the system (4.35). It is easy to see that $1 \leq T_t \leq k^{\tau n}$. A simple calculation shows that $M \subseteq \{\delta_{k^{\tau n}}^{a} : a \in \Gamma(\alpha)\}$, and,

$$\begin{cases} M \subseteq J_1(M), \\ J_\mu(M) = \Delta_{k^{\tau n}}, \end{cases}$$

holds for $\mu = T_t \leq k^{\tau n}$. By Lemma 4.7 and Lemma 4.8, $B_M > 0$ and $B_M^\mu > 0$. ∎

Finally, we give an illustrative example to show the effectiveness of the obtained results.

Example 4.6:

Consider the following delayed logical control network:

$$\begin{cases} x_1(t+1) = u(t) \wedge \{\neg x_1(t-1) \vee \neg x_2(t)\}, \\ x_2(t+1) = u(t) \wedge \{\neg x_1(t) \wedge \neg x_2(t)\}; \\ y_1(t) = x_1(t), \\ y_2(t) = x_2(t), \end{cases} \tag{4.36}$$

where x_1 and x_2 are state variables, $u(t)$ is control input, and $y_1(t)$ and $y_2(t)$ are output variables. Given the reference signal $y_r = (0,0)$. Our objective is to design a state feedback control (if any) such that the output of (4.36) tracks y_r.

Using the vector form of logical variables and setting $x(t) = x_1(t) \ltimes x_2(t)$, $z(t) = x(t-1) \ltimes x(t)$ and $y(t) = y_1(t) \ltimes y_2(t)$. By the semi-tensor product of matrices, we have the following algebraic form of (4.36):

$$\begin{cases} z(t+1) = Lu(t)z(t), \\ y(t) = Gz(t), \end{cases}$$

where:

$$\begin{aligned} L = \delta_{16}[&4\ 8\ 11\ 16\ 3\ 8\ 11\ 16 \\ &2\ 6\ 9\ 14\ 1\ 6\ 9\ 14 \\ &4\ 8\ 11\ 16\ 3\ 8\ 11\ 16 \\ &4\ 8\ 11\ 16\ 3\ 8\ 11\ 16], \end{aligned}$$

$G = \delta_4[1\ 2\ 3\ 4\ 1\ 2\ 3\ 4\ 1\ 2\ 3\ 4\ 1\ 2\ 3\ 4]$ and $y_r = \delta_2^2 \ltimes \delta_2^2 = \delta_4^4$.

A simple calculation gives $\Gamma(4) = \{4, 8, 12, 16\}$. Setting $M = \{\delta_{16}^{16}\} \subseteq \{\delta_{16}^4, \delta_{16}^8, \delta_{16}^{12}, \delta_{16}^{16}\}$ and $\mu = 5$. Since $B_M = 1 > 0$ and $B_M^\mu = [12\ 12\ 16\ 18\ 8\ 12\ 16\ 18\ 10\ 10\ 14\ 15\ 8\ 10\ 14\ 15] > 0$, by Theorem 4.4, the output of (4.36) tracks y_r under the following 256 state feedback gain matrices:

$$H = \delta_2[i_1\ \cdots\ i_8\ 2\ 2\ 1\ 2\ 1\ 2\ 1\ 2],$$

where $i_j \in \{1,2\}$, $j = 1, \cdots, 8$.

REFERENCES

1. Goodwin, B. (1963). Temporal Organization in Cell. New York, Academic Press.
2. Mu, Y. and Guo, L. (2009). Optimization and identification in a non-equilibrium dynamic game. Proc. IEEE Conf. Decis. Control, 5750C5755.
3. Cheng, D., Qi, H. and Li, Z. (2011). Analysis and Control of Boolean Networks: A Semi-tensor Product Approach. London, Springer.
4. Li, Z., Zhao, Y. and Cheng, D. (2011). Structure of higher order Boolean networks. Journal of the Graduate School of the Chinese Academy of Sciences, 28(4): 431–447.
5. Li, F. and Sun, J. (2012). Controllability of higher order Boolean control networks. Applied Mathematics and Computation, 219(1): 158–169.
6. Heidel, J., Maloney, J., Farrow, C. and Rogers, J. (2003). Finding cycles in synchronous Boolean networks with applications to biochemical systems. Int. J. Bifurcation and Chaos, 13(3): 535–552.
7. Zhang, K., Zhang, L. and Xie, L. (2015). Invertibility and nonsingularity of Boolean control networks. Automatica, 60: 155–164.
8. Hedlund, G. (1969). Endomorphisms and automorphisms of shift dynamical systems. Mathematical Systems Theory, 3: 320–375.
9. Lu, J., Zhong, J., Ho, D. W. C., Tang, Y. and Cao, J. (2016). On controllability of delayed Boolean control networks. SIAM J. Contr. Optim., 54(2): 475–494.
10. Li, H., Wang, Y. and Xie, L. (2015). Output tracking control of Boolean control networks via state feedback: Constant reference signal case. Automatica, 59: 54–59.
11. Li, H. and Wang, Y. (2016). Output tracking of switched Boolean networks under open-loop/closed-loop switching signals. Nonlinear Analysis: Hybrid Systems, 22: 137–146.
12. Li, H., Wang, Y. and Guo, P. (2016). State feedback based output tracking control of probabilistic Boolean networks. Information Sciences, 349-350: 1–11.
13. Li, H., Wang, Y. and Guo, P. (2017). Output reachability analysis and output regulation control design of Boolean control networks. Science China Information Sciences, 60(2): 022202.

5 Switched Logical Networks

5.1 INTRODUCTION TO SWITCHED LOGICAL NETWORKS

While typical Boolean networks are described by purely discrete dynamics, the dynamics of biological networks in practice is often governed by different switching models. A practical example is the cell's growth and division in a eukaryotic cell, which are usually described as a sequence of four processes triggered by a set of events [1]. Another typical example is the genetic switch in the bacteriophage λ, which contains two distinct models: lysis and lysogeny [2]. Besides, many switchings are generated by external interventions or control inputs that attempt to re-engineer a given network, and the logical switching phenomenon is often encountered. When modeling these networks as Boolean networks, we obtain switched Boolean networks (SBNs) [9–11]. In the past two decades, due to the great importance of switched systems, in both theoretical development and practical applications, the study of ordinary switched systems has drawn a great deal of attention [3–5].

The definition of switched Boolean networks is as follows:

DEFINITION 5.1

A logical network is called a switched Boolean network with n nodes and m models, if it is described as:

$$\begin{cases} x_1(t+1) = f_1^{\sigma(t)}(x_1(t),\cdots,x_n(t)), \\ x_2(t+1) = f_2^{\sigma(t)}(x_1(t),\cdots,x_n(t)), \\ \quad\vdots \\ x_n(t+1) = f_n^{\sigma(t)}(x_1(t),\cdots,x_n(t)), \end{cases} \tag{5.1}$$

where $\sigma : \mathbb{N} \mapsto \mathscr{W} = \{1,2,\cdots,w\}$ is the switching signal, $x_i \in \mathscr{D}$, $i = 1,2,\cdots,n$ are logical variables, $f_i^j : \mathscr{D}^n \mapsto \mathscr{D}$, $i = 1,\cdots,n$, $j = 1,2,\cdots,w$ are logical functions.

We give an example to illustrate these kind of systems [6].

Example 5.1: Apoptosis network

Consider the following apoptosis network (see Figure 5.1):

$$\begin{cases} x_1(t+1) & = & \neg x_2(t) \wedge u(t), \\ x_2(t+1) & = & \neg x_1(t) \wedge x_3(t), \\ x_3(t+1) & = & x_2(t) \vee u(t), \end{cases} \tag{5.2}$$

where the concentration level (high or low) of the inhibitor of apoptosis proteins (IAP) is denoted by x_1, the concentration level of the active caspase 3 (C3a) by x_2, and the concentration level of the active caspase 8 (C8a) by x_3; the concentration level of the tumor necrosis factor (TNF, a stimulus) is regarded as the control input u.

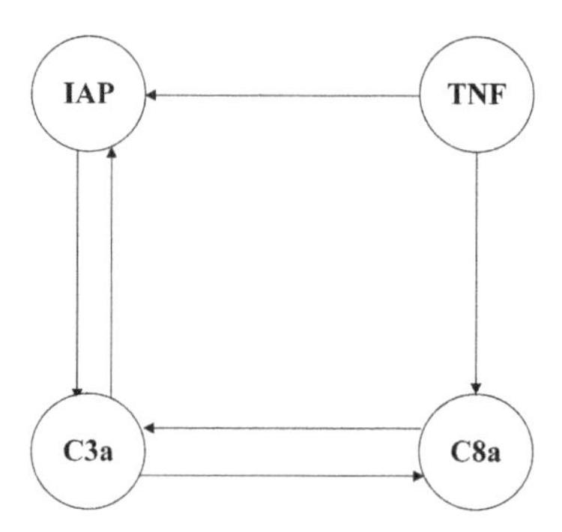

Figure 5.1 Apoptosis network in Example 5.1

By letting $u=1$ and $u=2$, respectively, the dynamics of the apoptosis network becomes the following switched Boolean network:

$$\begin{cases} x_1(t+1) = f_1^{\sigma(t)}(x_1(t), x_2(t), x_3(t)), \\ x_2(t+1) = f_2^{\sigma(t)}(x_1(t), x_2(t), x_3(t)), \\ x_3(t+1) = f_3^{\sigma(t)}(x_1(t), x_2(t), x_3(t)), \end{cases} \tag{5.3}$$

where $\sigma: \mathbb{N} \mapsto \mathscr{A} = \{1,2\}$ is the switching signal, $f_1^1 = \neg x_2$, $f_2^1 = \neg x_1 \wedge x_3$ and $f_3^1 = 1$ correspond to $u=1$, and $f_1^2 = 0$, $f_2^2 = \neg x_1 \wedge x_3$ and $f_3^2 = x_2$ correspond to $u=0$.

5.2 STABILITY ANALYSIS VIA SWITCHING-POINT REACHABILITY

DEFINITION 5.2

The system (5.1) is said to be globally stable at $X_e = (x_1^e, x_2^e, \cdots, x_n^e) \in \mathscr{D}^n$ under arbitrary switching signal, if the systems trajectory starting from any initial state converges to X_e under any switching signal.

The objective of this section is to establish some necessary and sufficient conditions for the global stability of the SBN (5.1) under arbitrary switching signal.

First, we introduce a concept of switching point reachability for SBNs, and then present some necessary and sufficient conditions for the switching point reachability and pointwise stabilizability, respectively.

DEFINITION 5.3

Consider the system (5.1). Let $X_0 = (x_1(0), \cdots, x_n(0)) \in \mathscr{D}^n$. Then, a point $X = (x_1, \cdots, x_n) \in \mathscr{D}^n$ is said to be switching reachable from X_0, if one can find an integer $k > 0$ and a switching signal σ such that under the switching signal, the trajectory of the system (5.1) starting from X_0 reaches X at time k.

To facilitate the analysis, we convert the system (5.1) into an equivalent algebraic form via the semi-tensor product method.

Using the vector form of logical variables and setting $x(t) = \ltimes_{i=1}^n x_i(t) \in \Delta_{2^n}$, by semi-tensor product method, the system (5.1) can be expressed as:

$$\begin{cases} x_1(t+1) = S_1^{\sigma(t)} x(t), \\ x_2(t+1) = S_2^{\sigma(t)} x(t), \\ \vdots \\ x_n(t+1) = S_n^{\sigma(t)} x(t), \end{cases} \tag{5.4}$$

where $S_i^{\sigma(t)} \in \mathscr{L}_{2\times 2^n}$, $i = 1, 2, \cdots, n$. Multiplying the equations in (5.4) together, yields the following algebraic form:

$$x(t+1) = L_{\sigma(t)} x(t), \tag{5.5}$$

where $L_{\sigma(t)} \in \mathscr{L}_{2^n \times 2^n}$, and,

$$Col_i(L_{\sigma(t)}) = \ltimes_{j=1}^n Col_i(S_j^{\sigma(t)}),\ i = 1, \cdots, 2^n.$$

Moreover, $X_e = (x_1^e, x_2^e, \cdots, x_n^e) \in \mathscr{D}^n$ becomes the vector form $x_e = \ltimes_{j=1}^n x_j^e = \delta_{2^n}^{i^*} \in \Delta_{2^n}, 1 \le i^* \le 2^n$.

Based on (5.5), we have the following result on the switching point reachability.

Theorem 5.1

Consider the system (5.1). Then,

1). $x = \delta_{2^n}^p$ is switching reachable from $x(0) = \delta_{2^n}^q$ at time k, if and only if:

$$\left(M^k\right)_{pq} > 0, \tag{5.6}$$

where $M = \sum_{i=1}^{w} L_i$, $\left(M^k\right)_{pq}$ denotes the (p,q)-th element of M^k;

2). $x = \delta_{2^n}^p$ is switching reachable from $x(0) = \delta_{2^n}^q$, if and only if:

$$\mathscr{C}_{pq} > 0, \tag{5.7}$$

where,

$$\mathscr{C} = \sum_{k=1}^{2^n} M^k. \tag{5.8}$$

■

Proof. First, we prove 1) by induction.

When $k = 1$, $x(1) = \delta_{2^n}^p$ is switching reachable from $x(0) = \delta_{2^n}^q$, if and only if there exists an integer $i \in \mathscr{A}$ such that $\delta_{2^n}^p = L_i \delta_{2^n}^q$. That is,

$$\sum_{i=1}^{w} (L_i)_{pq} = \left(\sum_{i=1}^{w} L_i\right)_{pq} = M_{pq} > 0,$$

which implies that (5.1) holds for $k = 1$.

Assume that 1) holds for $k = s$. Now, we consider the case of $k = s+1$. In this case, the trajectory from $x(0) = \delta_{2^n}^q$ to $x = \delta_{2^n}^p$ at $(s+1)$-th step can be decomposed into some trajectory from $x(0) = \delta_{2^n}^q$ to $x(s) = \delta_{2^n}^i$ at the s-th step and the trajectory from $x(s) = \delta_{2^n}^i$ to $x = \delta_{2^n}^p$ at one step. Noting that $x(s) = \delta_{2^n}^i$ has 2^n choices, we know that $x = \delta_{2^n}^p$ is switching reachable from $x(0) = \delta_{2^n}^q$ at time $s+1$, if and only if:

$$\sum_{i=1}^{2^n} M_{pi} \left(M^s\right)_{iq} = \left(M^{s+1}\right)_{pq} > 0,$$

which means that 1) is true for $k = s+1$.

By induction, Conclusion 1) holds for any positive integer k.

Next, we prove Conclusion 2).

From Definition 2 and Conclusion 1), $x = \delta_{2^n}^p$ is switching reachable from $x(0) = \delta_{2^n}^q$, if and only if $\sum_{k=1}^{\infty} \left(M^k\right)_{pq} > 0$.

Since M is a $2^n \times 2^n$ real square matrix, by Cayley-Hamilton theorem, it is easy to see that $\left(M^k\right)_{pq} = 0,\ \forall\, k \leq 2^n$, if and only if $\left(M^k\right)_{pq} = 0,\ \forall\, k \in \mathbb{Z}_+$. Therefore, we only need to consider $k \leq 2^n$ and thus Conclusion 2) is true. ■

In the following section, using the switching point reachability, we investigate the pointwise stabilizability of SBNs.

DEFINITION 5.4

The system (5.1) is said to be pointwise stabilizable to $X_e = (x_1^e, \cdots, x_n^e) \in \mathscr{D}^n$, if for any initial state $X_0 = (x_1(0), \cdots, x_n(0)) \in \mathscr{D}^n$, there exists a switching signal σ_{X_0}, under which the trajectory initialized at X_0 converges to X_e.

Note that if the system (5.1) is pointwise stabilizable to X_e, then X_e must be a fixed point of some sub-network(s) for the system (5.1). In fact, if X_e is not a fixed point of any sub-network, then under any switching signal, the next state of the trajectory starting from X_e is not X_e, which is a contradiction to the fact that the system (5.1) is pointwise stabilizable to X_e. Thus, X_e is a fixed point of at least one sub-network. Based on this, we give the following assumption.

Assumption 5.1: X_e is a fixed point of the k-th sub-network of the SBN (5.1).

Using the vector form of logical variables, we set the initial state $x(0) = \ltimes_{i=1}^n x_i(0) \in \Delta_{2^n}$ and the equilibrium $x_e = \ltimes_{i=1}^n x_i^e = \delta_{2^n}^{\mu}$. By Theorem 5.1, $x_e = \delta_{2^n}^{\mu}$ is switching reachable from the initial state $x(0) = \delta_{2^n}^j$, $1 \leq j \leq 2^n$, if and only if $\mathscr{C}_{\mu,j} > 0$. Moreover, when $\mathscr{C}_{\mu,j} > 0$, there exists a switching signal $\sigma_{x(0)}$, which forces $x(0)$ to x_e at some time s. From Assumption 5.1, we set $\sigma_{x(0)}(t) = k$, $\forall\, t \geq s$. Then, under $\sigma_{x(0)}$, the trajectory of the system (5.1) starting from $x(0) = \delta_{2^n}^j$ converges to x_e. Hence, if $Row_{\mu}(\mathscr{C}) > 0$, then for any initial state $x(0) = \delta_{2^n}^j$, $1 \leq j \leq 2^n$, there exists a switching signal $\sigma_{x(0)}$ such that the trajectory initialized at $x(0)$ converges to x_e under $\sigma_{x(0)}$.

Based on the above analysis, we have the following result.

Theorem 5.2

Suppose that the Assumption 5.1 holds. Then, the system (5.1) is pointwise stabilizable to $x_e = \delta_{2^n}^{\mu}$, if and only if:

$$Row_{\mu}(\mathscr{C}) > 0, \tag{5.9}$$

where $\mathscr{C}$ is given in (5.8) ■

Proof. From the above analysis, it is easy to see that the sufficiency holds. We need only to prove the necessity.

Suppose that the system (5.1) is pointwise stabilizable to $x_e = \delta_{2^n}^{\mu}$, but (5.9) does not hold. Then, there exists an integer $1 \leq j \leq 2^n$ such that $\mathscr{C}_{\mu,j} = 0$. By Theorem 5.1, $x_e = \delta_{2^n}^{\mu}$ is not switching reachable from $x(0) = \delta_{2^n}^j$, and thus one cannot find a switching signal $\sigma_{x(0)}$ which forces $x(0)$ to x_e. This is a contradiction to the fact that the system (5.1) is pointwise stabilizable to x_e. Thus, (5.9) holds. ∎

Next, based on the switching point reachability, we study the global stability of the system (5.1) under arbitrary switching signal. To this end, we first give the following two lemmas.

Lemma 5.1

Let $M = \sum\limits_{i=1}^{w} L_i$. Then,

$$\sum_{i=1}^{2^n} \left(M^k\right)_{ij} = w^k, \ \forall\, j = 1,2,\cdots,2^n,\ k \in \mathbb{Z}_+, \tag{5.10}$$

where w is the number of sub-networks of the system (5.1). ∎

Proof. We prove it by induction.

When $k=1$, since $L_p \in \mathscr{L}_{2^n \times 2^n}$, $p = 1,\cdots,w$, we have $\sum\limits_{i=1}^{2^n} (L_p)_{ij} = 1$. Thus,

$$\sum_{i=1}^{2^n} M_{ij} = \sum_{i=1}^{2^n}\sum_{p=1}^{w} (L_p)_{ij} = \sum_{p=1}^{w}\sum_{i=1}^{2^n} (L_p)_{ij} = w,$$

which implies that (5.10) holds for $k=1$.

Assume that (5.10) is true for $k=s$. Now, we consider the case of $k=s+1$. In this case, one can obtain:

$$\begin{aligned}
\sum_{i=1}^{2^n} \left(M^{s+1}\right)_{ij} &= \sum_{i=1}^{2^n}\sum_{p=1}^{2^n} \left(M^s\right)_{ip} M_{pj} \\
&= \sum_{i=1}^{2^n} \left(M^s\right)_{ip} \sum_{p=1}^{2^n} M_{pj} \\
&= w^s \sum_{p=1}^{2^n} M_{pj} = w^{s+1}.
\end{aligned}$$

Hence, (5.10) holds for $k = s+1$.

By induction, the conclusion holds for any $k \in \mathbb{Z}_+$. ∎

Lemma 5.2

Assume that the system (5.1) is globally stable at $x_e = \delta_{2^n}^{i^*}$, under arbitrary switching signal. Then, the trajectory starting from any initial state reaches x_e within time 2^n under any switching signal. ∎

Proof. If the conclusion is not true, then, one can find an initial state $x(0)$, an integer $T > 2^n$ and a switching signal $\sigma(t)$, such that under $\sigma(t)$, we have $x(t) \neq x_e$, $\forall\, 0 \leq t \leq T-1$, and $x(T) = x_e$. Since the number of different states for the system (5.1) is 2^n, there must exist two integers $0 \leq t_1 < t_2 \leq T-1$ such that $x(t_1) = x(t_2) \neq x_e$.

Now, for the initial state $\widehat{x}(0) = x(t_1)$, we set $\widehat{\sigma}(t) = \sigma(t_1+t)$, $t = 0, 1, \cdots, t_2 - t_1 - 1$. Then, under the switching signal $\widehat{\sigma}(t)$, one can see that $\widehat{x}(t_2 - t_1) = x(t_2) = x(t_1)$. Generally, for $k \in \mathbb{N}$, we construct the following switching signal:

$$\widehat{\sigma}(t) = \begin{cases} \sigma(t_1), & t = k(t_2 - t_1), \\ \sigma(t_1+1), & t = k(t_2-t_1)+1, \\ \vdots & \\ \sigma(t_2-1), & t = (k+1)(t_2-t_1)-1, \end{cases} \tag{5.11}$$

which is periodic. Then, under the switching signal $\widehat{\sigma}(t)$, the trajectory of the system (5.1) starting from the initial state $\widehat{x}(0) = x(t_1)$ forms a cycle $\{x(t_1), x(t_1+1), \cdots, x(t_2-1); x(t_1), x(t_1+1), \cdots, x(t_2-1); \cdots\}$, which is a contradiction to the global stability of the system under arbitrary switching signal. The conclusion follows. ∎

Based on Theorem 5.1 and the above two lemmas, we give the following result for the global stability of SBNs under arbitrary switching signal.

Theorem 5.3

The system (5.1) is globally stable at $x_e = \delta_{2^n}^{i^*}$ under arbitrary switching signal, if and only if there exists a positive integer $k^* \leq 2^n$ such that:

$$Row_{i^*}\left(M^{k^*}\right) = [\underbrace{w^{k^*} \cdots w^{k^*}}_{2^n}], \tag{5.12}$$

where $M = \sum_{i=1}^{w} L_i$, and w is the number of sub-networks for the system (5.1) ∎

Proof. (Necessity) Suppose that the system (5.1) is globally stable at $x_e = \delta_{2^n}^{i^*}$ under arbitrary switching signal. Then, Lemma 5.2 implies that under any switching signal $\sigma(t)$, the trajectory of the system (5.1) starting from any initial state $x_0 = \delta_{2^n}^{j}$ reaches $x_e = \delta_{2^n}^{i^*}$ in the time $k(x(0), \sigma(t)) \leq 2^n$, and then stays at x_e forever.

Set $k(x(0)) = \max_{\sigma(t)}\{k(x(0), \sigma(t))\} \leq 2^n$. Then, by Theorem 5.1 and Lemma 5.1, it is easy to see that for any integer $k \geq k(x(0))$, $\left(M^k\right)_{ij} = 0$, $\forall\ i \neq i^*$, and $\left(M^k\right)_{i^*j} = w^k$.

Letting $k^* = \max_{x(0)\in\Delta_{2^n}} \{k(x(0))\} \leq 2^n$, then from the above analysis one can obtain $\left(M^{k^*}\right)_{ij} = 0,\ \forall\, i \neq i^*,\ j = 1,\cdots,2^n$, and $\left(M^{k^*}\right)_{i^* j} = w^{k^*},\ j = 1,\cdots,2^n$, which imply that (5.12) holds.

(Sufficiency) Suppose that (5.12) holds. We need to prove that the system is globally stable at x_e under arbitrary switching signal.

In fact, from (5.12) it is easy to see that,

$$\left(M^{k^*}\right)_{ij} = 0,\ \forall\, i \neq i^*,\ j = 1,\cdots,2^n,$$

and,

$$\left(M^{k^*}\right)_{i^* j} = w^{k^*},\ j = 1,\cdots,2^n.$$

Thus, for any $k \geq k^*$ we have:

$$\left(M^{k}\right)_{ij} = 0,\ \forall\, i \neq i^*,\ j = 1,\cdots,2^n,$$

and,

$$\left(M^{k}\right)_{ij} = w^{k},\ j = 1,\cdots,2^n.$$

Hence, by Theorem 5.1 and Proposition 5.1, we know that under any switching signal $\sigma(t)$, the trajectory of the system starting from any initial state $x(0) = \delta_{2^n}^j$ will reach $x_e = \delta_{2^n}^{i^*}$ in the time $k^* \leq 2^n$, and then stays at x_e forever. Therefore, the system (5.1) is globally stable at x_e under arbitrary switching signal. ∎

Using the ASSR of SBNs, Theorem 5.3 provides a necessary and sufficient condition for the global stability of SBNs under arbitrary switching signal. Here, the matrix M^{k^*}, constructed from the algebraic form, plays a key role in establishing the condition. Note that one only needs to calculate some row of M^{k^*} to check the stability of SBNs under arbitrary switching signal. Thus, the condition is easily verified by using the MATLAB toolbox.

Example 5.2:

Recall Example 5.1. In the apoptosis network, $\{x_1 = 1,\ x_2 = 0\}$ stands for the cell survival [6]. We are interested in the following two problems.

Problem 1: Can the system (5.3) starting from the initial state $(0,0,0)$ achieve the cell survival state $X_e = (1,0,1)$ by a sequence of stimulus (a designed switching path)?

Problem 2: Is the system (5.3) pointwise stabilizable to the cell survival state X_e?

Using the vector form of logical variables and setting $x(t) = \ltimes_{i=1}^3 x_i(t)$, we have the following algebraic form for the system (5.3):

$$x(t+1) = L_{\sigma(t)} x(t), \tag{5.13}$$

where $L_1 = \delta_8[7\ 7\ 3\ 3\ 5\ 7\ 1\ 3]$, and $L_2 = \delta_8[7\ 7\ 8\ 8\ 5\ 7\ 6\ 8]$. Moreover, $X_e \sim x_e = \delta_8^3$, and $(0,0,0) \sim \delta_8^8$.

A simple calculation gives:

$$\mathscr{C} = \sum_{k=1}^{8} (L_1 + L_2)^k =$$
$$\begin{bmatrix} 170 & 170 & 0 & 0 & 0 & 170 & 85 & 0 \\ 0 & 0 & 0 & 0 & 0 & 0 & 0 & 0 \\ 0 & 0 & 255 & 255 & 0 & 0 & 0 & 255 \\ 0 & 0 & 0 & 0 & 0 & 0 & 0 & 0 \\ 0 & 0 & 0 & 0 & 510 & 0 & 0 & 0 \\ 170 & 170 & 0 & 0 & 0 & 170 & 85 & 0 \\ 170 & 170 & 0 & 0 & 0 & 170 & 340 & 0 \\ 0 & 0 & 255 & 255 & 0 & 0 & 0 & 255 \end{bmatrix}.$$

For Problem 1, since $\mathscr{C}_{3,8} = 255 > 0$, by Theorem 5.1, we conclude that the system (5.3) starting from the initial state $(0,0,0)$ can achieve the cell survival state $X_e = (1,0,1)$ by a sequence of stimulus.

For Problem 2, since:

$$Row_3(\mathscr{C}) = [0\ 0\ 255\ 255\ 0\ 0\ 0\ 255],$$

it is easy to see from Theorem 5.2 that the system (5.3) is not pointwise stabilizable to the cell survival state X_e.

Example 5.3:

Consider the following switched Boolean network:

$$\begin{cases} x_1(t+1) &= f_1^{\sigma(t)}(x_1(t), x_2(t)), \\ x_2(t+1) &= f_2^{\sigma(t)}(x_1(t), x_2(t)), \end{cases} \tag{5.14}$$

where $\sigma : \mathbb{N} \mapsto \mathscr{A} = \{1,2,3\}$ is the switching signal, and,

$$f_1^1(x_1,x_2) = \neg x_1 \wedge x_2,\ f_2^1(x_1,x_2) = 0,$$
$$f_1^2(x_1,x_2) = x_1 \bar{\vee} x_2,\ f_2^2(x_1,x_2) = \neg(x_1 \rightarrow x_2),$$
$$f_1^3(x_1,x_2) = x_1 \bar{\vee} x_2,\ f_2^3(x_1,x_2) = x_1 \bar{\vee} x_2.$$

Our objective is to check whether or not the system (5.14) is globally stable at $X_e = (0,0)$ under arbitrary switching signal.

Using the vector form of logical variables and setting $x(t) = \ltimes_{i=1}^{2} x_i(t)$, the system (5.14) can be expressed as:

$$x(t+1) = L_{\sigma(t)} x(t), \tag{5.15}$$

where $L_1 = \delta_4[4\ 4\ 2\ 4]$, $L_2 = \delta_4[4\ 1\ 2\ 4]$, $L_3 = \delta_4[4\ 1\ 1\ 4]$. In addition, $X_e \sim x_e = \delta_4^4$.

A simple calculation shows that:

$$M = L_1 + L_2 + L_3 = \begin{bmatrix} 0 & 2 & 1 & 0 \\ 0 & 0 & 2 & 0 \\ 0 & 0 & 0 & 0 \\ 3 & 1 & 0 & 3 \end{bmatrix},$$

$$M^2 = \begin{bmatrix} 0 & 0 & 4 & 0 \\ 0 & 0 & 0 & 0 \\ 0 & 0 & 0 & 0 \\ 9 & 9 & 5 & 9 \end{bmatrix},$$

$$M^3 = \begin{bmatrix} 0 & 0 & 0 & 0 \\ 0 & 0 & 0 & 0 \\ 0 & 0 & 0 & 0 \\ 27 & 27 & 27 & 27 \end{bmatrix}.$$

Thus, $Row_4\left(M^3\right) = [27\ \ 27\ \ 27\ \ 27]$. By Theorem 5.3, the system (5.14) is globally stable at $X_e = (0,0)$ under arbitrary switching signal.

5.3 CONTROLLABILITY ANALYSIS AND CONTROL DESIGN WITH STATE AND INPUT CONSTRAINTS

This section investigates the controllability analysis and the control design for switched Boolean networks (SBNs) with state and input constraints.

A SBN with state and input constraints can be described as:

$$\begin{cases} x_1(t+1) = f_1^{\sigma(t)}(x_1(t),\cdots,x_n(t),u_1(t),\cdots,u_m(t)), \\ x_2(t+1) = f_2^{\sigma(t)}(x_1(t),\cdots,x_n(t),u_1(t),\cdots,u_m(t)), \\ \vdots \\ x_n(t+1) = f_n^{\sigma(t)}(x_1(t),\cdots,x_n(t),u_1(t),\cdots,u_m(t)), \end{cases} \tag{5.16}$$

where $\sigma : \mathbb{N} \mapsto \mathscr{W} = \{1,2,\cdots,w\}$ is the switching signal, $X(t) := (x_1(t),\cdots,x_n(t)) \in C_x \subseteq \mathscr{D}^n$ is the logical state, $U(t) := (u_1(t),\cdots,u_m(t)) \in C_u \subseteq \mathscr{D}^m$ is the logical input, C_x $(1 \leq |C_x| \leq 2^n)$ is the state's constraint set, C_u $(1 \leq |C_u| \leq 2^m)$ is the input's constraint set. $f_i^j : C_x \times C_u \to \mathscr{D}$, $i = 1,\cdots,n$, $j = 1,\cdots,w$ are logical functions, $\mathbb{N}$ is the set of nonnegative integers, and $|C_x|$ and $|C_u|$ stand for the cardinalities of the sets C_x and C_u, respectively.

Given a positive integer l, denote by $\bar{l} := \{0,\cdots,l-1\}$. We allow $l = \infty$, which is defined as $\bar{\infty} := \mathbb{N}$. For a switching signal $\sigma : \bar{l} \mapsto W$, set $\sigma(k) = i_k$, $k \in \bar{l}$. Then, we obtain a switching sequence as:

$$\pi := \{(0,\ i_0),\ (1,\ i_1),\ \cdots,\ (l,\ i_l)\}. \tag{5.17}$$

Next, by the semi-tensor product of matrices, we express the system (5.16) into an algebraic form.

Using the vector form of logical variables and setting $x = \ltimes_{i=1}^n x_i$ and $u = \ltimes_{i=1}^m u_i$, the system (5.16) can be expressed as:

$$\begin{cases} x_1(t+1) &= Q_1^{\sigma(t)} u(t)x(t), \\ x_2(t+1) &= Q_2^{\sigma(t)} u(t)x(t), \\ &\vdots \\ x_n(t+1) &= Q_n^{\sigma(t)} u(t)x(t), \end{cases} \tag{5.18}$$

where $Q_i^{\sigma(t)} \in \mathcal{L}_{2\times 2^{m+n}}$. Multiplying the equations in (5.18) together yields the following algebraic form:

$$x(t+1) = L_{\sigma(t)}u(t)x(t), \tag{5.19}$$

where $Col_i(L_{\sigma(t)}) = \ltimes_{j=1}^{n} Col_i\left(Q_j^{\sigma(t)}\right)$, $i = 1,\cdots,2^{m+n}$, $L_{\sigma(t)} \in \mathcal{L}_{2^n\times 2^{m+n}}$.

Let $|C_x| = \alpha$ and $|C_u| = \beta$, then the state's constraint set and the input's constraint set can be expressed as $C_x = \{\delta_{2^n}^{i_k} : k = 1,\cdots,\alpha;\ i_1 < \cdots < i_\alpha\}$ and $C_u = \{\delta_{2^m}^{j_k} : k = 1,\cdots,\beta;\ j_1 < \cdots < j_\beta\}$, respectively. Now, we give an example to show how to convert a constrained SBN into its algebraic form.

Example 5.4: Algebraic form of constrained SBCN

Convert the following constrained SBN into its algebraic form:

$$\begin{cases} x_1(t+1) &= f_1^{\sigma(t)}(x_1(t),x_2(t),u(t)), \\ x_2(t+1) &= f_2^{\sigma(t)}(x_1(t),x_2(t),u(t)), \end{cases} \tag{5.20}$$

where $\sigma : \mathbb{N} \mapsto \mathcal{W} = \{1,2\}$ is the switching signal, $C_x = \{(1,1),(0,1),(0,0)\}$, $C_u = \{0\}$,

$$f_1^1 = x_1 \wedge x_2 \wedge u,\ f_2^1 = x_1 \to u,$$
$$f_1^2 = x_2 \leftrightarrow u,\ f_2^2 = x_1 \vee x_2 \vee u.$$

Using the vector form of logical variables and setting $x = x_1 \ltimes x_2$, we have:

$$\begin{cases} x_1(t+1) &= Q_1^{\sigma(t)}u(t)x(t), \\ x_2(t+1) &= Q_2^{\sigma(t)}u(t)x(t), \end{cases} \tag{5.21}$$

where $Q_1^1 = M_c^2 = \delta_2[1\ 2\ 2\ 2\ 2\ 2\ 2\ 2]$, $Q_2^1 = M_i E_d W_{[2,2]} W_{[2,4]} = \delta_2[1\ 1\ 1\ 1\ 2\ 2\ 1\ 1]$, $Q_1^2 = M_e E_d W_{[2,2]} = \delta_2[1\ 2\ 1\ 2\ 2\ 1\ 2\ 1]$, and $Q_2^2 = M_d^2 = \delta_2[1\ 1\ 1\ 1\ 1\ 1\ 1\ 2]$,

Multiplying the equations in (5.21) together yields:

$$x(t+1) = L_{\sigma(t)}u(t)x(t), \tag{5.22}$$

where $L_1 = \delta_4[1\ 3\ 3\ 3\ 4\ 4\ 3\ 3]$, $L_2 = \delta_4[1\ 3\ 1\ 3\ 3\ 1\ 3\ 2]$. Moreover $C_x = \{\delta_4^1, \delta_4^3, \delta_4^4\}$, $C_u = \{\delta_2^2\}$.

In the following section, we propose the so-called constrained incidence matrix for SBNs with state and input constraints, which is crucial to the analysis and control of the systems.

To begin with, we give a simple example to show the structure of the constrained incidence matrix.

Example 5.5: Constrained incidence matrix

Recall the system (5.20) with its algebraic form (5.22).

According to (5.22), we can draw the state transfer graph of the system (5.20), shown in Figure 5.2, where the solid line denotes the state transfer of the system (5.20) with $\sigma = 1$ and $u = \delta_2^2$, and the dotted line $\sigma = 2$ and $u = \delta_2^2$.

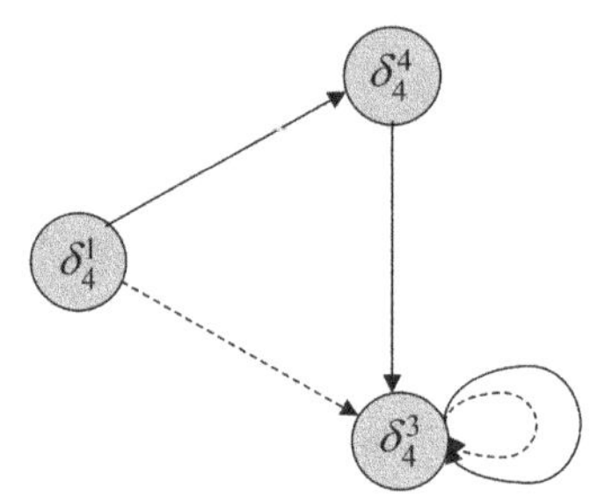

Figure 5.2 The state transfer graph of Example 5.5

Identifying the switching signal $\sigma = i \sim \delta_2^i$, $i \in \{1,2\}$ and using the vector form of logical variables, we set all the points formed by all possible switching signals, input, and state variables in the following order:

$$\begin{aligned} &P_1 = \delta_2^1 \times \delta_2^2 \times \delta_4^1,\ P_2 = \delta_2^1 \times \delta_2^2 \times \delta_4^3,\ P_3 = \delta_2^1 \times \delta_2^2 \times \delta_4^4, \\ &P_4 = \delta_2^2 \times \delta_2^2 \times \delta_4^1,\ P_5 = \delta_2^2 \times \delta_2^2 \times \delta_4^3,\ P_6 = \delta_2^2 \times \delta_2^2 \times \delta_4^4. \end{aligned} \tag{5.23}$$

$P_i = \delta_2^{i_1} \times \delta_2^{i_2} \times \delta_4^{i_3}$ is said to be reachable from $P_j = \delta_2^{j_1} \times \delta_2^{j_2} \times \delta_4^{j_3}$ in one step, denoted by $P_j \to P_i$, if $\delta_4^{i_3} = L_{j_1} \ltimes \delta_2^{j_2} \ltimes \delta_4^{j_3}$. P_i is said to be reachable from P_j with l steps, if there exist $l-1$ points P_{k_v}, $v = 1,\cdots,l-1$, such that $P_j \to P_{k_1} \to \cdots \to P_{k_{l-1}} \to P_i$. In this case, $P_j \to P_{k_1} \to \cdots \to P_{k_{l-1}} \to P_i$ is called a path from P_j to P_i.

Now, we construct a 6×6 matrix, $\mathscr{I} = \{\mathscr{I}_{ij}\}$, called the constrained incidence matrix of the system (5.20), in the following way:

$$\mathscr{I}_{ij} = \begin{cases} 1, & \text{if } P_j \to P_i, \\ 0, & \text{otherwise.} \end{cases} \tag{5.24}$$

Then, it is easy to obtain from Figure 5.2 that:

$$\mathscr{I} = \begin{bmatrix} 0 & 0 & 0 & \vdots & 0 & 0 & 0 \\ 0 & 1 & 1 & \vdots & 1 & 1 & 0 \\ 1 & 0 & 0 & \vdots & 0 & 0 & 0 \\ \cdots & \cdots & \cdots & \cdots & \cdots & \cdots & \cdots \\ 0 & 0 & 0 & \vdots & 0 & 0 & 0 \\ 0 & 1 & 1 & \vdots & 1 & 1 & 0 \\ 1 & 0 & 0 & \vdots & 0 & 0 & 0 \end{bmatrix}. \tag{5.25}$$

Obviously, $\mathscr{I}$ is a row-periodic matrix with period 3, and its basic block is:

$$\mathscr{I}_0 = \begin{bmatrix} 0 & 0 & 0 & \vdots & 0 & 0 & 0 \\ 0 & 1 & 1 & \vdots & 1 & 1 & 0 \\ 1 & 0 & 0 & \vdots & 0 & 0 & 0 \end{bmatrix}. \tag{5.26}$$

Next, we investigate the relation between $\mathscr{I}$ and L_i, $i = 1,2$.
Let $L_1 = [Blk_1(L_1)\ \ Blk_2(L_1)]$ and $L_2 = [Blk_1(L_2)\ \ Blk_2(L_2)]$, where:

$$Blk_1(L_1) = \begin{bmatrix} 1 & 0 & 0 & 0 \\ 0 & 0 & 0 & 0 \\ 0 & 1 & 1 & 1 \\ 0 & 0 & 0 & 0 \end{bmatrix}, \ Blk_2(L_1) = \begin{bmatrix} 0 & 0 & 0 & 0 \\ 0 & 0 & 0 & 0 \\ 0 & 0 & 1 & 1 \\ 1 & 1 & 0 & 0 \end{bmatrix},$$

$$Blk_1(L_2) = \begin{bmatrix} 1 & 0 & 1 & 0 \\ 0 & 0 & 0 & 0 \\ 0 & 1 & 0 & 1 \\ 0 & 0 & 0 & 0 \end{bmatrix}, \ Blk_2(L_2) = \begin{bmatrix} 0 & 1 & 0 & 0 \\ 0 & 0 & 0 & 1 \\ 1 & 0 & 1 & 0 \\ 0 & 0 & 0 & 0 \end{bmatrix},$$

then we can obtain the principal submatrices of $Blk_2(L_1)$ and $Blk_2(L_2)$ with the row and column indices $\{1,3,4\}$, respectively, as follows:

$$\widehat{Blk_2}(L_1) = \begin{bmatrix} 0 & 0 & 0 \\ 0 & 1 & 1 \\ 1 & 0 & 0 \end{bmatrix}, \ \widehat{Blk_2}(L_2) = \begin{bmatrix} 0 & 0 & 0 \\ 1 & 1 & 0 \\ 0 & 0 & 0 \end{bmatrix}. \tag{5.27}$$

Comparing (5.26) and (5.27), we find that:

$$\mathscr{I}_0 = [\widehat{Blk_2}(L_1)\ \ \widehat{Blk_2}(L_2)], \tag{5.28}$$

$$\mathscr{I} = \begin{bmatrix} \widehat{Blk_2}(L_1) & \widehat{Blk_2}(L_2) \\ \widehat{Blk_2}(L_1) & \widehat{Blk_2}(L_2) \end{bmatrix}. \tag{5.29}$$

One cannot help to ask why such a coincidence occurs. In fact, for any time $t \geq 0$, from (5.19) one can see that the kth block of L_i, $Blk_k(L_i)$, corresponds to $\sigma(t) = i$ and $u(t) = \delta_{2^m}^k$, and the jth column and the rth row of $Blk_k(L_i)$ correspond to $x(t) = \delta_{2^n}^j$ and $x(t+1) = \delta_{2^n}^r$, respectively. For the system (5.20), since $u(t) \in \{\delta_2^2\}$, one just studies $Blk_2(L_i)$, $i = 1,2$. In addition, since $x(t)$, $x(t+1) \in \{\delta_4^1, \delta_4^3, \delta_4^4\}$, we only need to consider the entries at (i,j), $i,j = 1,3,4$, in $Blk_2(L_1)$ and $Blk_2(L_2)$. Because $\sigma(t+1) \in \{\delta_2^1, \delta_2^2\}$ and $u(t+1) \in C_u$ are arbitrary, it follows that the constrained incidence matrix of the system (5.20) is (5.29), where the first block row corresponds to $\sigma(t+1) = \delta_2^1$ and $u(t+1) = \delta_2^2$, and the second block row corresponds to $\sigma(t+1) = \delta_2^2$ and $u(t+1) = \delta_2^2$.

Example 5.5, leads us to the definition of the constrained incidence matrix for the system (5.16).

Identifying the switching signal $\sigma = i \sim \delta_w^i \in \Delta_w,\ i \in \mathscr{W}$, and using the vector form of logical variables, we arrange all the $w\alpha\beta$ points formed by all possible switching signals, input, and state variables in the following order:

$$
\begin{aligned}
&P_1 = \delta_w^1 \times \delta_{2^m}^{j_1} \times \delta_{2^n}^{i_1}, \cdots, P_\alpha = \delta_w^1 \times \delta_{2^m}^{j_1} \times \delta_{2^n}^{i_\alpha}, \cdots, \\
&P_{\alpha\beta} = \delta_w^1 \times \delta_{2^m}^{j_\beta} \times \delta_{2^n}^{i_\alpha}, \cdots, P_{w\alpha\beta} = \delta_w^w \times \delta_{2^m}^{j_\beta} \times \delta_{2^n}^{i_\alpha},
\end{aligned} \tag{5.30}
$$

where the points are arranged according to the ordered multi-index $Id(k,q,r;w,\beta,\alpha)$. This ordered multi-index lets k, q, and r run from 1 to w, β, and α, respectively, and r runs first, q second, and k last.

DEFINITION 5.5

A $w\alpha\beta \times w\alpha\beta$ matrix, $\mathscr{I} = \{\mathscr{I}_{ij}\}$ is called the constrained incidence matrix of the system (5.16), if:

$$
\mathscr{I}_{ij} = \begin{cases} 1, \text{ if } P_j \to P_i, \\ 0, \text{ otherwise,} \end{cases} \tag{5.31}
$$

where P_i and P_j are give in (5.30), and $P_j \to P_i$ means that P_i is reachable from P_j in one step.

To give the structure of $\mathscr{I}$, we define the following block selection matrices:

$$
J_i^{(p,q)} := \underbrace{\begin{bmatrix} 0_{q\times q} & \cdots & 0_{q\times q} & \underbrace{I_q}_{i} & 0_{q\times q} & \cdots & 0_{q\times q} \end{bmatrix}}_{p} \in \mathbb{R}^{q\times pq}, \quad i = 1,2,\cdots,p, \tag{5.32}
$$

where $0_{q\times q}$ denotes the $q \times q$ zero matrix. The following proposition is obvious.

PROPOSITION 5.1

1). Given a matrix $A \in \mathbb{R}^{pq\times r}$, split A as:

$$
A = \begin{bmatrix} A_1 \\ \vdots \\ A_p \end{bmatrix},
$$

where $A_i \in \mathbb{R}^{q\times r}$. Then,

$$
J_i^{(p,q)} A = A_i. \tag{5.33}
$$

2). Given a matrix $B \in \mathbb{R}^{r\times pq}$, split B as:

$$B = \begin{bmatrix} B_1 & \cdots & B_p \end{bmatrix},$$

where $B_i \in \mathbb{R}^{r\times q}$. Then,

$$B\left(J_i^{(p,q)}\right)^T = B_i. \tag{5.34}$$

Consider the system (5.16). For any time $t \geq 0$, from (5.19) one can see that the kth block of L_i, $Blk_k(L_i)$, corresponds to $\sigma(t) = \delta_w^i$ and $u(t) = \delta_{2^m}^k$, and the jth column and the rth row of $Blk_k(L_i)$ correspond to $x(t) = \delta_{2^n}^j$ and $x(t+1) = \delta_{2^n}^r$, respectively. Since $x(t), x(t+1) \in C_x$, $u(t) \in C_u$, $\forall\, t \in \mathbb{N}$, letting $L_i := [Blk_1(L_i) \ \cdots \ Blk_{2^m}(L_i)]$ and using the block selection matrices, we can obtain a principal submatrix for each $Blk_s(L_i)$ with the row and column indexes $\{i_1, \ \cdots, \ i_\alpha\}$, and denote it by $\widehat{Blk}_s(L_i)$, where $s = 1, 2, \cdots, 2^m$, $i = 1, 2, \cdots, w$. Then, we select the blocks with indices $j_1, \ \cdots, \ j_\beta$ in $\left\{\widehat{Blk}_s(L_i) : \ s = 1, 2, \cdots, 2^m\right\}$. Using the block selection matrices, we thus obtain a new matrix for each i as follows:

$$\widehat{L}_i = \left[\widehat{Blk}_1(L_i) \ \cdots \ \widehat{Blk}_{2^m}(L_i)\right]\left[\left(J_{j_1}^{(2^m,\alpha)}\right)^T \ \cdots \ \left(J_{j_\beta}^{(2^m,\alpha)}\right)^T\right] \in \mathscr{B}_{\alpha\times\alpha\beta}, \tag{5.35}$$

where,

$$\widehat{Blk}_s(L_i) = \begin{bmatrix} J_{i_1}^{(2^n,1)} \\ \vdots \\ J_{i_\alpha}^{(2^n,1)} \end{bmatrix} Blk_s(L_i)\left[\left(J_{i_1}^{(2^n,1)}\right)^T \ \cdots \ \left(J_{i_\alpha}^{(2^n,1)}\right)^T\right] \in \mathscr{B}_{\alpha\times\alpha}. \tag{5.36}$$

Because $\sigma(t+1) \in \{\delta_w^1, \cdots, \delta_w^w\}$ and $u(t+1) \in C_u$ are arbitrary, with the order (5.30), we obtain the constrained incidence matrix as follows.

PROPOSITION 5.2

The constrained incidence matrix of the system (5.16) can be given as:

$$\mathscr{I} = \left.\begin{bmatrix} \widehat{L} \\ \widehat{L} \\ \vdots \\ \widehat{L} \end{bmatrix}\right\} w\beta \in \mathscr{B}_{w\alpha\beta\times w\alpha\beta}, \tag{5.37}$$

where, $\widehat{L} = [\widehat{L}_1 \ \cdots \ \widehat{L}_w] \in \mathscr{B}_{\alpha\times w\alpha\beta}$.

We have the following property of $\mathscr{I}$.

PROPOSITION 5.3

Consider the system (5.16) with its constrained incidence matrix (5.37). Then $\mathscr{I}$ is a row-periodic matrix with period α, and the basic block of $\mathscr{I}$ is $\mathscr{I}_0 = \widehat{L}$. In addition, $\mathscr{I}^l$, $l \in \mathbb{Z}_+$ is also a row-periodic matrix with period α, and the basic block of $\mathscr{I}^l$ is:

$$\mathscr{I}_0^l = \widehat{M}^{l-1}\widehat{L}, \tag{5.38}$$

where, $\widehat{M} = \sum_{i=1}^{w\beta} Blk_i(\widehat{L})$.

Proof. From (5.37), one can easily see that $\mathscr{I} \in \mathscr{B}_{w\alpha\beta \times w\alpha\beta}$ is a row-periodic matrix with period α. Thus, $\mathscr{I} = \mathbf{1}_{w\beta}\widehat{L}$, where $\mathbf{1}_{w\beta}$ is a $w\beta$-dimension column vector with all entries being 1. In addition, $\mathscr{I}^l$, $l \in \mathbb{Z}_+$ is a row-periodic matrix with period α.

Now, let us prove that (5.38) holds.

In fact,

$$\mathscr{I}_0^l = \widehat{L}\mathscr{I}^{l-1} = \widehat{L}\mathbf{1}_{w\beta}^T\mathscr{I}_0^{l-1} = \sum_{i=1}^{w\beta} Blk_i(\widehat{L})\mathscr{I}_0^{l-1} = \widehat{M}\mathscr{I}_0^{l-1} = \cdots = \widehat{M}^{l-1}\widehat{L},$$

which implies that (5.38) is true. ∎

Now, we give an example to show how to compute the constrained incidence matrix.

Example 5.6:

Recall the system (5.20) and use Proposition 5.2 to compute its constrained incidence matrix.

By Proposition 5.2, we need to calculate $\widehat{L}_1$ and $\widehat{L}_2$.

From (5.35) and (5.36), it is easy to obtain:

$$\begin{aligned}
\widehat{Blk}_1(L_1) &= \begin{bmatrix} J_1^{(4,1)} \\ J_3^{(4,1)} \\ J_4^{(4,1)} \end{bmatrix} Blk_1(L_1)\left[\left(J_1^{(4,1)}\right)^T \; \left(J_3^{(4,1)}\right)^T \; \left(J_4^{(4,1)}\right)^T\right] \\
&= \begin{bmatrix} 1 & 0 & 0 & 0 \\ 0 & 0 & 1 & 0 \\ 0 & 0 & 0 & 1 \end{bmatrix} \begin{bmatrix} 1 & 0 & 0 & 0 \\ 0 & 0 & 0 & 0 \\ 0 & 1 & 1 & 1 \\ 0 & 0 & 0 & 0 \end{bmatrix} \begin{bmatrix} 1 & 0 & 0 \\ 0 & 0 & 0 \\ 0 & 1 & 0 \\ 0 & 0 & 1 \end{bmatrix} \\
&= \begin{bmatrix} 1 & 0 & 0 \\ 0 & 1 & 1 \\ 0 & 0 & 0 \end{bmatrix},
\end{aligned}$$

$$\widehat{Blk}_2(L_1) = \begin{bmatrix} J_1^{(4,1)} \\ J_3^{(4,1)} \\ J_4^{(4,1)} \end{bmatrix} Blk_2(L_1)\left[\left(J_1^{(4,1)}\right)^T \left(J_3^{(4,1)}\right)^T \left(J_4^{(4,1)}\right)^T\right]$$
$$= \begin{bmatrix} 0 & 0 & 0 \\ 0 & 1 & 1 \\ 1 & 0 & 0 \end{bmatrix}.$$

Thus,

$$\widehat{L}_1 = \left[\widehat{Blk}_1(L_1)\ \widehat{Blk}_2(L_1)\right]\left(J_2^{(2,3)}\right)^T = \begin{bmatrix} 0 & 0 & 0 \\ 0 & 1 & 1 \\ 1 & 0 & 0 \end{bmatrix}.$$

Similarly, one can obtain:

$$\widehat{L}_2 = \begin{bmatrix} 0 & 0 & 0 \\ 1 & 1 & 0 \\ 0 & 0 & 0 \end{bmatrix}.$$

By Proposition 5.2, the constrained incidence matrix of the system (5.20) is given as:

$$\mathscr{I} = \begin{bmatrix} \widehat{L}_1 & \widehat{L}_2 \\ \widehat{L}_1 & \widehat{L}_2 \end{bmatrix},$$

which accords with (5.25).

Next, we study the physical meaning of the constrained incidence matrix. By the definition of $\mathscr{I}$, it is easy to see that $\mathscr{I}_{ij} > 0$ means that P_i is reachable from P_j in one step. Similarly, for the physical meaning of $(\mathscr{I}^l)_{ij} > 0$, $l \in \mathbb{Z}_+$, we have the following result.

Theorem 5.4

Consider the system (5.16). Assume that $(\mathscr{I}^l)_{ij} = c$, $l \in \mathbb{Z}_+$. Then, there are c paths such that P_i is reachable from P_j with l steps, where P_i and P_j are given in (5.30). ■

Proof. We prove it by induction. When $l = 1$, the conclusion follows from the definition of $\mathscr{I}$.

Assume that the conclusion holds for $l = k$, that is $(\mathscr{I}^k)_{ij}$ is the number of paths from P_j to P_i with k steps. Now, we consider the case of $l = k+1$. In this case, a path from P_j to P_i with $k+1$ steps can be decomposed into a path from P_j to P_v with k steps and a path from P_v to P_i. Noting that P_v has $w\alpha\beta$ choices, we know that the number of paths from P_j to P_i with $k+1$ steps is given as:

$$\sum_{v=1}^{w\alpha\beta} \mathscr{I}_{iv}(\mathscr{I}^k)_{vj} = \left(\mathscr{I}\mathscr{I}^k\right)_{ij} = \left(\mathscr{I}^{k+1}\right)_{ij} = c,$$

which means that the conclusion is true for $l = k+1$.

By induction, the conclusion holds for any $l \in \mathbb{Z}_+$. ∎

COROLLARY 5.1

P_i is reachable from P_j with l steps if and only if $(\mathscr{I}^l)_{ij} > 0$.

Theorem 5.4 and Corollary 5.1 imply that $\{\mathscr{I}^l : \forall\, l \in \mathbb{Z}_+\}$ contains the entire reachability information of the system (5.16) with state and input constraints. By the Cayley C Hamilton theorem, we only need to consider $\mathscr{I}_0^l,\ \forall\, l \leq w\alpha\beta$.

Next, based on the constrained incidence matrix, we investigate the controllability of the system (5.16) and present some necessary and sufficient conditions for the controllability of the system.

We first give the definition of the controllability for switched Boolean control networks with state and input constraints.

DEFINITION 5.6

Consider the system (5.16). Let $X_0 = (x_1(0), \cdots, x_n(0)) \in C_x$.

1. $X = (x_1, \cdots, x_n) \in C_x$ is said to be reachable from X_0 at time l if one can find a switching signal $\sigma : \{0, \cdots, l-1\} \mapsto \mathscr{A}$ and a sequence of controls $\{U(t) \in C_u : t = 0, \cdots, l-1\}$, under which the trajectory of the system starting from X_0 reaches X at time l, and $X(t) \in C_x,\ \forall\, t = 0, \cdots, l$. The reachable set of X_0 at time l is denoted by $R_l^c(X_0)$, and the reachable set of X_0 is given as: $R^c(X_0) = \bigcup\limits_{l=1}^{\infty} R_l^c(X_0)$.

2. The system (5.16) is said to be controllable at X_0 if $R^c(X_0) = C_x$. The system (5.16) is said to be controllable, if the system is controllable at any $X_0 \in C_x$.

For ease of expression, we now give a new definition. Let $A = (a_{ij}) \in \mathbb{R}^{n \times m}$ be a matrix; then we call $A > 0$ if and only if $a_{ij} > 0$ for any i and j.

Based on Theorem 5.4, and using the above definition, we have the following result for the controllability of the system (5.16).

Theorem 5.5

Consider the system (5.16). Suppose that $\delta_{2^n}^{i_p}, \delta_{2^n}^{i_q} \in C_x$. Then,

1). $x(l) = \delta_{2^n}^{i_p}$ is reachable from $x(0) = \delta_{2^n}^{i_q}$ at time l if and only if:

$$\sum_{i=1}^{w\beta} \left(Blk_i(\mathscr{I}_0^l)\right)_{pq} = \left(\widehat{M}^l\right)_{pq} > 0, \tag{5.39}$$

where $\mathscr{I}_0^l$ and $\widehat{M}$ are given in Proposition 5.3;

2). $x = \delta_{2^n}^{i_p}$ is reachable from $x(0) = \delta_{2^n}^{i_q}$ if and only if:

$$\sum_{l=1}^{w\alpha\beta} \left(\widehat{M}^l\right)_{pq} > 0; \tag{5.40}$$

3). the system is controllable at $x(0) = \delta_{2^n}^{i_q}$ if and only if:

$$\sum_{l=1}^{w\alpha\beta} Col_q\left(\widehat{M}^l\right) > 0; \tag{5.41}$$

4). the system is controllable if and only if:

$$\sum_{l=1}^{w\alpha\beta} \widehat{M}^l > 0. \tag{5.42}$$

■

Proof. Let us prove conclusion 1) first.

By Definition 5.6, $x(l) = \delta_{2^n}^{i_p}$ is reachable from $x(0) = \delta_{2^n}^{i_q}$ if and only if one can find a switching signal $\sigma : \{0,\cdots,l-1\} \mapsto \mathscr{W}$ and a control sequence $\{u(0),\cdots,u(l-1)\} \subseteq C_u$, with $\sigma(0) = \delta_w^{k_1}$, $u(0) = \delta_{2^m}^{j_{k_2}}$, such that the trajectory of the system (5.16) starting from $x(0)$ reaches $x(l)$ at time l, $x(t) \in C_x$, $\forall\ t \in \{0,\cdots,l-1\}$. Since $\sigma(l)$ and $u(l)$ are arbitrary, without loss of generality, we set $\sigma(l) = \delta_w^1$, $u(l) = \delta_{2^m}^{j_1}$. Then, we obtain two points $P_{k_0} = \delta_w^{k_1} \times \delta_{2^m}^{j_{k_2}} \times \delta_{2^n}^{i_q}$ and $P_p = \delta_w^1 \times \delta_{2^m}^{j_1} \times \delta_{2^n}^{i_p}$, where $k_0 = (k_1-1)\alpha\beta + (k_2-1)\alpha + q$. In this way, the reachability of the system (5.16) is converted to the reachability from P_{k_0} to P_p. From Corollary 5.1, P_p is reachable from P_{k_0} with l steps if and only if $(\mathscr{I}^l)_{p,k_0} > 0$. Set $k_3 = (k_1-1)\beta + k_2$, then $(\mathscr{I}^l)_{p,k_0} = (\mathscr{I}_0^l)_{p,k_0} = [Blk_{k_3}(\mathscr{I}_0^l)]_{pq}$. Now, since $1 \leq k_3 \leq w\beta$ has $w\beta$ choices, we know that $x(l) = \delta_{2^n}^{i_p}$ is reachable from $x(0) = \delta_{2^n}^{i_q}$ if and only if $\sum_{k_3=1}^{w\beta} [Blk_{k_3}(\mathscr{I}_0^l)]_{pq} > 0$. Noting that $\widehat{M} = \sum_{i=1}^{w\beta} Blk_i(\widehat{L})$, one can obtain:

$$\sum_{k_3=1}^{w\beta} Blk_{k_3}(\mathscr{I}_0^l) = \widehat{M}^{l-1} \sum_{k_3=1}^{w\beta} Blk_{k_3}(\widehat{L}) = \widehat{M}^l,$$

where $\widehat{L} = [\widehat{L}_1\ \cdots\ \widehat{L}_w]$. Therefore, conclusion 1) holds.

As for conclusions 2)–4), it is easy to see that they hold immediately from conclusion 1) and Definition 5.6, and thus the proof is completed. ■

When $w = 1$, the system (5.16) reduces to the nonswitching case. In this case, Theorem 5.5 can be used to check the controllability of Boolean control networks with state and input constraints.

A study of controllability of Boolean control networks was made earlier [7] with the state constraint and established some necessary and sufficient conditions via the Perron-Frobenius theory. Compared with [7], Theorem 5.5 has the following advantages: (i) Theorem 5.5 is applicable to the controllability analysis of Boolean control networks with both state and input constraints, while the results in [7] can be applied to the systems with only the state constraint; (ii) the computation complexity of Theorem 5.5 is $O(\alpha\beta)$, which is much less than the computation complexity of the results in [7], $O(2^{m+n})$, when $\alpha < 2^n$ and $\beta < 2^m$.

In the following section, we study the Mayer-type optimal control problem for switched Boolean control networks with state and input constraints and present a new design procedure for the problem.

Consider the system (5.16) with the initial state $x(0) = \delta_{2^n}^{i_q} \in C_x$. *The Mayer-type optimal control problem* can be stated as follows: find a switching signal σ : $\{0,\cdots,s-1\} \mapsto \mathscr{A}$ and a control sequence $\{u(t) \in C_u : t = 0,\cdots,s-1\}$ such that the cost function:

$$J(\sigma, u(0),\cdots,u(s-1);x(0)) = r^T x(s) \tag{5.43}$$

with $x(t) \in C_x$, $t = 0,\cdots,s$ is minimized, where $r = [r_1 \ \cdots \ r_{2^n}]^T \in \mathbb{R}^{2^n \times 1}$ is a fixed vector, and $s \geq 1$ is a fixed or designed shortest termination time.

We first give a straightforward lemma on the separation of logical variables, which will be used later.

Lemma 5.3

For any integer $1 \leq i \leq w\beta$, there exist unique positive integers i_1 and i_2 such that:

$$\delta_{w\beta}^{i} = \delta_{w}^{i_1} \ltimes \delta_{\beta}^{i_2}, \tag{5.44}$$

where,

$$i_1 = \begin{cases} k, \text{ if } i = k\beta, \ k = 1,\cdots,w; \\ \left[\frac{i}{\beta}\right] + 1, \text{ otherwise}, \end{cases} \tag{5.45}$$

$\left[\frac{i}{\beta}\right]$ denotes the largest integer less than or equal to $\frac{i}{\beta}$, and $i_2 = i - (i_1 - 1)\beta$. ∎

Now, we study the Mayer-type optimal control problem, which is divided into the following two cases.

Case I: $s \geq 1$ is a fixed termination time.

In this case, minimizing the cost functional (5.43) is equivalent to finding the minimum value of J under the following constraint:

$$x(s) \in R_s^C(x(0)) = \{\delta_{2^n}^{i_{k_1}}, \ \cdots, \ \delta_{2^n}^{i_{k_\gamma}}\}, \tag{5.46}$$

and meanwhile, we need to design a switching signal $\sigma : \{0,\cdots,s-1\} \mapsto \mathscr{W}$ and a control sequence $\{u(0), \cdots, u(s-1)\}$ to force $x(0)$ to the optimal terminal state $x^*(s)$, where $R_s^c(x(0))$ can be calculated by Theorem 5.17. Since $r^T \delta_{2^n}^k = Col_k(r^T) = r_k$, to find the minimum value of J under the constraint (5.46), we just calculate the minimum value of r_k, $k = i_{k_1}, \cdots, i_{k_\gamma}$.

Based on the above analysis, we have the following algorithm to design a switching signal $\sigma : \{0,\cdots,s-1\} \mapsto \mathscr{W}$ and a control sequence $\{u(0), \cdots, u(s-1)\}$ such that the cost functional (5.43) is minimized at the fixed termination time s.

ALGORITHM 5.1

Step 1: Calculate $R_s^c(x(0))$ by 5.5, denoted by $R_s^c(x(0)) = \{\delta_{2^n}^{i_{k_1}}, \cdots, \delta_{2^n}^{i_{k_\gamma}}\}$;
Step 2: Calculate the optimal value $J^* = \min\{r_k : k = i_{k_1}, \cdots, i_{k_\gamma}\} := r_{i_{k_{\nu^*}}}$;
Step 3: Divide $\mathscr{I}_0^s$ into $w\beta$ equal blocks as:

$$\mathscr{I}_0^s = \left[Blk_1(\mathscr{I}_0^s) \ \cdots \ Blk_{w\beta}(\mathscr{I}_0^s)\right],$$

and find a block, say, $Blk_\mu(\mathscr{I}_0^s)$, such that $\left[Blk_\mu(\mathscr{I}_0^s)\right]_{k_{\nu^*},q} > 0$, where $\mathscr{I}_0^s$ is given in Proposition 5.3. By Lemma 5.3, calculate μ_1 and μ_2 such that $\delta_w^{\mu_1} \ltimes \delta_\beta^{\mu_2} = \delta_{w\beta}^\mu$. Set $\sigma(0) = \delta_w^{\mu_1}$, $u(0) = \delta_{2^m}^{j_{\mu_2}}$, $x(s) = \delta_{2^n}^{i_{k_{\nu^*}}}$. If $s = 1$, end the calculation; else, go to the next step;

Step 4: Find k and ξ such that:

$$\left[Blk_\xi(\mathscr{I}_0)\right]_{k_{\nu^*},k} > 0, \ \left[Blk_\mu(\mathscr{I}_0^{s-1})\right]_{kq} > 0.$$

By Lemma 5.3, calculate ξ_1 and ξ_2 such that $\delta_w^{\xi_1} \ltimes \delta_\beta^{\xi_2} = \delta_{w\beta}^\xi$. Set $\sigma(s-1) = \delta_w^{\xi_1}$, $u(s-1) = \delta_{2^m}^{j_{\xi_2}}$, $x(s-1) = \delta_{2^n}^{i_k}$;

Step 5: If $s-1 = 1$, end the calculation; else, replace s and k_{ν^*} by $s-1$ and k, respectively, go to Step 4.

PROPOSITION 5.4

The switching signal $\sigma : \{0,\cdots,s-1\} \mapsto \mathscr{W}$ and control sequence $\{u(0), \cdots, u(s-1)\}$ obtained by Algorithm 5.1 can minimize the cost functional (5.43) at the fixed termination time s.

Proof. Since $x(s) = \delta_{2^n}^{i_{k_{v^*}}}$ can minimize J, we just need to prove that the switching signal $\sigma : \{0, \cdots, s-1\} \mapsto \mathscr{W}$ and control sequence $\{u(0), \cdots, u(s-1)\}$ can force the trajectory from $x(0)$ to $x(s)$, and $x(t) \in C_x$, $t = 0, \cdots, s$.

In fact, Since $x(s) \in R_s^c(x(0))$, by Theorem 5.5, we can find an integer μ such that $\left[Blk_\mu(\mathscr{I}_0^s)\right]_{k_{v^*},q} > 0$, which means that if $\sigma(0) = \delta_w^{\mu_1}$, $u(0) = \delta_{2^m}^{j_{\mu_2}}$ with $\delta_w^{\mu_1} \ltimes \delta_\beta^{\mu_2} = \delta_{w\beta}^{\mu}$, there exists at least one path from $x(0)$ to $x(s)$ with $x(t) \in C_x$, $\forall\, t = 0, \cdots, s$. Therefore, there must exist two integers k and ξ such that $x(0)$ can reach $\delta_{2^n}^{i_k} \in C_x$ at time $s-1$ under $\sigma(0) = \delta_w^{\mu_1}$ and $u(0) = \delta_{2^m}^{j_{\mu_2}}$, and $\delta_{2^n}^{i_k}$ can reach $x(s)$ under $\sigma(s-1) = \delta_w^{\xi_1}$ and $u(s-1) = \delta_{2^m}^{j_{\xi_2}}$ in one step, where $\delta_w^{\xi_1} \ltimes \delta_\beta^{\xi_2} = \delta_{w\beta}^{\xi}$. Thus,

$$\left[Blk_\xi(\mathscr{I}_0)\right]_{k_{v^*},k} > 0, \ \left[Blk_\mu(\mathscr{I}_0^{s-1})\right]_{kq} > 0.$$

Similarly, we can find two integers k' and ξ' such that $x(0)$ can reach $\delta_{2^n}^{i_{k'}} \in C_x$ at time $s-2$ under $\sigma(0) = \delta_w^{\mu_1}$ and $u(0) = \delta_{2^m}^{j_{\mu_2}}$, and $\delta_{2^n}^{i_{k'}}$ can reach $x(s-1)$ under $\sigma(s-2) = \delta_w^{\xi'_1}$ and $u(s-2) = \delta_{2^m}^{j_{\xi'_2}}$ in one step, where $\delta_w^{\xi'_1} \ltimes \delta_\beta^{\xi'_2} = \delta_{w\beta}^{\xi'}$.

Continuing this process, the switching signal $\sigma : \{0, \cdots, s-1\} \mapsto \mathscr{A}$ and the sequence of controls $\{u(0), \cdots, u(s-1)\}$ can be determined, under which the trajectory starting from $x(0)$ can reach $x(s)$, and $x(t) \in C_x$, $t = 0, \cdots, s$.

Case II: $s \geq 1$ is a designed shortest termination time.

In this case, we take by the following two steps: 1) find an $x_d \in R^c(x(0))$ to minimize the cost functional (5.43) in the shortest time s to be designed, and 2) design the shortest termination time s, a switching signal $\sigma : \{0, \cdots, s-1\} \mapsto \mathscr{W}$ and a control sequence $\{u(0), \cdots, u(s-1)\}$ such that, under the switching signal and control sequence, $x(0)$ can reach x_d at time s. Based on the above analysis, we have the following algorithm for Case II. ∎

ALGORITHM 5.2

Step 1: By Theorem 5.5, calculate $R^c(x(0))$, denoted by $R^c(x(0)) = \{\delta_{2^n}^{i_{k_1}}, \cdots, \delta_{2^n}^{i_{k_\gamma}}\}$;

Step 2: Calculate the optimal value $J^* = \min\{r_k : k = i_{k_1}, \cdots, i_{k_\gamma}\} := r_{i_{k_{v^*}}}$;

Step 3: Choose the smallest positive integer s such that in

$$\mathscr{I}_0^s = \left[Blk_1(\mathscr{I}_0^s) \ \cdots \ Blk_{w\beta}(\mathscr{I}_0^s)\right],$$

we find a block, say, $Blk_\mu(\mathscr{I}_0^s)$, satisfying $\left[Blk_\mu(\mathscr{I}_0^s)\right]_{k_{v^*},q} > 0$. By Lemma 5.3, calculate μ_1 and μ_2 such that $\delta_w^{\mu_1} \ltimes \delta_\beta^{\mu_2} = \delta_{w\beta}^{\mu}$. Set $\sigma(0) = \delta_w^{\mu_1}$, $u(0) = \delta_{2^m}^{j_{\mu_2}}$, $x(s) =$

$\delta_{2^n}^{i_{k_{v^*}}}$. If $s=1$, end the calculation; else, go to the next step;

Step 4: Find k and ξ such that

$$\left[Blk_{\xi}(\mathscr{I}_0)\right]_{k_{v^*},k}>0,\ \left[Blk_{\mu}(\mathscr{I}_0^{s-1})\right]_{kq}>0.$$

By Lemma 5.3, calculate ξ_1 and ξ_2 such that $\delta_w^{\xi_1}\ltimes\delta_{\beta}^{\xi_2}=\delta_{w\beta}^{\xi}$. Set $\sigma(s-1)=\delta_w^{\xi_1}$, $u(s-1)=\delta_{2^m}^{j_{\xi_2}}$, $x(s-1)=\delta_{2^n}^{i_k}$;

Step 5: If $s-1=1$, end the calculation; else, replace s and k_{v^*} by $s-1$ and k, respectively, and go to Step 4.

PROPOSITION 5.5

The termination time s obtained in Algorithm 5.2 is the shortest one. Moreover, the switching signal $\sigma:\{0,\cdots,s-1\}\mapsto\mathscr{W}$ and the control sequence $\{u(0),\ \cdots,\ u(s-1)\}$ generated by Algorithm 5.2 can minimize the cost functional (5.43) at time s.

Proof. We just need to show that the termination time s obtained in Algorithm 5.2 is the shortest one, and the remainder can be proved by an argument similar to the proof of Proposition 5.5.

In fact, since $\delta_{2^n}^{i_{k_{v^*}}}\in R^c(x(0))$, by Theorem 5.5, there exists the smallest positive integer s such that $[Blk_{\mu}(\mathscr{I}_0^s)]_{k_{v^*},q}>0$, that is to say, the termination time s obtained in Algorithm 5.2 is the shortest one. ∎

In the following, we investigate the stabilization problem for the system (5.16). The constrained system is converted into an unconstrained equivalent one, based on which we present some necessary and sufficient conditions for the stabilization with open-loop and closed-loop controls, respectively.

We first, give the concept of stabilization for the switched Boolean control network with state and input constraints.

DEFINITION 5.7

The system (5.16) is said to be stabilizable to $X_e=(x_1^e,\cdots,x_n^e)\in C_x$, if there exist a switching signal $\sigma:\mathbb{N}\mapsto\mathscr{W}$ and a control sequence $\{U(t)=(u_1(t),\cdots,u_m(t)):t\in\mathbb{N}\}\subseteq C_u$, under which the trajectory initialized at any $X_0=(x_1(0),\cdots,x_n(0))\in C_x$ converges to X_e, and all $X(t)=(x_1(t),\cdots,x_n(t))\in C_x$, $\forall\, t\in\mathbb{N}$.

Next, to facilitate the analysis, we convert the system (5.16) into an unconstrained equivalent one.

Recall that, by the vector form of logical variables, $C_x = \{\delta_{2^n}^{i_k} : k = 1, \cdots, \alpha;\ i_1 < \cdots < i_\alpha\}$, $C_u = \{\delta_{2^m}^{j_k} : k = 1, \cdots, \beta;\ j_1 < \cdots < j_\beta\}$. Letting $x_e = \ltimes_{i=1}^n x_i^e = \delta_{2^n}^{i_p} \in C_x$ in the vector form of logical variables, and identifying $\sigma(t) = i \sim \delta_w^i$, $\delta_{2^n}^{i_k} \sim \delta_\alpha^k$, $k = 1, \cdots, \alpha$, and $\delta_{2^m}^{j_k} \sim \delta_\beta^k$, $k = 1, \cdots, \beta$, one can convert (5.19) into the following form:

$$z(t+1) = \widehat{L}\sigma(t)v(t)z(t), \tag{5.47}$$

where $z(t) \in \Delta_\alpha$ is the state variable, $\sigma(t) \in \Delta_w$ and $v(t) \in \Delta_\beta$ are the input variables, and $\widehat{L} \in \mathscr{B}_{\alpha \times w\alpha\beta}$ is given in Proposition 5.2. In this case, $z(t) = \delta_\alpha^k \sim x(t) = \delta_{2^n}^{i_k}$, $v(t) = \delta_\beta^k \sim u(t) = \delta_{2^m}^{j_k}$. Obviously, the system (5.47) is an unconstrained equivalent form of the system (5.16).

The following result can be obtained by a straightforward computation.

PROPOSITION 5.6

The system (5.16) is stabilizable to $x_e = \delta_{2^n}^{i_p}$, if and only if the system (5.47) is stabilizable to $z_e = \delta_\alpha^p$.

In what follows, based on the unconstrained system (5.47) and Proposition 5.6, we study the stabilization for the system (5.16).

We first study the stabilization of the system (5.16) by an open-loop control sequence $\{u(t) \in C_u : t \in \mathbb{N}\}$ and a free-form switching sequence $\{(0, \sigma(0)), \cdots, (l, \sigma(l)), \cdots\}$. In this case, $\{u(t) \in C_u : t \in \mathbb{N}\} \sim \{v(t) \in \Delta_\beta : t \in \mathbb{N}\}$. From (5.47), for any $\tau \in \mathbb{Z}_+$, we have:

$$\begin{aligned} z(\tau) &= \widehat{L}\sigma(\tau-1)v(\tau-1)z(\tau-1) \\ &= \widehat{L}(\sigma(\tau-1)v(\tau-1)) \cdots \widehat{L}(\sigma(0)v(0))z(0) \\ &= \widehat{L}(I_{w\beta} \otimes \widehat{L}) \cdots (I_{(w\beta)^{\tau-1}} \otimes \widehat{L}) \ltimes_{i=\tau-1}^{0} (\sigma(i)v(i))z(0) \\ &:= \widehat{\widehat{L}} \ltimes_{i=\tau-1}^{0} (\sigma(i)v(i))z(0), \end{aligned}$$

where $\ltimes_{i=\tau-1}^{0}(\sigma(i)v(i)) = (\sigma(\tau-1)v(\tau-1)) \ltimes \cdots \ltimes (\sigma(0)v(0))$,

$$\widehat{\widehat{L}} = \widehat{L}(I_{w\beta} \otimes \widehat{L}) \cdots (I_{(w\beta)^{\tau-1}} \otimes \widehat{L}) \in \mathscr{B}_{\alpha \times (w\beta)^\tau \alpha}. \tag{5.48}$$

Split $\widehat{\widehat{L}}$ into $(w\beta)^\tau$ equal blocks as:

$$\widehat{\widehat{L}} = \left[Blk_1(\widehat{\widehat{L}}) \ \cdots \ Blk_{(w\beta)^\tau}(\widehat{\widehat{L}}) \right]. \tag{5.49}$$

Then, for $\ltimes_{i=\tau-1}^{0}(\sigma(i)v(i)) = \delta_{(w\beta)^\tau}^\mu$, it is easy to see that:

$$z(\tau) = Blk_\mu(\widehat{\widehat{L}})z(0). \tag{5.50}$$

Theorem 5.6

The system (5.16) is stabilized to $x_e = \delta_{2^n}^{i_p}$ by an open-loop control sequence and a free-form switching sequence if and only if:

1). there exists an integer $1 \leq \eta \leq w\beta$ such that:

$$Col_p\left(Blk_\eta(\widehat{L})\right) = \delta_\alpha^p; \tag{5.51}$$

2). there exist integers $1 \leq \tau \leq w\beta\alpha^\alpha$ and $1 \leq \mu \leq (w\beta)^\tau$ such that:

$$Blk_\mu(\widehat{\overline{L}}) = \delta_\alpha[\underbrace{p \cdots p}_{\alpha}]. \tag{5.52}$$

Moreover, if (5.51) and (5.52) hold, then the open-loop control sequence and the free-form switching sequence can be designed respectively as:

$$u(t) = \begin{cases} u^*(t), \ 0 \leq t \leq \tau - 1, \\ \delta_{2^m}^{j_{\eta_1}}, \ t \geq \tau, \end{cases} \tag{5.53}$$

$$\sigma(t) = \begin{cases} \sigma^*(t), \ 0 \leq t \leq \tau - 1, \\ \delta_w^{\eta_2}, \ t \geq \tau, \end{cases} \tag{5.54}$$

where $\delta_w^{\eta_2} \ltimes \delta_\beta^{\eta_1} = \delta_{w\beta}^\eta$, $u^*(t) \sim v^*(t)$, $0 \leq t \leq \tau - 1$, v^* and σ^* are determined by $\ltimes_{i=\tau-1}^{0}(\sigma^*(i)v^*(i)) = \delta_{(w\beta)^\tau}^\mu$. ■

Proof. By Proposition 5.6, we only need to prove that the system (5.47) is stabilized to $z_e = \delta_\alpha^p$ by open-loop control sequences $\{v(t) \in \Delta_\beta : v(t) \sim u(t), t \in \mathbb{N}\}$ and $\{\sigma(t) : t \in \mathbb{N}\}$, if and only if conditions 1) and 2) hold.

(Sufficiency) Suppose that conditions 1) and 2) hold. Then, for $\ltimes_{i=\tau-1}^{0}(\sigma^*(i)v^*(i)) = \delta_{(w\beta)^\tau}^\mu$, (5.50) and (5.52) imply that:

$$z(\tau) = Blk_\mu(\widehat{\overline{L}})z(0) = \delta_\alpha[\underbrace{p \cdots p}_{\alpha}]z(0) = \delta_\alpha^p, \ \forall\, z(0) \in \Delta_\alpha.$$

On the other hand, for $v(t) = \delta_\beta^{\eta_1}$ and $\sigma(t) = \delta_w^{\eta_2}$, $\forall\, t \geq \tau$, from (5.51) we have

$$\begin{aligned} z(t+1) &= \widehat{L}\sigma(t)v(t)z(t) = \widehat{L}\delta_{w\beta}^\eta z(t) = Blk_\eta(\widehat{L})z(t) \\ &= [Blk_\eta(\widehat{L})]^{t+1-\tau}z(\tau) = \delta_\alpha^p. \end{aligned}$$

Thus, the system (5.47) is stabilized to $z_e = \delta_\alpha^p$ by open-loop control sequences $\{v(t) \in \Delta_\beta : v(t) \sim u(t), t \in \mathbb{N}\}$ and $\{\sigma(t) : t \in \mathbb{N}\}$, where $u(t)$ and $\sigma(t)$ are given in (5.53) and (5.54).

(Necessity) Assume that the system (5.47) is stabilized to $z_e = \delta_\alpha^p$ by open-loop control sequences $\{v(t) \in \Delta_\beta : v(t) \sim u(t), t \in \mathbb{N}\}$ and $\{\sigma(t) : t \in \mathbb{N}\}$.

We first prove that condition 1) holds.

If condition 1) does not hold, then setting $z(t) = \delta_\alpha^p$, for any $v(t) \in \Delta_\beta$ and $\sigma(t) \in \Delta_w$ with $\sigma(t) \ltimes v(t) = \delta_{w\beta}^\eta$, we have:

$$z(t+1) = \widehat{L}\sigma(t)v(t)z(t) = Blk_\eta(\widehat{L})z(t) = Col_p\left(Blk_\eta(\widehat{\overline{L}})\right) \neq \delta_\alpha^p,$$

which is a contradiction to the fact that the system (5.47) is stabilized to $z_e = \delta_\alpha^p$ by open-loop control sequences $\{v(t) \in \Delta_\beta : v(t) \sim u(t), t \in \mathbb{N}\}$ and $\{\sigma(t) : t \in \mathbb{N}\}$. Hence, condition 1) holds.

Now, we prove that condition 2) holds. We divide the proof into two steps.

Step 1. We prove the conclusion that there exist open-loop control sequences $\{v(t) \in \Delta_\beta : t \in \mathbb{N}\}$ and $\{\sigma(t) : t \in \mathbb{N}\}$ under which the trajectory of the system (5.47) initialized at any initial state $z(0) \in \Delta_\alpha$ converges to z_e within time $\omega\beta\alpha^\alpha$.

Given open-loop control sequences $\{v(t) \in \Delta_\beta : t \in \mathbb{N}\}$ and $\{\sigma(t) : t \in \mathbb{N}\}$ that stabilize the system (5.47) to z_e, denote by $T_{v,\sigma}$ the shortest time among the ones within which the trajectories starting from any initial state converge to z_e under v and σ. If $T_{v,\sigma} \leq \omega\beta\alpha^\alpha$, the proof is completed. Otherwise, if $T_{v,\sigma} > \omega\beta\alpha^\alpha$, for $z^i(0) = \delta_\alpha^i$, $i = 1, 2, \cdots, \alpha$, since the total number of choices for $(z^1(t), \cdots, z^\alpha(t), v(t), (t))$, $t \in \mathbb{N}$, is $\omega\beta\alpha^\alpha$, there exist two integers $0 \leq t_1^0 < t_2^0 \leq T_{v,\sigma} - 1$ such that $z^i(t_1^0) = z^i(t_2^0)$, $i = 1, \cdots, \alpha$, $v(t_1^0) = v(t_2^0)$, and $\sigma(t_1^0) = \sigma(t_2^0)$. We construct the following open-loop control sequences:

$$v_1(t) = \begin{cases} v(t), \ 0 \leq t < t_1^0, \\ v(t + t_2^0 - t_1^0), \ t \geq t_1^0, \end{cases} \tag{5.55}$$

and

$$\sigma_1(t) = \begin{cases} \sigma(t), \ 0 \leq t < t_1^0, \\ \sigma(t + t_2^0 - t_1^0), \ t \geq t_1^0. \end{cases} \tag{5.56}$$

Then, it is easy to see that the shortest time among the ones within which the trajectories starting from any initial state converge to z_e under v_1 and σ_1 is:

$$T_{v_1,\sigma_1} = T_{v,\sigma} - (t_2^0 - t_1^0) < T_{v,\sigma}. \tag{5.57}$$

If $T_{v_1,\sigma_1} \leq \omega\beta\alpha^\alpha$, the proof is completed. Otherwise, one can find integers $0 \leq t_1^1 < t_2^1 \leq T_{v_1,\sigma_1} - 1$ such that $z^i(t_1^1) = z^i(t_2^1)$, $i = 1, \cdots, \alpha$, $v_1(t_1^1) = v_1(t_2^1)$, and $\sigma_1(t_1^1) = \sigma_1(t_2^1)$. We construct the following open-loop control sequences:

$$v_2(t) = \begin{cases} v_1(t), \ 0 \leq t < t_1^0, \\ v_1(t + t_2^0 - t_1^0), \ t \geq t_1^0, \end{cases} \quad \text{and } \sigma_2(t) = \begin{cases} \sigma_1(t), \ 0 \leq t < t_1^0, \\ \sigma_1(t + t_2^0 - t_1^0), \ t \geq t_1^0. \end{cases}$$

Then, the shortest time taken by the trajectories starting from any initial state converge to z_e under v_2 and σ_2 is $T_{v_2,\sigma_2} = T_{v_1,\sigma_1} - (t_2^1 - t_1^1) < T_{v_1,\sigma_1}$.

Keep this procedure going, then we obtain open-loop control sequences $v_q(t)$ and $\sigma_q(t)$. If $T_{v_q,\sigma_q} \leq \omega\beta\alpha^{\alpha}$, the proof is completed. Otherwise, one can find integers $0 \leq t_1^q < t_2^q \leq T_{v_q,\sigma_q} - 1$ such that $z^i(t_1^q) = z^i(t_2^q)$, $i = 1, \cdots, \alpha$, $v_q(t_1^q) = v_q(t_2^q)$, and $\sigma_q(t_1^q) = \sigma_q(t_2^q)$. We construct the following open-loop control sequences:

$$v_{q+1}(t) = \begin{cases} v_q(t),\ 0 \leq t < t_1^0, \\ v_q(t + t_2^0 - t_1^0),\ t \geq t_1^0, \end{cases} \text{ and } \sigma_{q+1}(t) = \begin{cases} \sigma_q(t),\ 0 \leq t < t_1^0, \\ \sigma_q(t + t_2^0 - t_1^0),\ t \geq t_1^0. \end{cases}$$

Then, the shortest time taken by the trajectories starting from any initial state converge to z_e under v_{q+1} and σ_{q+1} is:

$$T_{v_{q+1},\sigma_{q+1}} = T_{v_q,\sigma_q} - (t_2^q - t_1^q) < T_{v_q,\sigma_q}.$$

Since $t_2^j - t_1^j \geq 1$, $j = 0, 1, \cdots, q$, there must exist an integer $0 \leq q \leq T_{v,\sigma} - \omega\beta\alpha^{\alpha}$ such that $\sum_{j=0}^{q} (t_2^j - t_1^j) \geq T_{v,\sigma} - \omega\beta\alpha^{\alpha}$. In this case, $T_{v_{q+1},\sigma_{q+1}} \leq \omega\beta\alpha^{\alpha}$, which implies that the conclusion in Step 1 holds.

Step 2. By Step 1, there exist open-loop control sequences $v^*(t)$ and $\sigma^*(t)$ under which the trajectory initialized at any initial state $z(0) \in \Delta_\alpha$ converges to z_e in time $1 \leq \tau \leq \omega\beta\alpha^{\alpha}$. Set $\ltimes_{i=\tau-1}^{0}(\sigma^*(i)v^*(i)) = \delta_{(w\beta)^\tau}^{\mu}$. Then, one can see from (5.50) that:

$$z(\tau) = Blk_\mu\left(\widehat{\widehat{L}}\right)z(0) = z_e = \delta_\alpha^p,\ \forall\, z(0) \in \Delta_\alpha,$$

which implies that condition 2) is true. ∎

Finally, we investigate the stabilization of the system (5.16) by a closed-loop control $u(t) = G_1x(t)$ and a state feedback switching signal $\sigma(t) = G_2x(t)$, where $G_1 \in \mathcal{L}_{2^m \times 2^n}$, $G_2 \in \mathcal{L}_{w \times 2^n}$.

In this case, for any time $t \geq 0$, one can see that the kth column and the rth row of G_1 correspond to $x(t) = \delta_{2^n}^k$ and $u(t) = \delta_{2^m}^r$, and the kth column of G_2 correspond to $x(t) = \delta_{2^n}^k$. Since $x(t) \in C_x$ and $u(t) \in C_u$, we consider the entries at (r,k), $r = j_1, \cdots, j_\beta, k = i_1, \cdots, i_\alpha$ in G_1, and columns $i_1, \cdots, i_\alpha$ in G_2. Then, using the block selection matrices, we obtain two new matrices as:

$$\widehat{G}_1 = \begin{bmatrix} J_{j_1}^{(2^m,1)} \\ \vdots \\ J_{j_\beta}^{(2^m,1)} \end{bmatrix}, G_1\left[\left(J_{i_1}^{(2^n,1)}\right)^T \cdots \left(J_{i_\alpha}^{(2^n,1)}\right)^T\right] \in \mathcal{B}_{\beta\times\alpha}, \tag{5.58}$$

$$\widehat{G}_2 = G_2\left[\left(J_{i_1}^{(2^n,1)}\right)^T \cdots \left(J_{i_\alpha}^{(2^n,1)}\right)^T\right] \in \mathcal{L}_{w\times\alpha}. \tag{5.59}$$

Substituting $v(t) = \widehat{G}_1z(t)$ and $\sigma(t) = \widehat{G}_2z(t)$ into (5.47), we have:

$$\begin{aligned} z(t+1) &= \widehat{L}\widehat{G}_2z(t)\widehat{G}_1z(t)z(t) = \widehat{L}\widehat{G}_2(I_\alpha \otimes \widehat{G}_1)z(t)z(t)z(t) \\ &= \widehat{L}\widehat{G}_2(I_\alpha \otimes \widehat{G}_1)M_{r,\alpha}M_{r,\alpha}z(t) := \widehat{\widehat{L}}z(t), \end{aligned}$$

where,

$$\widehat{\widehat{L}} = \widehat{L}\widehat{G}_2(I_\alpha \otimes \widehat{G}_1)M_{r,\alpha}M_{r,\alpha}, \tag{5.60}$$

$$M_{r,\alpha} = diag\{\delta_\alpha^1, \cdots, \delta_\alpha^\alpha\}. \tag{5.61}$$

Theorem 5.7

The system (5.16) is stabilized to $x_e = \delta_{2^n}^{i_p}$ by a closed-loop control $u(t) = G_1x(t)$ and a state feedback switching signal $\sigma(t) = G_2x(t)$ if and only if there exists an integer $1 \leq \tau \leq \alpha$ such that:

$$\left(\widehat{\widehat{L}}\right)^\tau = \delta_\alpha[\underbrace{p \cdots p}_{\alpha}]. \tag{5.62}$$

■

Proof. By Proposition 5.6 holds, we only need to prove that the system (5.47) is stabilized to $z_e = \delta_\alpha^p$ by $v(t) = \widehat{G}_1z(t)$ and $\sigma(t) = \widehat{G}_2z(t)$.

(Sufficiency) Assuming that (5.62) holds, we prove that the system (5.47) is stabilized to z_e by $v(t) = \widehat{G}_1z(t)$ and $\sigma(t) = \widehat{G}_2z(t)$.

In fact, from (5.62), under the controls $v(t) = \widehat{G}_1z(t)$ and $\sigma(t) = \widehat{G}_2z(t)$, along the trajectory starting from any initial state $z(0) \in \Delta_\alpha$, we have:

$$z(\tau) = \left(\widehat{\widehat{L}}\right)^\tau z(0) = \delta_\alpha^p = z_e,$$

and,

$$z(t) = \left(\widehat{\widehat{L}}\right)^t z(0) = \left(\widehat{\widehat{L}}\right)^\tau \left[\left(\widehat{\widehat{L}}\right)^{t-\tau} z(0)\right] = z_e, \ \forall\, t \geq \tau.$$

Thus, the system (5.47) is stabilized to z_e by $v(t) = \widehat{G}_1z(t)$ and $\sigma(t) = \widehat{G}_2z(t)$.

(Necessity) Suppose that the system (5.47) is stabilized to z_e by $v(t) = \widehat{G}_1z(t)$ and $\sigma(t) = \widehat{G}_2z(t)$. Since $z(t) \in \Delta_\alpha$ has α choices, one can see that under $v(t) = \widehat{G}_1z(t)$ and $\sigma(t) = \widehat{G}_2z(t)$, the trajectory of the system (5.47) initialized at any initial state $z(0) \in \Delta_\alpha$ convergences to z_e within time α. Therefore, one can find an integer $1 \leq \tau \leq \alpha$ such that:

$$z(\tau) = \left(\widehat{\widehat{L}}\right)^\tau z(0) = z_e = \delta_\alpha^p, \ \forall\, z(0) \in \Delta_\alpha,$$

which implies that (5.62) holds. ■

It is noted that the stabilization by a closed-loop control $u(t) = G_1x(t)$ and a state feedback switching signal $\sigma(t) = G_2x(t)$ is quite different from the one by an open-loop control sequence and a free-form switching sequence. As for how to determine G_1 and G_2, since the choices of $G_1 \in \mathscr{L}_{2^m \times 2^n}$ and $G_2 \in \mathscr{L}_{w \times 2^n}$ are finite, one can choose them by a trial-and-error procedure.

Example 5.7:

Recall the apoptosis network in Example 5.1. When modeling the system (5.2) as the deterministic asynchronous Boolean network, one can convert it into the following switched Boolean control network:

$$\begin{cases} x_1(t+1) = f_1^{\sigma(t)}(x_1(t), x_2(t), x_3(t), u(t)), \\ x_2(t+1) = f_2^{\sigma(t)}(x_1(t), x_2(t), x_3(t), u(t)), \\ x_3(t+1) = f_3^{\sigma(t)}(x_1(t), x_2(t), x_3(t), u(t)), \end{cases} \tag{5.63}$$

where $\sigma : \mathbb{N} \mapsto \mathscr{A} = \{1, 2, \cdots, 8\}$ is the switching signal, and

$$\begin{aligned}
&f_1^1 = x_1,\ f_2^1 = x_2,\ f_3^1 = x_3; \\
&f_1^2 = \neg x_2 \wedge u,\ f_2^2 = x_2,\ f_3^2 = x_3; \\
&f_1^3 = x_1,\ f_2^3 = \neg x_1 \wedge x_3,\ f_3^3 = x_3; \\
&f_1^4 = x_1,\ f_2^4 = x_2,\ f_3^4 = x_2 \vee u; \\
&f_1^5 = \neg x_2 \wedge u,\ f_2^5 = \neg x_1 \wedge x_3,\ f_3^5 = x_3; \\
&f_1^6 = \neg x_2 \wedge u,\ f_2^6 = x_2,\ f_3^6 = x_2 \vee u; \\
&f_1^7 = x_1,\ f_2^7 = \neg x_1 \wedge x_3,\ f_3^7 = x_2 \vee u; \\
&f_1^8 = \neg x_2 \wedge u,\ f_2^8 = \neg x_1 \wedge x_3,\ f_3^8 = x_2 \vee u.
\end{aligned}$$

Using the vector form of logical variables and setting $x(t) = \ltimes_{i=1}^3 x_i(t)$, we have:

$$x(t+1) = L_{\sigma(t)} u(t) x(t), \tag{5.64}$$

where,

$$\begin{aligned}
L_1 &= \delta_8[1\,2\,3\,4\,5\,6\,7\,8\,1\,2\,3\,4\,5\,6\,7\,8], \\
L_2 &= \delta_8[5\,6\,3\,4\,5\,6\,3\,4\,5\,6\,7\,8\,5\,6\,7\,8], \\
L_3 &= \delta_8[3\,4\,3\,4\,5\,6\,7\,8\,3\,4\,3\,4\,5\,6\,7\,8], \\
L_4 &= \delta_8[1\,1\,3\,3\,5\,5\,7\,7\,1\,1\,4\,4\,5\,5\,8\,8], \\
L_5 &= \delta_8[7\,8\,3\,4\,5\,8\,1\,4\,7\,8\,7\,8\,5\,8\,5\,8], \\
L_6 &= \delta_8[5\,5\,3\,3\,5\,5\,3\,3\,5\,5\,8\,8\,5\,5\,8\,8], \\
L_7 &= \delta_8[3\,3\,3\,3\,5\,7\,5\,7\,3\,3\,4\,4\,5\,7\,6\,8], \\
L_8 &= \delta_8[7\,7\,3\,3\,5\,7\,1\,3\,7\,7\,8\,8\,5\,7\,6\,8].
\end{aligned}$$

In the following section, we study two kinds of control problems.

Kind I: Find a control strategy and a switching path such that the stimulus is applied as little as possible, and meanwhile, the cell survival is achieved.

Since $\{x_1 = 1,\ x_2 = 0\}$ stands for the cell survival, and $u = 0$ implies that the stimulus is not applied to the system, we set $C_x = \Delta_8$, $C_u = \{\delta_2^2\}$. Our goal is to find a switching path and a control strategy to minimize the following cost function:

$$J(\sigma, u(0), \cdots, u(s-1); x(0)) = [1\,1\,0\,0\,2\,2\,1\,1] x(s), \tag{5.65}$$

where $s=2$ and $x(0)=\delta_8^1$ is an initial cell state. Moreover, $x_1(s)\bar{\vee}1+x_2(s)\bar{\vee}0$ denotes the Hamming distance between the state $(x_1(s),x_2(s))$ and the cell survival state $(1,0)$.

Now, we use Algorithm 5.1 to solve this problem.

Step 1: Since $\left(\widehat{M}^2\right)_{v1}>0$, $v=1,3,4,5,6,7,8$, we obtain:

$$R_2^c(x(0))=\{\delta_8^1,\delta_8^3,\delta_8^4,\delta_8^5,\delta_8^6,\delta_8^7,\delta_8^8\},$$

where $\widehat{M}=\sum_{i=1}^{8}Blk_i(\widehat{L})$, $\widehat{L}=[\widehat{L}_1\ \cdots\ \widehat{L}_8]$ can be calculated by (5.35) and (5.36).

Step 2: $J^*=\min\{1,0,0,2,2,1,1\}=0$, $k_{v^*}=3$ or 4.

Step 3: For $k_{v^*}=3$, one can easily see that $\left[Blk_1(\mathscr{I}_0^2)\right]_{31}>0$, where $\mathscr{I}_0^2=\widehat{M}\widehat{L}$. Set $\sigma(0)=\delta_8^1\sim1$, $u(0)=\delta_2^2\sim0$, $x(2)=\delta_8^3$.

Step 4: Since:

$$\left[Blk_3(\mathscr{I}_0)\right]_{31}>0,\ \left[Blk_1(\mathscr{I}_0)\right]_{11}>0,$$

we set $\sigma(1)=\delta_8^3\sim3$, $u(1)=\delta_2^2\sim0$, $x(1)=\delta_8^1$. Since $s=1$ here, we end the calculation.

Thus, we obtain a switching sequence $\pi=\{(0,1),(1,3)\}$ and a sequence of controls $\{u(0)=0,u(1)=0\}$ that minimize J in time $s=2$. Since $J^*=0$, we know that the cell survival is achieved.

Kind II: When the cell survival is always achieved, design a control strategy and a switching signal, under which the system (5.63) is stabilized to $x_e=\delta_8^3$.

In this case, $C_x=\{\delta_8^3,\delta_8^4\}$, $C_u=\{\delta_2^1,\delta_2^2\}$, we have:

$$\begin{aligned}\widehat{L}\ =\ &\delta_2[1\,2\,\vdots\,1\,2\,\vdots\,1\,2\,\vdots\,0\,0\,\vdots\,1\,2\,\vdots\,1\,2\,\vdots\,1\,1\,\vdots\,2\,2\\ &\vdots\,1\,2\,\vdots\,0\,0\,\vdots\,1\,1\,\vdots\,0\,0\,\vdots\,1\,1\,\vdots\,2\,2\,\vdots\,1\,1\,\vdots\,0\,0],\end{aligned}$$

where $\delta_2^0:=\begin{bmatrix}0\\0\end{bmatrix}$.

We first apply Theorem 5.6 to design an open-loop control strategy such that the system (5.63) is stabilized to $x_e=\delta_8^3$.

Setting $\tau=1$, one can see that $Col_1(Blk_2(\widehat{L}))=\delta_2^1$, $Blk_7(\widehat{\widehat{L}})=Blk_7(\widehat{L})=\delta_2[1\ 1]$. Hence, by Theorem 5.6, we get the following open-loop control sequence and free-form switching sequence:

$$u(t)=\begin{cases}\delta_2^1\sim1,\ t=0,\\ \delta_2^2\sim0,\ t\geq1,\end{cases}\ \text{and}\ \sigma(t)=\begin{cases}\delta_8^4\sim4,\ t=0,\\ \delta_8^1\sim1,\ t\geq1,\end{cases}$$

Second, we apply Theorem 5.7 to design a closed-loop control strategy such that the system (5.63) is stabilized to $x_e=\delta_8^3$.

According to Theorem 5.7, after a trial-and-error procedure, we set $G_1=\delta_2[2\ 2\ 1\ 1\ 1\ 2\ 1\ 2]$, $G_2=\delta_8[2\ 2\ 1\ 4\ 3\ 3\ 5\ 7]$. From (5.58)-(5.60), it is easy to see that $\widehat{G}_1=\delta_2[1\ 1]$, $\widehat{G}_2=\delta_8[1\ 4]$, $\widehat{\widehat{L}}=\delta_2[1\ 1]$. Thus, by Theorem 5.7, the system (5.63) can be stabilized to $x_e=\delta_8^3$ by $u(t)=\delta_2[2\ 2\ 1\ 1\ 1\ 2\ 1\ 2]x(t)$ and $\sigma(t)=\delta_8[2\ 2\ 1\ 4\ 3\ 3\ 5\ 7]x(t)$.

5.4 DISTURBANCE DECOUPLING CONTROLLER DESIGN

This section investigates a kind of disturbance decoupling problems (DDPs) for switched Boolean control networks (BCNs) and gives a constructive procedure to design all possible state feedback and output feedback disturbance decoupling controllers.

Consider a switched BCN with n nodes, m control inputs, p outputs, q disturbance inputs and w sub-networks as:

$$\begin{cases} x_1(t+1) = f_1^{\sigma(t)}(X(t),U(t),\xi(t)), \\ x_2(t+1) = f_2^{\sigma(t)}(X(t),U(t),\xi(t)), \\ \quad \vdots \\ x_n(t+1) = f_n^{\sigma(t)}(X(t),U(t),\xi(t)), \\ y_i(t) = h_i(x_1(t),\cdots,x_r(t)),\ i = 1,\cdots,p, \end{cases} \tag{5.66}$$

where $\sigma : \mathbb{N} \mapsto \mathscr{A} = \{1,2,\cdots,w\}$ is the switching signal, $X(t) = (x_1(t),\cdots,x_n(t))$, $U(t) = (u_1(t),\cdots,u_m(t))$, $\xi(t) = (\xi_1(t),\cdots,\xi_q(t))$, $x_i \in \mathscr{D}$, $i = 1,\cdots,n$ are logical variables, $u_i \in \mathscr{D}$, $i = 1,\cdots,m$ are control inputs, $y_i \in \mathscr{D}$, $i = 1,\cdots,p$ are outputs, and $f_i^j : \mathscr{D}^{m+n+q} \mapsto \mathscr{D}$, $i = 1,\cdots,n$, $j = 1,\cdots,w$ and $h_i : \mathscr{D}^n \mapsto \mathscr{D}$, $i = 1,\cdots,p$ are logical functions.

Now, we give the definition for the DDP of switched BCNs studied in this section.

DEFINITION 5.8

Consider the system (5.66). The disturbance-independent output decoupling problem is solvable, if one can find a state feedback control:

$$u_i(t) = \varphi_i(x_1(t),\cdots,x_n(t)),\ i = 1,\cdots,m, \tag{5.67}$$

such that the closed-loop system consisting of (5.66) and (5.67) becomes:

$$\begin{cases} x_i(t+1) = \widehat{f_i}^{\sigma(t)}(x_1(t),\cdots,x_r(t)), i = 1,\cdots,r, \\ x_k(t+1) = \widehat{f_k}^{\sigma(t)}(x_1(t),\cdots,x_n(t),\xi(t)), k = r+1,\cdots,n, \\ y_j(t) = \widehat{h}_j(x_1(t),\cdots,x_r(t)),\ j = 1,\cdots,p, \end{cases} \tag{5.68}$$

where $\widehat{f_i^j} : \mathscr{D}^r \mapsto \mathscr{D}, i = 1,\cdots,r, j = 1,\cdots,w$, $\widehat{f_k^j} : \mathscr{D}^{n+q} \mapsto \mathscr{D}$, $k = r+1,\cdots,n$, $j = 1,\cdots,w$ and $\widehat{h}_j : \mathscr{D}^r \mapsto \mathscr{D}, j = 1,\cdots,p$ are logical functions.

It should be pointed out that when a switched BCN does not directly have the form of (5.66), one can construct a coordinate transformation to convert such a switched BCN into the form of (5.66) by using the technique proposed in [8]. This is also the main difficulty in solving the DDP of BCNs. However, this difficulty was not completely solved in [8]. How to find a proper subspace which contains the disturbance inputs and is in the kernel of outputs, still needs to be further investigated. In the following, we will study the DDP and design controllers such that the DDP is solvable.

Consider the system (5.66). First, we convert the state feedback control (5.67) and the system (5.66) into algebraic forms.

Using the vector form of logical variables and setting $x^1(t) = \ltimes_{i=1}^{r} x_i(t)$, $x^2(t) = \ltimes_{i=r+1}^{n} x_i(t)$, $u(t) = \ltimes_{i=1}^{m} u_i(t)$, $x(t) = x^1(t) \ltimes x^2(t)$ and $\xi(t) = \ltimes_{i=1}^{q} \xi_i(t)$, one can convert (5.67) and the dynamics of $\{x_1, \cdots, x_r\}$ in (5.66) into:

$$u(t) = Kx(t), \tag{5.69}$$

and:

$$z^1(t+1) = L_{\sigma(t)} u(t) z^1(t) z^2(t) \xi(t), \tag{5.70}$$

respectively, where $K \in \mathcal{L}_{2^m \times 2^n}$ and $L_i \in \mathcal{L}_{2^r \times 2^{m+n+q}}, i \in \mathscr{A}$. Noting that the output of the system (5.66) is only dependent on $\{x_1, \cdots, x_r\}$, to make the disturbance-independent output decoupling problem solvable, we should design a control $u(t) = Kx(t)$ such that for any $\sigma \in \mathscr{A}$, x^2 and ξ are redundant variables in (5.70).

Next, to use the redundant variable separation technique, we partition L_σ into 2^m equal blocks as:

$$L_\sigma = \left[\left(L_\sigma\right)_1 \ \left(L_\sigma\right)_2 \ \cdots \ \left(L_\sigma\right)_{2^m} \right],$$

where $\left(L_\sigma\right)_i \in \mathcal{L}_{2^r \times 2^{n+q}}, i = 1, 2, \cdots, 2^m$. For each integer $1 \leq i \leq 2^m$, partition $\left(L_\sigma\right)_i$ into 2^n equal blocks as:

$$\left(L_\sigma\right)_i = \left[\left(L_\sigma\right)_{i,1} \ \left(L_\sigma\right)_{i,2} \ \cdots \ \left(L_\sigma\right)_{i,2^n} \right],$$

where $\left(L_\sigma\right)_{i,j} \in \mathcal{L}_{2^r \times 2^q}, j = 1, 2, \cdots, 2^n$.

Then, for a given state feedback gain matrix $K = \delta_{2^m}[v_1 \ v_2 \ \cdots \ v_{2^n}]$, it is easy to obtain that:

$$\begin{aligned} x^1(t+1) &= L_\sigma K x(t) x^1(t) x^2(t) \xi(t) \\ &= \left[\left(L_\sigma\right)_{v_1,1} \ \left(L_\sigma\right)_{v_2,2} \ \cdots \ \left(L_\sigma\right)_{v_{2^n},2^n} \right] x^1(t) x^2(t) \xi(t). \end{aligned} \tag{5.71}$$

Thus, the disturbance-independent output decoupling problem is solvable by $u(t) = Kx(t)$, if and only if for any $\sigma \in \mathscr{A}$, x^2 and ξ are redundant variables in (5.71).

Lemma 5.4

For the given logical mapping, $G, x_{r+1}, \cdots, x_n$ are redundant variables, if and only if:

$$rank(M_i) = 1, \ i = 1, \cdots, 2^r,$$

where $M_G = [M_1\ M_2\ \cdots M_{2^r}]$ is the structural matrix of G. ■

By Lemma 5.4, for any $\sigma \in \mathscr{A}$, x^2 and ξ are redundant variables in (5.71) if and only if for any $s = 1, \cdots, 2^r$, $\sigma = 1, \cdots, w$,

$$rank\left(\left[\left(L_\sigma\right)_{v_{s(1)},s(1)} \cdots \left(L_\sigma\right)_{v_{s(2^{n-r})},s(2^{n-r})}\right]\right) = 1 \tag{5.72}$$

where $s(l) := (s-1)2^{n-r} + l$, $\forall l = 1, \cdots, 2^{n-r}$.

Now, for any integer $1 \le j \le 2^n$, define the following sets:

$$\begin{aligned}\Gamma_j = \Big\{&(k_1, \cdots, k_w) \in \{1, 2, \cdots, 2^r\}^w : \text{ there exists an integer } 1 \le i \le 2^m \\ &\text{such that } \left(L_\sigma\right)_{i,j} = \delta_{2^r}[k_\sigma \ \cdots \ k_\sigma], \sigma = 1, \cdots, w\Big\};\end{aligned} \tag{5.73}$$

$$\begin{aligned}\Pi_j^{(k_1,\cdots,k_w)} = \Big\{&i \in \{1, 2, \cdots, 2^m\} : \left(L_\sigma\right)_{i,j} = \delta_{2^r}[k_\sigma \ \cdots \ k_\sigma], \\ &\sigma = 1, \cdots, w\Big\}, \ \forall\ (k_1, \cdots, k_w) \in \Gamma_j.\end{aligned} \tag{5.74}$$

Based on the above analysis, we have the following result.

Theorem 5.8

The disturbance-independent output decoupling problem is solvable, if and only if:

$$\Lambda_s := \bigcap_{l=1}^{2^{n-r}} \Gamma_{(s-1)2^{n-r}+l} \neq \emptyset, \ \forall\ s = 1, 2, \cdots, 2^r. \tag{5.75}$$

Moreover, if (5.75) holds, then all the state feedback gain matrices are designed as:

$$K = \delta_{2^m}[v_1\ v_2\ \cdots\ v_{2^n}],$$

with,

$$v_j \in \Upsilon_j := \bigcup_{(k_{1,s},\cdots,k_{w,s}) \in \Lambda_s} \Pi_j^{(k_{1,s},\cdots,k_{w,s})}, \ j = 1, 2, \cdots, 2^n, \tag{5.76}$$

where $1 \le s \le 2^r$ is the unique integer such that $1 \le j - (s-1)2^{n-r} \le 2^{n-r}$. ■

Proof. We just prove that all the state feedback gain matrices are $K = \delta_{2^m}[v_1\ v_2\ \cdots\ v_{2^n}]$ with v_j being given in (5.76). In fact, given a logical matrix $K = \delta_{2^m}[v_1\ v_2\ \cdots\ v_{2^n}]$, one can see that the disturbance-independent output decoupling problem is solvable under the control $u(t) = Kx(t)$, if and only if (5.72) holds, say,

$$\left[\left(L_\sigma\right)_{v_{s(1)},s(1)}\left(L_\sigma\right)_{v_{s(2)},s(2)}\cdots\left(L_\sigma\right)_{v_{s(2^{n-r})},s(2^{n-r})}\right]$$

$$=\delta_{2^r}[k_{\sigma,s}\ \cdots\ k_{\sigma,s}],\ \forall\ s=1,2,\cdots,2^r,\ \forall\ \sigma=1,\cdots,w$$

hold for some $k_{\sigma,s}\in\{1,2,\cdots,2^r\}$, which implies that $(k_{1,s},\cdots,k_{w,s})\in\Lambda_s$, $\forall\ s=1,2,\cdots,2^r$. Therefore,

$$v_j\in\bigcup_{(k_{1,s},\cdots,k_{w,s})\in^{\xi}\Lambda_s}\Pi_j^{(k_{1,s},\cdots,k_{w,s})}$$

hold for $\forall\ j=(s-1)2^{n-r}+1,(s-1)2^{n-r}+2,\cdots,s2^{n-r}$, $\forall\ s=1,2,\cdots,2^r$. This completes the proof. ∎

It should be pointed out that all the state feedback controllers obtained in Theorem 5.8 are mode-independent. Next, we study how to design a mode-dependent state feedback control:

$$u(t)=K_{\sigma(t)}x(t) \tag{5.77}$$

for the DDP, where $K_\sigma=\delta_{2^m}[v_1^\sigma\ \cdots\ v_{2^n}^\sigma]\in\mathscr{L}_{2^m\times 2^n}$.

For any integer $1\le j\le 2^n$, define:

$$\begin{aligned}\widehat{\Gamma}_j=&\Big\{(k_1,\cdots,k_w)\in\{1,\cdots,2^r\}^w:\ \text{for any }\sigma=1,\cdots,w,\text{ there exists an integer}\\&1\le i_\sigma\le 2^m\text{ such that}\left(L_\sigma\right)_{i_\sigma,j}=\delta_{2^r}[k_\sigma\ \cdots\ k_\sigma]\Big\};\end{aligned} \tag{5.78}$$

$$\begin{aligned}\widehat{\Pi}_j^{(k_1,\cdots,k_w)}=&\Big\{(i_1,\cdots,i_w)\in\{1,2,\cdots,2^m\}^w:\left(L_\sigma\right)_{i_\sigma,j}=\delta_{2^r}[k_\sigma\ \cdots\ k_\sigma],\\&\sigma=1,\cdots,w\Big\},\ \forall\ (k_1,\cdots,k_w)\in\widehat{\Gamma}_j.\end{aligned} \tag{5.79}$$

We have the following result on the design of mode-dependent state feedback controllers.

Theorem 5.9

The disturbance-independent output decoupling problem is solvable under a state feedback control in the form of (5.77), if and only if:

$$\widehat{\Lambda}_s:=\bigcap_{l=1}^{2^{n-r}}\widehat{\Gamma}_{(s-1)2^{n-r}+l}\neq\emptyset,\ \forall\ s=1,2,\cdots,2^r. \tag{5.80}$$

Moreover, if (5.80) holds, then all the mode-dependent state feedback gain matrices are designed as:

$$K_\sigma=\delta_{2^m}[v_1^\sigma\ v_2^\sigma\ \cdots\ v_{2^n}^\sigma],$$

with:

$$(v_j^1, \cdots, v_j^w) \in \widehat{\Upsilon}_j := \bigcup_{(k_{1,s}, \cdots, k_{w,s}) \in \widehat{\Lambda}_s} \widehat{\Pi}_j^{(k_{1,s}, \cdots, k_{w,s})}, \ j = 1, 2, \cdots, 2^n, \tag{5.81}$$

where $1 \leq s \leq 2^r$ is the unique integer such that $1 \leq j - (s-1)2^{n-r} \leq 2^{n-r}$. ■

Proof. The proof of this theorem is similar to that of Theorem 5.8. We omit it. ▮

Finally, based on the results obtained in Theorem 5.8, we study how to design output feedback disturbance decoupling controllers in the form of:

$$u_i(t) = \psi_i(y_1(t), \cdots, y_p(t)), \ i = 1, \cdots, m, \tag{5.82}$$

where $\psi_i : \mathscr{D}^p \mapsto \mathscr{D}, i = 1, \cdots, m$ are logical functions to be determined.

Using the vector form of logical variables and setting $x(t) = \ltimes_{i=1}^n x_i(t)$, $u(t) = \ltimes_{i=1}^m u_i(t)$ and $y(t) = \ltimes_{i=1}^p y_i(t)$, by semi-tensor product method, one can convert (5.82) and the output equations in (5.66) into:

$$u(t) = Gy(t) \tag{5.83}$$

and,

$$y(t) = Hz(t), \tag{5.84}$$

where $G = \delta_{2^m}[\alpha_1 \ \alpha_2 \ \cdots \ \alpha_{2^p}] \in \mathscr{L}_{2^m \times 2^p}, H = \delta_{2^p}[h_1 \ h_2 \ \cdots \ h_{2^n}] \in \mathscr{L}_{2^p \times 2^n}$.

For each integer $1 \leq k \leq 2^p$, $\mathscr{O}(k)$ denotes the set of states whose output is $\delta_{2^p}^k$. Noting that $\widehat{H}\delta_{2^n}^i = Col_i(\widehat{H})$, we have:

$$\mathscr{O}(k) = \{\delta_{2^n}^i : Col_i(H) = \delta_{2^p}^k\}. \tag{5.85}$$

It is easy to see that $\mathscr{O}(k_1) \bigcap \mathscr{O}(k_2) = \emptyset$, $\forall \ k_1 \neq k_2$, and $\bigcup_{k=1}^{2^p} \mathscr{O}(k) = \Delta_{2^n}$.

Now, for each set $\mathscr{O}(k), k = 1, 2, \cdots, 2^p$, we construct a set, denoted by I(k), as:

$$I(k) = \begin{cases} \bigcap_{\delta_{2^n}^i \in \mathscr{O}(k)} \Upsilon_i, \text{ if } \mathscr{O}(k) \neq \emptyset, \\ \{1, 2, \cdots, 2^m\}, \text{ if } \mathscr{O}(k) = \emptyset, \end{cases} \tag{5.86}$$

where $\Upsilon_i, i = 1, 2, \cdots, 2^n$ are defined in (5.76). Then, we have the following result.

Theorem 5.10

Suppose that (5.75) holds. Then, the disturbance-independent output decoupling problem is solvable by an output feedback control, if and only if:

$$I(k) \neq \emptyset, \ \forall \ k = 1, 2, \cdots, 2^p. \tag{5.87}$$

Moreover, if (5.87) holds, then all the output feedback gain matrices are:

$$G = \delta_{2^m}[\alpha_1\ \alpha_2\ \cdots\ \alpha_{2^p}],\ \alpha_k \in I(k). \tag{5.88}$$

■

Proof. (Sufficiency). Assume that (5.87) holds. We construct an output feedback control in the form of (5.88). Then, for $H = \delta_{2^p}[h_1\ h_2\ \cdots\ h_{2^n}]$, it is easy to see that:

$$GH = \delta_{2^m}[\alpha_{h_1}\ \alpha_{h_2}\ \cdots\ \alpha_{h_{2^n}}]. \tag{5.89}$$

Since $\alpha_{h_i} \in I(h_i) \subseteq \Upsilon_i$, $\forall\ i = 1,2,\cdots,2^n$, from Theorem 5.8, we conclude that the disturbance-independent output decoupling problem is solvable by the state feedback control $u(t) = (GH)x(t)$. Therefore, the disturbance-independent output decoupling problem is solvable by the output feedback control $u(t) = Gy(t)$, which implies that the sufficiency holds.

(Necessity) Suppose that the disturbance-independent output decoupling problem is solvable by an output feedback control, say, $u(t) = Gy(t) = \delta_{2^m}[\alpha_1\ \alpha_2\ \cdots\ \alpha_{2^p}]y(t)$. We prove that (5.87) holds.

In fact, if (5.87) does not hold, then there exists an integer $1 \leq k \leq 2^p$ such that $I(k) = \emptyset$. In this case, one can see from (5.86) that $\mathscr{O}(k) \neq \emptyset$ and $\bigcap_{\delta_{2^n}^i \in \mathscr{O}(k)} \Upsilon_i = \emptyset$.

Assume that $\mathscr{O}(k) = \{\delta_{2^n}^{i_1},\cdots,\delta_{2^n}^{i_q}\}$.

On the other hand, since the disturbance-independent output decoupling problem is solvable by $u(t) = Gy(t)$, one can see that $u(t) = (GH)x(t)$ is a state feedback disturbance-independent output decoupling controller.

For $H = \delta_{2^p}[h_1\ h_2\ \cdots\ h_{2^n}]$, from (5.89) and the fact that $h_{i_1} = \cdots = h_{i_q} = k$, we have:

$$\alpha_{h_{i_1}} = \cdots = \alpha_{h_{i_q}} = \alpha_k.$$

Hence, $\alpha_k \in \bigcap_{\delta_{2^n}^i \in \mathscr{O}(k)} \Upsilon_i$, which is a contradiction to $\bigcap_{\delta_{2^n}^i \in \mathscr{O}(k)} \Upsilon_i = \emptyset$.

Thus, $I(k) \neq \emptyset$, $\forall\ k = 1,2,\cdots,2^p$. ■

Example 5.8:

Consider the following switched BCN:

$$\begin{cases} x_1(t+1) = f_1^{\sigma(t)}(x_1(t),x_2(t),x_3(t),u_1(t),u_2(t),\xi(t)), \\ x_2(t+1) = f_2^{\sigma(t)}(x_1(t),x_2(t),x_3(t),u_1(t),u_2(t),\xi(t)), \\ x_3(t+1) = f_3^{\sigma(t)}(x_1(t),x_2(t),x_3(t),u_1(t),u_2(t),\xi(t)), \\ y(t) = x_1(t) \wedge x_2(t), \end{cases} \tag{5.90}$$

where $\sigma : \mathbb{N} \mapsto \mathscr{A} = \{1,2\}$ is the switching signal, and,

$$f_1^1 = x_2 \wedge [(x_3 \to \xi) \vee u_1],\ f_2^1 = \neg x_1 \vee (\xi \wedge u_2),\ f_3^1 = x_1 \vee u_1,$$
$$f_1^2 = x_2 \wedge \xi \wedge u_2,\ f_2^2 = x_1 \vee x_3 \vee \xi \vee u_1,\ f_3^2 = (x_2 \leftrightarrow \xi) \wedge u_1.$$

Our objective is to design mode-independent state feedback controllers for the disturbance-independent output decoupling of the system (5.90).

For this example, we have the following structure matrix:

$$\begin{aligned} L_1 \quad = \quad & \delta_4[1\,2\,1\,2\,3\,4\,3\,4\,1\,1\,1\,1\,3\,3\,3\,3 \\ & 2\,2\,2\,2\,4\,4\,4\,4\,1\,1\,1\,1\,3\,3\,3\,3 \\ & 1\,4\,1\,2\,3\,4\,3\,4\,1\,3\,1\,1\,3\,3\,3\,3 \\ & 2\,4\,2\,2\,4\,4\,4\,4\,1\,3\,1\,1\,3\,3\,3\,3], \end{aligned}$$

and,

$$\begin{aligned} L_2 \quad = \quad & \delta_4[1\,3\,1\,3\,3\,3\,3\,3\,1\,3\,1\,3\,3\,3\,3\,3 \\ & 3\,3\,3\,3\,3\,3\,3\,3\,3\,3\,3\,3\,3\,3\,3\,3 \\ & 1\,3\,1\,3\,3\,3\,3\,3\,1\,3\,1\,4\,3\,3\,3\,4 \\ & 3\,3\,3\,3\,3\,3\,3\,3\,3\,3\,3\,4\,3\,3\,3\,4]. \end{aligned}$$

From (5.73) and (5.74), we have $\Gamma_1 = \{(2,3)\}$, $\Gamma_2 = \{(2,3)\}$, $\Gamma_3 = \{(4,3)\}$, $\Gamma_4 = \{(4,3)\}$, $\Gamma_5 = \{(1,3)\}$, $\Gamma_6 = \{(1,3)\}$, $\Gamma_7 = \{(3,3)\}$, $\Gamma_8 = \{(3,3)\}$. Since $\Lambda_1 = \Gamma_1 \bigcap \Gamma_2 = \{(2,3)\}$, $\Lambda_2 = \Gamma_3 \bigcap \Gamma_4 = \{(4,3)\}$, $\Lambda_3 = \Gamma_5 \bigcap \Gamma_6 = \{(1,3)\}$, $\Lambda_4 = \Gamma_7 \bigcap \Gamma_8 = \{(3,3)\}$. By Theorem 5.8, the disturbance-independent output decoupling problem is solvable, and all the mode-independent state feedback gain matrices are:

$$K = \delta_4[v_1\ v_2\ v_3\ v_4\ v_5\ v_6\ v_7\ v_8],$$

where $v_1 \in \Upsilon_1 = \Pi_1^{(2,3)} = \{2\}$, $v_2 \in \Upsilon_2 = \Pi_2^{(2,3)} = \{2,4\}$, $v_3 \in \Upsilon_3 = \Pi_3^{(4,3)} = \{2,4\}$, $v_4 \in \Upsilon_4 = \Pi_4^{(4,3)} = \{2,4\}$, $v_5 \in \Upsilon_5 = \Pi_5^{(1,3)} = \{2\}$, $v_6 \in \Upsilon_6 = \Pi_6^{(1,3)} = \{2\}$, $v_7 \in \Upsilon_7 = \Pi_7^{(3,3)} = \{1,2,3,4\}$, $v_8 \in \Upsilon_8 = \Pi_8^{(3,3)} = \{1,2\}$. For example, letting $v_1 = v_2 = v_3 = v_4 = v_5 = v_6 = 2$, and $v_7 = v_8 = 1$, the corresponding state feedback control is: $u_1(t) = 1$, $u_2(t) = \neg x_1(t) \wedge \neg x_2(t)$.

Example 5.9:

Consider the following Boolean control network, which is a reduced model of the lac operon in the Escherichia coli [12]:

$$\begin{cases} x_1(t+1) = \neg u_1(t) \wedge (x_2(t) \vee x_3(t)), \\ x_2(t+1) = \neg u_1(t) \wedge u_2(t) \wedge x_1(t) \wedge \xi(t), \\ x_3(t+1) = \neg u_1(t) \wedge (u_2(t) \vee (u_3(t) \wedge x_1(t))), \end{cases} \tag{5.91}$$

where x_1, x_2 and x_3 are state variables which denote the lac mRNA, the lactose in high concentrations, and the lactose in medium concentrations, respectively; u_1, u_2 and u_3 are control inputs which represent the extracellular glucose, the high extracellular lactose, and the medium extracellular lactose, respectively; ξ is an external disturbance.

In this example, we are interested in observing the genes x_1 and x_2 of the system (5.91). Then, the output equation is:

$$\begin{cases} y_1(t+1) = x_1(t), \\ y_2(t+1) = x_2(t), \end{cases} \tag{5.92}$$

The objective is to design all possible output feedback controllers for the DDP of the system (5.91) with the output equation (5.92). From (5.70), for this example, we have:

$$\begin{aligned} L_1 \quad = \quad & \delta_4[4\,4\,4\,4\,4\,4\,4\,4\,4\,4\,4\,4\,4\,4\,4\,4 \\ & 4\,4\,4\,4\,4\,4\,4\,4\,4\,4\,4\,4\,4\,4\,4\,4 \\ & 4\,4\,4\,4\,4\,4\,4\,4\,4\,4\,4\,4\,4\,4\,4\,4 \\ & 4\,4\,4\,4\,4\,4\,4\,4\,4\,4\,4\,4\,4\,4\,4\,4 \\ & 1\,2\,1\,2\,1\,2\,3\,4\,2\,2\,2\,2\,2\,2\,4\,4 \\ & 1\,2\,1\,2\,1\,2\,3\,4\,2\,2\,2\,2\,2\,2\,4\,4 \\ & 2\,2\,2\,2\,2\,2\,4\,4\,2\,2\,2\,2\,2\,2\,4\,4 \\ & 2\,2\,2\,2\,2\,2\,4\,4\,2\,2\,2\,2\,2\,2\,4\,4]. \end{aligned}$$

Then, from (5.73)–(5.76), for this example, one can obtain $\Upsilon_1 = \{1,2,3,4,7,8\}$, $\Upsilon_2 = \{1,2,3,4,7,8\}$, $\Upsilon_3 = \{1,2,3,4\}$, $\Upsilon_4 = \{1,2,3,4,7,8\}$, $\Upsilon_5 = \{1,2,3,4,5,6,7,8\}$, $\Upsilon_6 = \{1,2,3,4,5,6,7,8\}$, $\Upsilon_7 = \{1,2,3,4\}$, $\Upsilon_8 = \{1,2,3,4,5,6,7,8\}$.

From (5.84) and (5.85), for this example, it is easy to see that $H = \delta_4[1\ 1\ 2\ 2\ 3\ 3\ 4\ 4]$, $\mathcal{O}(1) = \{\delta_8^1, \delta_8^2\}$, $\mathcal{O}(2) = \{\delta_8^3, \delta_8^4\}$, $\mathcal{O}(3) = \{\delta_8^5, \delta_8^6\}$ and $\mathcal{O}(4) = \{\delta_8^7, \delta_8^8\}$.

Then, one can obtain from (5.86) that $I(1) = \Upsilon_1 \bigcap \Upsilon_2 = \{1,2,3,4,7,8\}$, $I(2) = \Upsilon_3 \bigcap \Upsilon_4 = \{1,2,3,4\}$, $I(3) = \Upsilon_5 \bigcap \Upsilon_6 = \{1,2,3,4,5,6,7,8\}$, $I(4) = \Upsilon_7 \bigcap \Upsilon_8 = \{1,2,3,4\}$.

By Theorem 5.10, the DDP is solvable, and all the output feedback gain matrices are:

$$G = \delta_8[\alpha_1\ \alpha_2\ \alpha_3\ \alpha_4]$$

with $\alpha_i \in I(i)$, $i = 1,2,3,4$. For example, letting $\alpha_1 = 8$, $\alpha_2 = 4$, $\alpha_3 = 8$ and $\alpha_4 = 2$, the corresponding output feedback control is:

$$\begin{cases} u_1(t+1) = \neg y_2(t), \\ u_2(t+1) = \neg y_1(t) \wedge \neg y_2(t), \\ u_3(t) = 0. \end{cases} \tag{5.93}$$

REFERENCES

1. Lewin, B. (2000). Genes VII. Cambridge, Oxford Univ. Press.
2. El-Farra, N. H., Gani, A. and Christofides, P. D. (2005). Analysis of mode transitions in biological networks. AIChE J., 51(8): 2220–2234.
3. Agrachev, A. A. and Liberzon, D. (2001). Lie-algebraic stability criteria for switched systems. SIAM J. Control Optim., 40: 253–269.
4. Sun, Z. (2012). Robust switching of discrete-time switched linear systems. Automatica, 48(1): 239–242.
5. Dehghan, M. and Ong, C. J. (2012). Discrete-time switching linear system with constraints: Characterization and computation of invariant sets under dwell-time consideration. Automatica, 48: 964–969.

6. Kobayashi, K. and Hiraishi, K. (2011). Optimal control of asynchronous Boolean networks modeled by petri nets. Proc. 2nd Int. Workshop Biol. Process Petri Nets, 7–20.
7. Laschov, D. and Margaliot, M. (2012). Controllability of Boolean control networks via the Perron-Frobenius theory. Automatica, 48: 1218–1223.
8. Cheng, D. (2011). Disturbance decoupling of Boolean control networks. IEEE Trans. Aut. Contr., 56(1): 2–10.
9. Li, H., Wang, Y. and Liu, Z. (2014). Stability analysis for switched Boolean networks under arbitrary switching signals. IEEE Trans. Aut. Contr., 59(7): 1978–1982.
10. Li, H. and Wang, Y. (2015). Controllability analysis and control design for switched Boolean networks with state and input constraints. SIAM J. Contr. Optim., 53(5): 2955–2979.
11. Li, H., Wang, Y., Xie, L. and Cheng, D. (2014). Disturbance decoupling control design for switched Boolean control networks. Systems and Control Letters, 72: 1–6.
12. Li, H., Wang, Y. and Liu, Z. (2013). Simultaneous stabilization for a set of Boolean control networks. Systems and Control Letters, 62(12): 1168–1174.

6 Probabilistic Logical Networks

6.1 INTRODUCTION TO PROBABILISTIC LOGICAL NETWORKS

This section introduces some basic concepts of probabilistic logical networks.

Consider a mix-valued logical control network as follows:

$$\begin{cases} x_1(t+1) = f_1(u_1(t), \cdots, u_m(t), x_1(t) \cdots, x_n(t)), \\ x_2(t+1) = f_2(u_1(t), \cdots, u_m(t), x_1(t) \cdots, x_n(t)), \\ \vdots \\ x_n(t+1) = f_n(u_1(t), \cdots, u_m(t), x_1(t) \cdots, x_n(t)), \end{cases} \tag{6.1}$$

where $x_i(t) \in \Delta_{k_i}$, $i = 1, 2, \cdots, n$, are the states, $u_j(t) \in \Delta_{q_j}$, $j = 1, 2, \cdots, m$, are the input controls and

$$f_i : \Delta_{q_1} \times \cdots \times \Delta_{q_m} \times \Delta_{k_1} \times \cdots \times \Delta_{k_n} \to \Delta_{k_i}, \quad i = 1, 2, \cdots, n$$

are mix-valued logical functions.

Using the semi-tensor product (refer to [1]), system (6.1) can be expressed in an algebraic form as:

$$x(t+1) = Lu(t)x(t), \tag{6.2}$$

where $x(t) = x_1(t) \ltimes x_2(t) \ltimes \cdots \ltimes x_n(t) \in \Delta_k$, $u(t) = u_1(t) \ltimes u_2(t) \ltimes \cdots \ltimes u_n(t) \in \Delta_q$, $k = k_1 k_2 \cdots k_n$, $q = q_1 q_2 \cdots q_m$, and $L \in \mathscr{L}_{k \times qk}$ is called the structural matrix of (6.1).

Consider the mix-valued logical control network (6.1). It is called *a probabilistic mix-valued logical control network (MLCN)* if:

$$f_i \in \{f_i^j,\ j = 1, 2, \cdots, l_i\}$$

with the probability:

$$P\{f_i = f_i^j\} = p_i^j,\ j = 1, 2, \cdots, l_i,\ i = 1, 2, \cdots, n.$$

Define an index set [2] of possible models for the probabilistic MLCN as follows:

$$K = \begin{bmatrix} 1 & 1 & \cdots & 1 & 1 \\ 1 & 1 & \cdots & 1 & 2 \\ \vdots & \vdots & \ddots & \vdots & \vdots \\ 1 & 1 & \cdots & 1 & l_n \\ 1 & 1 & \cdots & 2 & 1 \\ \vdots & \vdots & \ddots & \vdots & \vdots \\ 1 & 1 & \cdots & 2 & l_n \\ 1 & 1 & \cdots & 3 & 1 \\ \vdots & \vdots & \ddots & \vdots & \vdots \\ l_1 & l_2 & \cdots & l_{n-1} & l_n \end{bmatrix} \in \mathbb{R}^{N\times n},$$

where $N = \prod_{i=1}^{n} l_i$ is the number of all possible models. Denote each model by Σ_λ, $\lambda = 1,2,\cdots,N$. Then, each model Σ_λ consists of the logical functions $\left\{f_j^{K_{\lambda j}} : j = 1,2,\cdots,n\right\}$ and has the probability:

$$p_\lambda = P\{\text{the network } \Sigma_\lambda \text{ is selected}\} = \prod_{j=1}^{n} P_j^{K_{\lambda j}},$$

where $K_{\lambda j}$ denotes the (λ, j)-th element of the matrix K, $\lambda = 1,2,\cdots,N$.

Similar to (6.2), using the semi-tensor product, each model of the probabilistic MLCN can be converted into:

$$x(t+1) = L_\lambda u(t)x(t),\ \lambda = 1,2,\cdots,N. \tag{6.3}$$

Hence, the overall expected value of $x(t+1)$ is:

$$Ex(t+1) = \sum_{\lambda=1}^{N} p_\lambda L_\lambda u(t)Ex(t) := Lu(t)Ex(t), \tag{6.4}$$

where,

$$L = \sum_{\lambda=1}^{N} p_\lambda L_\lambda \tag{6.5}$$

is called the transition matrix.

To explain the meaning of the transition matrix L, we now split L into q equal blocks, that is, $L = [L^1\ L^2\ \cdots L^q]$, where $L^s \in \mathcal{L}_{k\times k}$, $s = 1,2,\cdots,q$. Consider the control $u(t) = \delta_q^s$ and the (i,j)-th element, L_{ij}^s, of L^s. It is easy to see that L_{ij}^s is just the probability of transition from the state $x(t) = \delta_k^j$ to the state $x(t+1) = \delta_k^i$ with the control $u(t) = \delta_q^s$. That is, $P\left\{x(t+1) = \delta_k^i \,|\, x(t) = \delta_k^j, u(t) = \delta_q^s\right\} = L_{ij}^s$.

6.2 CONTROLLABILITY OF CONTEXT-SENSITIVE PROBABILISTIC LOGICAL CONTROL NETWORKS

This section first introduces two algebraic forms for context-sensitive probabilistic MLCNs with/without constrains respectively. Then, controllability of probabilistic MLCNs with constrains is analyzed. After that, we investigate optimal finite-horizon control of the probabilistic mix-valued logical networks and apply the results, obtained in this section, to first-passage model.

6.2.1 AN ALGEBRAIC FORM FOR CONTEXT-SENSITIVE PROBABILISTIC MLCNS

Consider the probabilistic MLCN (6.4). Assume that at each time, the choice of the present context Σ_λ is probability-based on the previous one, and this random switching is governed by a random logical variable with a small probability r. That is, if the context Σ_α is active at time t, then it is still active at time $t+1$ with probability $1-r$, or it transits to another context Σ_β at time $t+1$ with probability $r\frac{p_\beta}{1-p_\alpha}$, where $\beta = 1,2,\cdots,N$ and $\beta \neq \alpha$. In this situation, the system (6.4) is called *a context-sensitive probabilistic MLCN*.

Now, we derive the transition probability matrix of the system.

Theorem 6.1

Consider the context-sensitive probabilistic MLCN (6.3). Then, its transition matrix is expressed as:

$$\tilde{L} = (1-r)\cdot L + r\cdot \hat{L}, \tag{6.6}$$

where $L = \sum_{\lambda=1}^{N} p_\lambda L_\lambda$ is defined as (6.5), and

$$\hat{L} = \sum_{\alpha=1}^{N}\sum_{\beta\neq\alpha}\frac{p_\alpha}{1-p_\alpha}p_\beta L_\beta = \sum_{\alpha=1}^{N}\left(\sum_{\beta=1,\beta\neq\alpha}^{N}\frac{p_\beta}{1-p_\beta}\right)p_\alpha L_\alpha. \tag{6.7}$$

is called the switching transition matrix. ■

Proof. Assume that the present state is $x(t)$ and the next state is $x(t+1)$. The transition probability from $x(t)$ to $x(t+1)$ is:

$$\begin{aligned} & P\Big\{x(t+1) \mid x(t), u(t)\Big\} \\ = & \sum_{\lambda=1}^{N} P\Big\{x(t+1) \mid x(t) \text{ is governed by the } \lambda\text{-th context}\Big\}\cdot p_\lambda. \end{aligned} \tag{6.8}$$

For the next context, that is, the dynamics of $x(t+1)$, we consider the following two cases:

Case 1: the context remains the same as the previous one, that is, there is no switching. Assume that the same context is α-th context Σ_α. In this case, the probability of transition from $x(t)$ to $x(t+1)$ is:

$$(1-r)\left[x^T(t+1)L_\alpha u(t)x(t)\right]. \tag{6.9}$$

Case 2: the new context is selected as Σ_β, $\beta \neq \alpha$. In this case, the probability of transition from $x(t)$ to $x(t+1)$ is:

$$r \cdot \frac{p_\beta}{1-p_\alpha}\left[x^T(t+1)L_\beta u(t)x(t)\right]. \tag{6.10}$$

Hence, when $x(t)$ is governed by the α-th context, from (6.9) and (6.10), the overall probability of transition from $x(t)$ to $x(t+1)$ is:

$$\begin{aligned} & P\{x(t+1)\,|\,x(t) \text{ is governed by the } \alpha\text{-th context}\} \\ = \;& x^T(t+1)\left[(1-r)L_\alpha + r\sum_{\beta\neq\alpha}\frac{p_\beta L_\beta}{1-p_\alpha}\right]u(t)x(t). \end{aligned} \tag{6.11}$$

According to (6.8) and (6.11), we have:

$$\begin{aligned} & P\left\{x(t+1)\mid x(t),u(t)\right\} \\ = \;& x^T(t+1)\left[(1-r)\sum_{\alpha=1}^{N}p_\alpha L_\alpha + r\sum_{\alpha=1}^{N}\sum_{\beta\neq\alpha}\frac{p_\alpha}{1-p_\alpha}p_\beta L_\beta\right]u(t)x(t). \end{aligned}$$

Thus, the proof is completed. ∎

COROLLARY 6.1

Consider the context-sensitive probabilistic MLCN (6.3). It can be expressed as an algebraical form:

$$Ex(t+1) = \tilde{L}u(t)Ex(t). \tag{6.12}$$

It is clear that the transition matrix $\tilde{L}$ of the context-sensitive probabilistic MLCN consists of two parts: the general transition matrix L and the switching transition matrix $\hat{L}$. To explain the meaning of the transition matrix $\tilde{L}$, we now split $\tilde{L}$ into q equal blocks, that is, $\tilde{L} = [L^1\ L^2\ \cdots L^q]$, where $L^s \in \mathscr{L}_{k\times k}$, $s = 1,2,\cdots,q$. Consider the control $u(t) = \delta_q^s$ and the (i,j)-th element, L_{ij}^s, of L^s. It is easy to see that L_{ij}^s is just the probability of transition from the state $x(t) = \delta_k^j$ to the state $x(t+1) = \delta_k^i$ under the control $u(t) = \delta_q^s$, that is,

$$P\left\{x(t+1) = \delta_k^j\,|\,x(t) = \delta_k^i,\ u(t) = \delta_q^s\right\} = x^T(t+1)\tilde{L}u(t)x(t) = L_{ij}^s. \tag{6.13}$$

6.2.2 AN ALGEBRAIC FORM OF PROBABILISTIC MLCNS WITH CONSTRAINTS

In the following section, we consider the context-sensitive probabilistic MLCN (6.12) with constraints on states and controls.

Denote the sets of states and controls by C_x and C_u, respectively. Let:

$$C_x = \left\{\delta_k^{i_1}, \delta_k^{i_2}, \cdots, \delta_k^{i_\alpha}\right\} \subset \Delta_k \ (i_1 < i_2 < \cdots < i_\alpha)$$

and,

$$C_u = \left\{\delta_q^{j_1}, \delta_q^{j_2}, \cdots, \delta_q^{j_\beta}\right\} \subset \Delta_q \ (j_1 < j_2 < \cdots < j_\beta),$$

which are proper subsets of Δ_k and Δ_q respectively, where $x = x_1 \ltimes x_2 \ltimes \cdots \ltimes x_n$ and $u = u_1 \ltimes u_2 \ltimes \cdots \ltimes u_m$.

It is obvious that $P\{x(t+1) \mid u(t), x(t)\} = 0$ if $u(t) \notin C_u$. Thus, we set $L_{ij}^s = 0$ for all i, j and $s \neq j_1, \cdots, j_\beta$. Since $x(t) \in C_x$, let $L_{i,j}^{\lambda_s} = 0$ for all s and $i \neq i_1, \cdots, i_\alpha$ or $j \neq i_1, \cdots, i_\alpha$.

Next, we further study the transition matrix of the system with constraints on states and controls. Denote by Π_{C_u} the matrix with $\delta_q^{j_i}$ as its j_i-th column ($i = 1, 2, \cdots, \beta$) and other columns being zero, and by Π_{C_x} the matrix with $\delta_k^{i_j}$ as its i_j-th column ($j = 1, 2, \cdots, \alpha$) and other columns being zero.

Set,

$$\check{L} = \tilde{L} \ltimes \Pi_{C_u}. \tag{6.14}$$

Then, we have the following results.

PROPOSITION 6.1

If $u(t) \in C_u$, then $\check{L}u(t) = \tilde{L}u(t)$. If $u(t) \notin C_u$, then $\check{L}u(t) = 0_{k\times k}$, where $0_{k\times k}$ denotes the $k \times k$ zero matrix.

PROPOSITION 6.2

If $x(t) \in C_x$, then $\Pi_{C_x} \ltimes x(t) = x(t)$. If $x(t) \notin C_x$, then $\Pi_{C_x} \ltimes x(t) = 0_k$ and $\Pi_{C_x}^T \ltimes x(t) = 0_k$, where 0_k denotes the zero vector of dimension k.

It is noted that when $x(t) \notin C_x$, $u(t) \notin C_u$ or $x(t+1) \notin C_x$, the transition probability is regarded as zero, that is,

$$P\{x(t+1) \,|\, x(t), u(t)\} = 0. \tag{6.15}$$

Therefore, we obtain the following results.

Theorem 6.2

The transition matrix of the context-sensitive probabilistic MLCN (6.12) with constraints on states and controls is given as:

$$\Theta = \left[\Pi_{C_x}^T \ltimes \left(\check{L} \ltimes W_{[k,q]} \ltimes \Pi_{C_x}\right)\right] \ltimes W_{[q,k]}. \tag{6.16}$$

■

Proof. For any $x(t)$, $x(t+1)$ and $u(t)$, by (6.14) and (6.16) we obtain:

$$\begin{aligned} x^T(t+1)\Theta u(t)x(t) &= x^T(t+1)\left[\Pi_{C_x}^T \ltimes \left(\check{L} \ltimes W_{[k,q]} \ltimes \Pi_{C_x}\right)\right] \ltimes W_{[q,k]}u(t)x(t) \\ &= x^T(t+1)\Pi_{C_x}^T \ltimes \check{L} \ltimes W_{[k,q]} \ltimes \Pi_{C_x}x(t)u(t) \\ &= [\Pi_{C_x}x(t+1)]^T \ltimes \tilde{L} \ltimes [\Pi_{C_u}u(t)] \ltimes [\Pi_{C_x}x(t)]. \end{aligned}$$

If $x(t), x(t+1) \in C_x$ and $u(t) \in C_u$, from Propositions 6.1 and 6.2 we have $\Pi_{C_x}x(t+1) = x(t+1)$, $\Pi_{C_x}x(t) = x(t)$ and $\Pi_{C_u}u(t) = u(t)$. Therefore,

$$x^T(t+1)\Theta u(t)x(t) = x^T(t+1) \ltimes \tilde{L} \ltimes u(t)x(t) = P\{x(t+1)\,|\,x(t),u(t)\}.$$

If $x(t) \notin C_x$, $u(t) \notin C_u$ or $x(t+1) \notin C_x$, we have $\Pi_{C_x}x(t+1) = 0_k$, $\Pi_{C_x}x(t) = 0_k$ or $\tilde{L}\Pi_{C_u}u(t) = 0_{k\times k}$. Thus, by (6.15):

$$x^T(t+1)\Theta u(t)x(t) = 0 = P\{x(t+1)\,|\,x(t),u(t)\}.$$

Summarizing the above analysis,

$$x^T(t+1)\Theta u(t)x(t) = P\{x(t+1)\,|\,x(t),u(t)\}$$

holds for all $x(t)$, $x(t+1)$ and $u(t)$.

Thus, according to (6.13), Θ is a transition matrix of the context-sensitive probabilistic MLCN with constraints, and the proof is completed. ■

COROLLARY 6.2

The context-sensitive probabilistic MLCN with constraints can be expressed in an algebraic form as:

$$Ex(t+1) = \Theta u(t)Ex(t). \tag{6.17}$$

6.2.3 CONTROLLABILITY OF PROBABILISTIC MLCNS WITH CONSTRAINTS

First, we give the definition of controllability for the context-sensitive probabilistic MLCNs with constraints.

DEFINITION 6.1: [3]

Consider the context-sensitive constrained probabilistic MLCN (6.17).

1. Given an initial state $x(0) = x_0 \in C_x$ and a destination state $x_d \in C_x$, x_d is said to be reachable with probability 1 from the initial state x_0, if there exists an integer $s > 0$ and a sequence of control $\{u(0), u(1), \cdots, u(s-1)\} \subset C_u$ such that:

$$P\{x_d = x(s) \mid x(0) = x_0, u(0), \cdots, u(s-1)\} = 1.$$

2. The constrained probabilistic MLCN (6.17) is said to be controllable (with probability 1) at $x(0) = x_0 \in C_x$, if for any $x_d \in C_x$, x_d is reachable from x_0 (with probability 1).
3. The constrained probabilistic MLCN (6.17) is said to be controllable (with the probability 1), if for any $x_0 \in C_x$, the MLCN is controllable at x_0.

Now, we study the reachability of the constrained probabilistic MLCN (6.17) at the s-th step.

Theorem 6.3

Consider the constrained probabilistic MLCN (6.17). Then, $x_d \in C_x$ is reachable at time s from the initial state $x(0) = x_0 \in C_x$ with probability 1 under a control sequence $\{u(0), u(1), \cdots, u(s-1)\} \subset C_u$, if and only if:

$$x_d \in Col_{C_x}\{\Theta_*^s x_0\},$$

where $\Theta_* = \Theta W_{[k,q]}$ and $Col_{C_x}\{\Theta_*^s x_0\} = Col\{C_x\} \bigcap Col\{\Theta_*^s x_0\}$. ■

Proof. Using the properties of the swap matrix [1], we can rewrite (6.17) as:

$$Ex(t+1) = \Theta W_{[k,q]} Ex(t)u(t) := \Theta_* Ex(t)u(t).$$

A straightforward computation shows that:

$$\begin{aligned} Ex(1) &= \Theta_* Ex(0)u(1) = \Theta_* x(0)u(0), \\ Ex(2) &= \Theta_* Ex(1)u(1) = \Theta_*^2 x(0)u(0)u(1). \end{aligned}$$

Applying mathematical induction, we can show:

$$Ex(s) = \Theta_*^s x(0)u(0)u(1)\cdots u(s-1).$$

Thus, the proof is completed. ■

From Theorem 6.3, we can obtain the reachable set from the initial state $x_0 \in C_x$, denoted by:

$$R(x_0) = \bigcup_{s=1}^{\infty} Col_{C_x}\{\Theta_*^s x_0\}.$$

About $R(x_0)$, we have the following result.

Theorem 6.4

Consider the constrained probabilistic MLCN (6.17). Then, the state $x_0 \in C_x$ is controllable if and only if:

$$R(x_0) = C_x.$$

■

According to Theorem 3.2 in [3], if there exists an integer r such that $Col\{\Theta_*^{r+1}\} \subset Col\{\Theta_*^i | i = 1,2,\cdots,r\}$, then $R(x_0) = \bigcup_{s=1}^{r} Col_{C_x}\{\Theta_*^s x_0\}$. For Boolean networks, the existence of r was proved in [4]. Unfortunately, we find that "r" does not necessarily exist for the probabilistic MLCNs.

Based on Theorems 6.1, 6.2 and 6.4, in order to study the controllability of the context-sensitive probabilistic MLCN with constrains, we can take the following steps:

S_1: Compute the matrices L_λ and the probabilities p_λ, $\lambda = 1,2,\cdots,N$, and then obtain the transition matrix $L = \sum_{\lambda=1}^{N} p_\lambda L_\lambda$;
S_2: Compute the switching transition matrix $\hat{L}$ by (6.7), and then obtain the transition matrix of the context-sensitive probabilistic MLCN $\tilde{L}$ by (6.6);
S_3: Compute the transition matrix Θ of the constrained probabilistic MLCN by (6.14) and (6.16);
S_4: Check whether $x_d \in Col_{C_x}\{\Theta_*^s x_0\}$ for all $x_0, x_d \in C_x$.

In the following section, we study the reachability of the constrained probabilistic MLCN (6.17) with a kind of practicable control inputs [3,4].

As we know, in the medical field, the state of a constrained probabilistic MLCN is used to denote human cells, and the corresponding control input represents therapeutic measures in the treatment of diseases such as cancer. During the treatment, the past measure is often considered in the present treatment, i.e., the present control is influenced by the previous ones. Thus, the following control form is useful in the medical field:

$$\begin{cases} u_1(t+1) = & g_1(u_1(t),\cdots,u_m(t)), \\ u_2(t+1) = & g_2(u_1(t),\cdots,u_m(t)), \\ & \vdots \\ u_m(t+1) = & g_m(u_1(t),\cdots,u_m(t)), \end{cases} \tag{6.18}$$

where $g_i : \Delta_{q_1} \times \Delta_{q_2} \times \cdots \times \Delta_{q_m} \to \Delta_{q_i}$, $i = 1,2,\cdots,m$, are mix-valued logical functions.

Convert (6.18) into an algebraic form as $u(t+1) = Gu(t)$, where G is the transition matrix of (6.18). Since $u \in C_u$, we obtain a new algebraic form of (6.18) as:

$$u(t+1) = \Pi_{C_u}^T G \Pi_{C_u} u(t) := \widetilde{G}u(t). \tag{6.19}$$

For the reachability of the context-sensitive constrained probabilistic MLCN (6.17) with (6.18), we have the following result.

Theorem 6.5

Consider the context-sensitive constrained probabilistic MLCN (6.17) with the control (6.19). If:

$$x_d \in Col_{C_x}\left\{\Phi_{\widetilde{G}}(s)x(0)\right\},$$

then, $x_d \in C_x$ is s-th step reachable from the initial state $x(0) \in C_x$ with probability 1 under the control (6.19), where:

$$\Phi_{\widetilde{G}}(s) = \Theta_*^s \ltimes_{j=1}^{s-1}\left\{\left[I_{kq} \otimes \widetilde{G}^j\right][I_k \otimes M_{r,q}]\right\}.$$

■

Proof. Through a straightforward computation, we obtain:

$$Ex(1) = \Theta_* x(0)u(0), \tag{6.20}$$

From (6.19) and (6.20),

$$Ex(2) = \Theta_* Ex(1)u(1) = \Theta_*^2 x(0)u(0)\widetilde{G}u(0) = \Theta_*^2\left[I_{kq} \otimes \widetilde{G}\right][I_k \otimes M_{r,q}]x(0)u(0).$$

Then,

$$\begin{aligned} Ex(3) &= \Theta_* Ex(2)u(2) = \Theta_*^3\left[I_{kq} \otimes \widetilde{G}\right][I_k \otimes M_{r,q}]x(0)u(0)\widetilde{G}^2u(0) \\ &= \Theta_*^3\left[I_{kq} \otimes \widetilde{G}\right][I_k \otimes M_{r,q}]\left[I_{kq} \otimes \widetilde{G}^2\right][I_k \otimes M_{r,q}]x(0)u(0). \end{aligned}$$

Using the mathematical induction, we can prove that:

$$Ex(t) = \Theta_*^t \ltimes_{j=1}^{t-1}\left\{\left[I_{kq} \otimes \widetilde{G}^j\right][I_k \otimes M_{r,q}]\right\}x(0)u(0) =: \Phi_{\widetilde{G}}(t)x(0)u(0).$$

Assume that $x_d = Col_i\left[\Phi_{\widetilde{G}}(s)x(0)\right] \in C_x$. We can choose $u(0) = \delta_q^i$ such that

$$x_d = x(s) = Col_i\left[\Phi_{\widetilde{G}}(s)x(0)\right] = \Phi_{\widetilde{G}}(s)x(0)u(0).$$

Thus, the proof is completed. ■

COROLLARY 6.3

Consider the constrained probabilistic MLCN (6.17). If:

$$\bigcup_{s=1}^{\infty} Col_{C_x}\left\{\Phi_{\tilde{G}}(s)x(0)\right\} = C_x,$$

then, the state $x_0 \in C_x$ is controllable.

Theorems 6.3 and 6.5 can be applied to the system without constraints. But it is noted that θ_*, $\tilde{G}$ and Col_{C_x} need to be placed by $\tilde{L}$ in (6.6), G and Col, respectively.

The following example will illustrate the effectiveness of the results obtained in this section.

Example 6.1: A gene regulatory network containing four genes

Consider a gene regulatory network containing four genes: WNT5A, pirin, S100P and STC3. The network was used to study metastatic melanoma in [5].

These genes can take value "1" or "0", if the gene can be expressed or unexpressed respectively. When the gene is not obvious to be expressed, we assume that the gene takes value "1/2". Then, we can apply the procedure given in [5] to generate a mix-valued logical networks. Assume that WNT5A and pirin are states (denoted by x_1 and x_2, respectively), and both S100P and STC3 are controls (denoted by u_1 and u_2, respectively). Set $x_1, u_1, u_2 \in \Delta = \{0,1\}$ and $x_2 \in \Delta_3 = \{0, \frac{1}{2}, 1\}$. It is noted that x_1 is desirable if WNT5A=0 and undesirable if WNT5A=1. Therefore, the desirable state set is $C_x = \{(x_1,x_2) \mid (0,1),(0,0.5),(0,0)\}$.

It is easy to see that the dynamics of the gene regulatory network can be described as:

$$\begin{cases} x_1(t) &= f_1(x_1(t),x_2(t),u_1(t),u_2(t)), \\ x_2(t) &= f_2(x_1(t),x_2(t),u_1(t),u_2(t)). \end{cases} \tag{6.21}$$

Assume that $f_1 \in \{f_1^1, f_1^2\}$ with probability $p_1^1 = 0.6$ and $p_1^2 = 0.4$, and the switching probability $r = 0.01$. Using the vector form of logical variables and the semi-tensor product, the algebraic form of the system (6.21) is given as:

$$\begin{cases} x_1(t) &= M_1 u(t)x(t), \\ x_2(t) &= M_2 u(t)x(t), \end{cases}$$

where, $x(t) = x_1(t) \ltimes x_2(t)$, $u(t) = u_1(t) \ltimes u_2(t)$, $M_1 \in \{M_1^1, M_1^2\}$,

$$\begin{aligned} M_1^1 &= \delta_2[1\ 2\ 1\ 1\ 2\ 1\ 2\ 2\ 1\ 2\ 1\ 2\ 1\ 2\ 2\ 1\ 3\ 3\ 2\ 1\ 2\ 2\ 2\ 2], \\ M_1^2 &= \delta_2[1\ 1\ 1\ 1\ 2\ 2\ 1\ 1\ 2\ 2\ 1\ 2\ 1\ 1\ 2\ 1\ 3\ 3\ 1\ 2\ 2\ 2\ 2\ 2], \\ M_2 &= \delta_3[1\ 1\ 3\ 2\ 2\ 1\ 1\ 1\ 2\ 3\ 3\ 1\ 1\ 1\ 1\ 3\ 2\ 2\ 2\ 2\ 3\ 2\ 1\ 3], \end{aligned}$$

and M_1^j and M_2 are the structural matrices of functions f_1^j and f_2, respectively.

Hence, the two contexts corresponding to (M_1^1, M_2) and (M_1^2, M_2) are given as:

$$x(t+1) = L_i u(t)x(t), i = 1,2,$$

with the probabilities $p_1 = 0.6$ and $p_2 = 0.4$, respectively, where:

$$L_1 = \delta_6[1\ 4\ 3\ 4\ 6\ 1\ 4\ 4\ 2\ 6\ 3\ 4\ 1\ 4\ 4\ 3\ 5\ 5\ 5\ 2\ 6\ 5\ 4\ 6]$$

and,

$$L_2 = \delta_6[1\ 1\ 3\ 4\ 6\ 4\ 1\ 1\ 5\ 6\ 3\ 4\ 1\ 1\ 4\ 3\ 5\ 5\ 2\ 5\ 6\ 5\ 4\ 6].$$

Thus, we obtain

$$L = \begin{bmatrix} 1 & 0.4 & 0 & 0 & 0 & 0.6 & 0.4 & 0.4 & 0 & 0 & 0 & 0 & 1 & 0.4 & 0 & 0 & 0 & 0 & 0 & 0 & 0 & 0 & 0 & 0 \\ 0 & 0 & 0 & 0 & 0 & 0 & 0 & 0 & 0.6 & 0 & 0 & 0 & 0 & 0 & 0 & 0 & 0 & 0 & 0.4 & 0.6 & 0 & 0 & 0 & 0 \\ 0 & 0 & 1 & 0 & 0 & 0 & 0 & 0 & 0 & 0 & 1 & 0 & 0 & 0.6 & 0 & 1 & 0 & 0 & 0 & 0 & 0 & 0 & 0 & 0 \\ 0 & 0.6 & 0 & 1 & 0 & 0.4 & 0.6 & 0.6 & & 0 & 0 & 1 & 0 & 0 & 1 & 0 & 0 & 0 & 0 & 0 & 0 & 0 & 1 & 0 \\ 0 & 0 & 0 & 0 & 0 & 0 & 0 & 0 & 0.4 & 0 & 0 & 0 & 0 & 0 & 0 & 0 & 1 & 1 & 0.6 & 0.4 & 0 & 1 & 0 & 0 \\ 0 & 0 & 0 & 0 & 1 & 0 & 0 & 0 & 0 & 1 & 0 & 0 & 0 & 0 & 0 & 0 & 0 & 0 & 0 & 0 & 1 & 0 & 0 & 1 \end{bmatrix}$$

From (6.7), the switching transition matrix of the system is given as:

$$\begin{aligned} \hat{L} &= \frac{p_1 p_2}{1-p_2} \cdot L_1 + \frac{p_1 p_2}{1-p_1} L_2 = \frac{0.4 \times 0.6}{1-0.6} L_1 + \frac{0.4 \times 0.6}{1-0.4} L_2 \\ &= \begin{bmatrix} 1 & 0.6 & 0 & 0 & 0 & 0.4 & 0.6 & 0.6 & 0 & 0 & 0 & 0 & 1 & 0.6 & 0 & 0 & 0 & 0 & 0 & 0 & 0 & 0 & 0 & 0 \\ 0 & 0 & 0 & 0 & 0 & 0 & 0 & 0 & 0.4 & 0 & 0 & 0 & 0 & 0 & 0 & 0 & 0 & 0 & 0.6 & 0.4 & 0 & 0 & 0 & 0 \\ 0 & 0 & 1 & 0 & 0 & 0 & 0 & 0 & 0 & 0 & 1 & 0 & 0 & 0.4 & 0 & 1 & 0 & 0 & 0 & 0 & 0 & 0 & 0 & 0 \\ 0 & 0.4 & 0 & 1 & 0 & 0.6 & 0.4 & 0.4 & & 0 & 0 & 1 & 0 & 0 & 1 & 0 & 0 & 0 & 0 & 0 & 0 & 0 & 1 & 0 \\ 0 & 0 & 0 & 0 & 0 & 0 & 0 & 0 & 0.6 & 0 & 0 & 0 & 0 & 0 & 0 & 0 & 1 & 1 & 0.4 & 0.6 & 0 & 1 & 0 & 0 \\ 0 & 0 & 0 & 0 & 1 & 0 & 0 & 0 & 0 & 1 & 0 & 0 & 0 & 0 & 0 & 0 & 0 & 0 & 0 & 0 & 1 & 0 & 0 & 1 \end{bmatrix} \end{aligned} \tag{6.22}$$

With (6.6) and (6.22), we have:

$$\tilde{L} = (1-0.01) \cdot L + 0.01 \cdot \hat{L}.$$

On the other hand, using the vector form of logical variables, C_x and C_u can be expressed as:

$$C_x = \left\{ \delta_6^4, \delta_6^5, \delta_6^6 \right\}, \quad C_u = \Delta_4.$$

Thus, we have $\Pi_{C_u} = I_4$, $\Pi_{C_x} = \left[0_6\ 0_6\ 0_6\ \delta_6^4\ \delta_6^5\ \delta_6^6 \right]$, and,

$$Ex(t+1) = \Theta u(t) Ex(t),$$

where,

$$\begin{aligned} &\Theta \\ = &\left[\Pi_{C_x}^T \ltimes \left(\tilde{L} \ltimes \Pi_{C_u} \ltimes W_{[6,4]} \ltimes \Pi_{C_x} \right) \right] \ltimes W_{[4,6]} \\ = &\begin{bmatrix} 1 & 0.598 & 0 & 0 & 0 & 0.402 & 0.598 & 0.598 & 0 & 0 & 0 & 0 & 1 & 0.598 & 0 & 0 & 0 & 0 & 0 & 0 & 0 & 0 & 0 & 0 \\ 0 & 0 & 0 & 0 & 0 & 0 & 0 & 0 & 0.402 & 0 & 0 & 0 & 0 & 0 & 0 & 0 & 0 & 0 & 0.598 & 0.402 & 0 & 0 & 0 & 0 \\ 0 & 0 & 1 & 0 & 0 & 0 & 0 & 0 & 0 & 0 & 1 & 0 & 0 & 0.402 & 0 & 1 & 0 & 0 & 0 & 0 & 0 & 0 & 0 & 0 \\ 0 & 0.402 & 0 & 1 & 0 & 0.598 & 0.402 & 0.403 & & 0 & 0 & 1 & 0 & 0 & 1 & 0 & 0 & 0 & 0 & 0 & 0 & 0 & 1 & 0 \\ 0 & 0 & 0 & 0 & 0 & 0 & 0 & 0 & 0.598 & 0 & 0 & 0 & 0 & 0 & 0 & 0 & 1 & 1 & 0.402 & 0.598 & 0 & 1 & 0 & 0 \\ 0 & 0 & 0 & 0 & 1 & 0 & 0 & 0 & 0 & 1 & 0 & 0 & 0 & 0 & 0 & 0 & 0 & 0 & 0 & 0 & 1 & 0 & 0 & 1 \end{bmatrix} \end{aligned}$$

Now, we study the controllability of the system (6.21) at state $x(0) = \delta_6^4$. Through computation, we have:

$$Ex(1) = \Theta W_{[6,4]} \delta_6^4 u(0) = \begin{bmatrix} 0 & 0 & 0 & 0 \\ 0 & 0 & 0 & 0 \\ 0 & 0 & 0 & 0 \\ 1 & 0 & 0 & 0 \\ 0 & 0 & 0 & 0 \\ 0 & 1 & 0 & 1 \end{bmatrix} u(0).$$

According to Theorem 6.3, the states δ_6^4, δ_6^5 and δ_6^6 are reachable from $x_0 = \delta_6^4$ at the first step under the control δ_4^1, δ_4^4 and δ_4^2, respectively. Thus, δ_6^4 is controllable. Similarly, we can prove that δ_6^5 and δ_6^6 are also controllable. Therefore, the constrained system (6.21) is controllable, that is, when WNT5A is unexpressed (WNT5A=0), the state can remain unexpressed through some suitable controls in the evolution of the gene regulatory network.

6.3 OPTIMAL CONTROL FOR PROBABILISTIC LOGICAL CONTROL NETWORKS

This section studies the optimal control problems of the probabilistic mix-valued logical network. First, the probabilistic mix-valued logical network is converted into an algebraic form by using the semi-tensor product. Second, using the algebraic form and the dynamic programming, a new algorithm is proposed to design the optimal strategy for the optimal finite-horizon control problem. Third, the first-passage model based optimal control problem is investigated for the probabilistic mix-valued logical network, and an algorithm is designed to find the optimal policy for the control problem.

6.3.1 OPTIMAL FINITE-HORIZON CONTROL OF THE PROBABILISTIC MIX-VALUED LOGICAL NETWORKS

Consider system (6.4). Define the expected cost of control over the finite-horizon M as:

$$E\left[\sum_{t=0}^{M-1} C_t(x(t),u(t)) + C_M(x(M)) \,|\, x(0)\right],$$

where $x(0)$ is an initial value, $C_t(u(t),x(t))$ is the cost at the t-th step of applying the control input $u(t)$ when the state is $x(t)$, $t = 0,1,\cdots,M-1$, $C_M(x(M))$ is a penalty or terminal cost, and M is a finite number of steps over which the control input is applied. Let us assume that $u(t) = G_t x(t)$ is the control input at each step, where $G_t \in \mathscr{L}_{q\times k}$, $t = 0,1,\cdots,M-1$.

The optimal finite-horizon control problem of the probabilistic mix-valued logical network (6.4) can be stated as follows: Given the initial value $x(0)$, how to find a control sequence $\pi = \{G_0, G_1, \cdots, G_{M-1}\}$ to minimize the cost functional:

$$J_\pi(x(0)) = E\left[\sum_{t=0}^{M-1} C_t(x(t), G_t x(t)) + C_M(x(M))\right]. \tag{6.23}$$

In order to find the optimal solution by using dynamic programming, we need the following Lemma 6.1.

Lemma 6.1

Let $J^*(x(0))$ be the minimum value of the cost functional (6.23). Then:

$$J^*(x(0)) = J_0(x(0)),$$

where the function J_0 is given by the last step of the following dynamic programming algorithm which proceeds backward in time from step $M-1$ to step 0:

$$\begin{aligned} J_M(x_M) &= C_M(x(M)), \\ J_t(x(t)) &= \min_{u(t)}\{E[C_t(x(t),u(t)) + J_{t+1}(x(t+1))]\}, \quad t = M-1,\cdots,1,0. \end{aligned} \tag{6.24}$$

Furthermore, if $u^*(t) = G_t^* x(t)$ minimizes $E[C_t(x(t),u(t)) + J_{t+1}(x(t+1))]$ at time t, the sequence $\pi^* = \{G_0^*, G_1^*, \cdots, G_{M-1}^*\}$ is the optimal control sequence. ■

If a state is not reachable at the first step, then $P\{x(1) \mid x(0),u(0)\} = 0$. Therefore, $J^*(x(0))$ cannot be changed, nor can the optimal control sequence be affected, no matter whether all terminal states are reachable. This is an assumption of Lemma 6.1. Moreover, in the trivial algorithm of "testing all possible cases", to complete the optimal value at the t-th step needs to consider the former states and controllers of all the $t-1$ steps. Whereas, in our algorithm (see below), only the optimal values of the $(t+1)$-th step are considered. This will undoubtedly make the computational load much less than that of "testing all possible cases".

To simplify the description, we give some notations first. Denote the structural matrix of the cost function $C_t(x(t),u(t))$ by $\mathscr{C}_t$, i.e., $C_t(x(t),u(t)) = \mathscr{C}_t u(t)x(t)$ and let,

$$D_t := \begin{bmatrix} J_t(\delta_k^1) & J_t(\delta_k^2) & \cdots & J_t(\delta_k^k) \end{bmatrix}.$$

Set $F_t = \mathscr{C}_t + D_{t+1}L$ and split $F_t W_{[k,q]}$ into k equal blocks, that is,

$$F_t W_{[k,q]} = [F_t^1 \quad F_t^2 \quad \cdots \quad F_t^k],$$

where $F_t^j \in \mathbb{R}^{1\times q}$, $j = 1,2,\cdots,k$, $t = M-1,\cdots,1,0$. Denote by $\left(F_t^j\right)_{i_j}$ the minimum element of F_t^j, where i_j is the column index of the minimum element.

Using the above notations, we have the following result based on Lemma 6.1.

Theorem 6.6

Consider probabilistic mix-valued logical network (6.4) with the cost function (6.23), and assume that the initial condition is $x(0) = \delta_k^{j_0}$. Then the optimal solution of (6.23) is:

$$J^*(x(0)) = J_0(\delta_k^{j_0}) = \left(F_0^{j_0}\right)_{i_{j_0}},$$

where J_0 is the last one of $J_t(x(t))$ defined in (6.24). ■

Proof. Note that the expectation $E\{J_{t+1}(x(t+1))\}$ on the right-hand side of (6.24) is conditioned on $x(t)$ and $u(t)$. Hence, it follows that:

$$\begin{aligned} E\{J_{t+1}(x(t+1))\,|\,x(t),u(t)\} &= [J_{t+1}(\delta_k^1)\ J_{t+1}(\delta_k^2)\cdots J_{t+1}(\delta_k^k)]\cdot Lu(t)x(t) \\ &= D_{t+1}Lu(t)x(t). \end{aligned}$$

Then,

$$\begin{aligned} J_t(x(t)) &= \min_{u(t)} E[C_t(x(t),u(t)) + J_{t+1}(x(t+1))] \\ &= \min_{u(t)}[\mathscr{C}_t + D_{t+1}L]u(t)x(t) = \min_{u(t)}\{F_t u(t)x(t)\} \\ &= \min_{u(t)}\left\{F_t W_{[k,q]}x(t)u(t)\right\}. \end{aligned}$$

For each $x(t) = \delta_k^j$, we have,

$$J_t(x(t)) = \min_{u(t)}\left\{F_t^j u(t)\right\} = \left(F_t^j\right)_{i_j}.$$

Furthermore, we obtain the optimal control $u^*(t) = \delta_q^{i_j}$ corresponding to $x(t) = \delta_k^j$. From $t = M-1$ to $t = 0$, we can calculate $J_t(x(t))$, F_t and D_t, respectively. By Lemma 6.1, we have:

$$J^*(x(0)) = J_0(\delta_k^{j_0}) = \left(F_0^{j_0}\right)_{i_{j_0}}.$$

Hence, the proof is completed. ■

Based on Theorem 6.6, we propose an algorithm to find the optimal value $J^*(x(0))$ and the optimal control sequence of the probabilistic mix-valued logical network.

ALGORITHM 6.1

Consider the probabilistic mix-valued logical network (6.4). Given the initial value $x(0)$, to find the optimal control sequence $\pi^* = \left\{G_0^*,\cdots,G_{M-1}^*\right\}$ that minimizes the cost functional (6.23), the algorithm can be stated as:

1) Set $D_M = \left[J_M(\delta_k^1), J_M(\delta_k^2), \cdots, J_M(\delta_k^k)\right]$.
2) Compute $F_{M-1} = \mathscr{C}_{M-1} + D_M L$, and split $F_{M-1}W_{[k,q]}$ into k equal blocks:

$$\left[F_{M-1}^1 \; F_{M-1}^2 \; \cdots \; F_{M-1}^k\right].$$

For each state $x(M-1) = \delta_k^j$, find the minimum element of F_{M-1}^j, denoted by $\left(F_{M-1}^j\right)_{i_j}$, $j = 1, 2, \cdots, k$. Then, set,

$$D_{M-1} = \left[\left(F_{M-1}^1\right)_{i_1} \quad \left(F_{M-1}^2\right)_{i_2} \quad \cdots \quad \left(F_{M-1}^k\right)_{i_k}\right]$$

and,

$$G_{M-1}^* = \left[\delta_q^{i_1} \; \delta_q^{i_2} \cdots \; \delta_q^{i_k}\right].$$

3) If $M-1 = 0$, stop. Else, set $M = M-1$ and go back to 2.
4) The optimal control sequence is:

$$\pi^* = \left\{G_0^*, G_1^*, \cdots, G_{M-1}^*\right\}.$$

5) For the different initial value $x(0) = \delta_k^{j_0}$, the optimal value $J^*(x(0))$ equals the j_0-th element of D_0 obtained from 3.

6.3.2 FIRST-PASSAGE MODEL BASED OPTIMAL CONTROL

In this section, we give the first-passage model for the probabilistic mix-valued logical network, and also study the optimal control problem based on this model.

Consider the probabilistic mix-valued logical network (6.4) with the initial condition $x(0) \in \Delta_k$. Let $S_0 \subset \Delta_k$ be the set of all undesirable states and let $S_1 \subset \Delta_k$ denote the set of all desirable states. Obviously, $S_0 \bigcup S_1 = \Delta_k$ and $S_0 \bigcap S_1 = \phi$.

Now, we give the concept of "the first-arrive time" before introducing optimality criterion. Let τ be the minimum integer of the system's first visiting the set S_1, that is,

$$\tau = \inf\{t : \; x(t) \in S_1\}.$$

Then, for any $s > 0$, we define a probability criterion $J(x(0), s)$ as:

$$J(x(0), s) = P\{\tau \leq s \,|\, x(0), v(x(0), s), v(y, s-i), y \in \Delta_k, \; i = s-1, \cdots, 2, 1\}, \quad (6.25)$$

where s is called a threshold value, $v(y, s-i)$ is a control chosen at the first step about the state y and the threshold value $s-i$, $i = s-1, \cdots, 1, 0$ and $J(x(0), s)$ stands for the probability of the system jumping to the set of desirable states from $x(0)$ within s time units. Thus, the first-passage model can be formulated as system (6.4) with both the first-arrive time τ and the probability criterion $J(x(0), s)$.

The first-passage-model based optimal control problem of the probabilistic mix-valued logical network (6.4) can be stated as follows: Given the initial value $x(0)$, for any $s > 0$, find a control sequence $\{v(x(0),s), v(y,s-i), y \in \Delta_k,\ i = s-1, \cdots, 2, 1\}$ to maximize probability criterion (6.25).

It is easy to obtain the following results.

PROPOSITION 6.3

If $x(t) \in S_1$, then $J(x(t),s) = 1$ for any threshold value s.

PROPOSITION 6.4

Consider system (6.4). Then, for any $x(t) \in S_0$ and any threshold value $s \geq 1$, the optimal value $J^*(x(t),s)$ of the probability criterion (6.25) can be given by:

$$\begin{aligned} J^*(x(t),s) &= \max_{v(x(t),s)} \left\{ \sum_{x(t+1)\in S_1} x^T(t+1) + \sum_{x(t+1)\in S_0} x^T(t+1) J^*(x(t+1), s-1) \right\} \\ &\quad \cdot LW_{[k,q]} x(t) v(x(t),s). \end{aligned}$$

Proof. The state $x(t+1)$ may be either in S_0 or S_1, with the control $v(x(t),s)$. If $x(t+1) \in S_1$, the probability of the system jumping to the set of desirable states is:

$$\sum_{x(t+1)\in S_1} P(x(t+1) \mid x(t), v(x(t),s)).$$

If $x(t+1) \in S_0$, then the threshold value is $s-1$; that is, the system should try to jump to the set of desirable states from $x(t+1)$ within $s-1$ time units. Thus, if $x(t+1) \in S_0$, the probability of the system jumping to the set of desirable states is:

$$\sum_{x(t+1)\in S_0} P(x(t+1) \mid x(t), v(x(t),s)) J(x(t+1), s-1).$$

Summarizing the above analysis, the probability of the system jumping to the set of desirable states from the initial value $x(t) \in S_0$ is:

$$\begin{aligned} J(x(t),s) &= \sum_{x(t+1)\in S_1} P(x(t+1) \mid x(t), v(x(t),s)) \\ &\quad + \sum_{x(t+1)\in S_0} P(x(t+1) \mid x(t), v(x(t),s)) J(x(t+1), s-1). \end{aligned}$$

Hence,

$$\begin{aligned}
& J^*(x(t),s) \\
= & \max_{v(x(t),s)} \Big\{ \sum_{x(t+1)\in S_1} P(x(t+1)\,|\,x(t),v(x(t),s)) \\
& + \sum_{x(t+1)\in S_0} P(x(t+1)\,|\,x(t),v(x(t),s))J^*(x(t+1),s-1) \Big\} \\
= & \max_{v(x(t),s)} \Big\{ \sum_{x(t+1)\in S_1} x^T(t+1)Lv(x(t),s)x(t) \\
& + \sum_{x(t+1)\in S_0} x^T(t+1)Lv(x(t),s)x(t)J^*(x(t+1),s-1) \Big\} \\
= & \max_{v(x(t),s)} \Big\{ \sum_{x(t+1)\in S_1} x^T(t+1) + \sum_{x(t+1)\in S_0} x^T(t+1)J^*(x(t+1),s-1) \Big\} \\
& \cdot LW_{[k,q]}x(t)v(x(t),s).
\end{aligned}$$

Thus, the proof is completed. ∎

Now, we propose an algorithm to find the optimal policy for the first-passage-model based optimal control of the probabilistic mix-valued logical network.

ALGORITHM 6.2

Consider probabilistic mix-valued logical network (6.4) with initial condition $x(0)$. Assume that the expected maximum threshold value is s_0. In order to find the optimal control sequence and the optimal value $J^*(x(0),s_0)$, the algorithm can be stated as:

1) Set $J^*(x(t),0) = 0$ for all $x(t) \in S_0$, and $J^*(x(t),s) = 1$ for all $x(t) \in S_1$ and any value s.
2) For any $x(t) \in S_0$, compute $J^*(x(t),s)$ by:

$$\max_{v(x(t),s)} \Big\{ \sum_{x(t+1)\in S_1} x^T(t+1) + \sum_{x(t+1)\in S_0} x^T(t+1)J^*(x(t+1),s-1) \Big\} \cdot LW_{[k,q]}x(t)v(x(t),s).$$

Suppose that the w-th element of:

$$\Big\{ \sum_{x(t+1)\in S_1} x^T(t+1) + \sum_{x(t+1)\in S_0} x^T(t+1)J^*(x(t+1),s-1) \Big\} LW_{[k,q]}x(t) \in \mathbb{R}^{1\times q}$$

is maximum. Then, set $v^*(x(t),s) = \delta_q^w$ and,

$$\begin{aligned}
J^*(x(t),s) \;=\; & Col_w\{[\sum_{x(t+1)\in S_1} x^T(t+1) \\
& + \sum_{x(t+1)\in S_0} x^T(t+1)J^*(x(t+1),s-1)]LW_{[k,q]}x(t)\}.
\end{aligned}$$

3) If $s = s_0$ or $J^* = 1$, stop. Else, set $s = s+1$ and go back to 2).

It is noted that, for the given initial value $x(0)$ and threshold value s_0 in Algorithm 6.2, the optimal value and the optimal control sequence can be obtained by finding the maximum column of the matrix:

$$\left\{ \sum_{x(1)\in S_1} x^T(1) + \sum_{x(1)\in S_0} x^T(1) J^*(x(1), s_0 - 1) \right\} LW_{[k,q]}x(0) \in \mathbb{R}^{1\times q}$$

and the corresponding column index, respectively.

Since the probabilistic Boolean network is a special case of the probabilistic mix-valued logical network, Algorithms 6.1 and 6.2 can be used to deal with the corresponding optimal control problems for probabilistic Boolean networks.

In the following section, we give an example to illustrate the effectiveness of the results/algorithms obtained in this section.

Example 6.2:

Consider the following probabilistic mix-valued logical network system:

$$\begin{cases} x_1(t+1) &= f_1(u(t), x_1(t), x_2(t)), \\ x_2(t+1) &= f_2(u(t), x_1(t), x_2(t)), \end{cases} \tag{6.26}$$

where $u(t), x_1(t) \in \mathscr{D} = \{1, 0\}$, $x_2(t) \in \mathscr{D}_3 = \{1, 0.5, 0\}$, $f_1 \in \{f_1^1\}$ and $f_2 \in \{f_2^1, f_2^2\}$ with probabilities P_2^j, $j = 1,2$, shown in Table 6.1.

Table 6.1: Truth table for Example 6.2

u	x_1	x_2	f_1^1	f_2^1	f_2^2
1	1	1	1	0	1
1	1	0.5	0	1	0
1	1	0	1	0.5	0
1	0	1	0	1	0.5
1	0	0.5	0	1	0.5
1	0	0	0	1	1
0	1	1	1	0.5	0.5
0	1	0.5	1	0	1
0	1	0	0	0	1
0	0	1	1	0	0
0	0	0.5	1	0.5	0
0	0	0	0	1	0.5
P_i^j			1	0.5	0.5

Based on the semi-tensor product method, from Table 6.1 we obtain the algebraic form of system (6.26) as

$$\begin{cases} x_1(t+1) &= M_1 u(t) x_1(t) x_2(t), \\ x_2(t+1) &= M_2 u(t) x_1(t) x_2(t), \end{cases}$$

where $u(t), x_1(t) \in \Delta$, $x_2(t) \in \Delta_3$, $M_2 \in \{M_2^1, M_2^2\}$,

$$M_1 = \delta_2[1\ 2\ 1\ 2\ 2\ 2\ 1\ 1\ 2\ 1\ 1\ 2],$$

$$M_2^1 = \delta_3[3\ 1\ 2\ 1\ 1\ 1\ 2\ 3\ 3\ 3\ 2\ 1],$$

and

$$M_2^2 = \delta_3[1\ 3\ 3\ 2\ 2\ 1\ 2\ 1\ 1\ 3\ 3\ 2].$$

Hence, the probabilistic mix-valued logical network has two possible networks, that is,

$$x(t+1) = L_i u(t) x(t),$$

with,

$$p_i = P\{\text{ the } i\text{-th network is selected }\} = 0.5,\ i = 1,2,$$

where $x(t) = x_1(t) \ltimes x_2(t)$,

$$L_1 = \delta_6[3\ 4\ 2\ 4\ 4\ 4\ 2\ 3\ 6\ 3\ 2\ 4],$$

$$L_2 = \delta_6[1\ 6\ 3\ 5\ 5\ 4\ 2\ 1\ 4\ 3\ 3\ 5].$$

Thus,

$$Ex(t+1) = Lu(t)Ex(t),$$

where,

$$L = p_1 L_1 + p_2 L_2 = \begin{bmatrix} 0.5 & 0 & 0 & 0 & 0 & 0 & 0 & 0.5 & 0 & 0 & 0 & 0 \\ 0 & 0 & 0.5 & 0 & 0 & 0 & 1 & 0 & 0 & 0 & 0.5 & 0 \\ 0.5 & 0 & 0.5 & 0 & 0 & 0 & 0 & 0.5 & 0 & 1 & 0.5 & 0 \\ 0 & 0.5 & 0 & 0.5 & 0.5 & 1 & 0 & 0 & 0.5 & 0 & 0 & 0.5 \\ 0 & 0 & 0 & 0.5 & 0.5 & 0 & 0 & 0 & 0 & 0 & 0 & 0.5 \\ 0 & 0.5 & 0 & 0 & 0 & 0 & 0 & 0 & 0.5 & 0 & 0 & 0 \end{bmatrix}.$$

Assume that the desirable state's set is $S_1 = \{(1,1),(1,0.5)\}$, that is, $S_1 = \{(\delta_2^1, \delta_3^1), (\delta_2^1, \delta_3^2)\}$. Then, the undesirable state's set is $S_0 = \{(\delta_2^1, \delta_3^3), (\delta_2^2, \delta_3^1), (\delta_2^2, \delta_3^2), (\delta_2^2, \delta_3^3)\}$. Now, we find a control sequence which minimizes the probability criterion (6.25) by using Algorithm 6.2.

For $s = 1$, by Algorithm 6.2, we obtain,

$$\begin{aligned} J^*(\delta_6^3, 1) &= \max_{v(\delta_6^3,1)} \left\{ (\delta_6^1)^T + (\delta_6^2)^T \right\} LW_{[6,2]} \delta_6^3 v(\delta_6^3, 1) \\ &= \max_{v(\delta_6^3,1)} \left\{ [1\ 1\ 0\ 0\ 0\ 0] \times \begin{bmatrix} 0 & 0.5 & 0.5 & 0 & 0 & 0 \\ 0 & 0 & 0 & 0.5 & 0 & 0.5 \end{bmatrix}^T \right\} v(\delta_6^3, 1) \\ &= \max_{v(\delta_6^3,1)} [0.5\ \ 0] v(\delta_6^3, 1) = 0.5. \end{aligned}$$

It is easy to see that the optimal policy is $v^*(\delta_6^3, 1) = \delta_2^1$. Similarly, we can obtain,

$$\begin{aligned} J^*(\delta_6^4, 1) &= 0, & v^*(\delta_6^4, 1) &= \delta_2^1 \text{ or } \delta_2^2, \\ J^*(\delta_6^5, 1) &= 0.5, & v^*(\delta_6^5, 1) &= \delta_2^2 \end{aligned}$$

and,

$$J^*(\delta_6^6, 1) = 0,\ \ v^*(\delta_6^6, 1) = \delta_2^1 \text{ or } \delta_2^2.$$

If $s=2$, we can compute $J^*(\delta_6^3,2)$ by:

$$\begin{aligned} J^*(\delta_6^3,2) &= \max_{v(\delta_6^3,2)} \left\{ (\delta_6^1)^T + (\delta_6^2)^T + \sum_{i=3}^{6} (\delta_6^i)^T J^*(\delta_6^i,1) \right\} LW_{[6,2]} \delta_6^3 v(\delta_6^3,2) \\ &= \max_{v(\delta_6^3,2)} \left\{ [1\ \ 1\ \ 0.5\ \ 0\ \ 0.5\ \ 0] \times \begin{bmatrix} 0 & 0.5 & 0.5 & 0 & 0 & 0 \\ 0 & 0 & 0 & 0.5 & 0 & 0.5 \end{bmatrix}^T \right\} v(\delta_6^3,2) \\ &= \max_{v(\delta_6^3,2)} [0.75\ \ 0] v(\delta_6^3,2) = 0.75. \end{aligned}$$

Obviously, the optimal policy $v^*(\delta_6^3,2)$ is δ_2^1. Moreover, we have,

$$J^*(\delta_6^4,2) = 0.5, \quad v^*(\delta_6^4,1) = \delta_2^2,$$

$$J^*(\delta_6^5,2) = 0.75, \quad v^*(\delta_6^5,1) = \delta_2^2$$

and

$$J^*(\delta_6^6,2) = 0.25, \quad v^*(\delta_6^6,1) = \delta_2^2.$$

By the same argument, for any $s \geq 3$, we can proceed until the expected threshold value is reached, and obtain the optimal value J^* and the optimal control v^*.

6.4 OUTPUT TRACKING CONTROL OF PROBABILISTIC LOGICAL CONTROL NETWORKS

In this section, we study output tracking control of probabilistic logical control networks. Consider the following probabilistic logical control network:

$$\begin{cases} X(t+1) = f(X(t),U(t)), \\ Y(t) = h(X(t)), t \in \mathbb{N}, \end{cases} \tag{6.27}$$

where $X(t) = (x_1(t),x_2(t),\cdots,x_n(t)) \in \Delta_k^n$, $U(t) = (u_1(t),\cdots,u_m(t)) \in \Delta_k^m$ and $Y(t) = (y_1(t),\cdots,y_p(t)) \in \Delta_k^p$ are the state, the control input and the output of the system (6.27), respectively. $f : \Delta_k^{m+n} \mapsto \Delta_k^n$ is chosen from the set $\{f_1, f_2, \cdots, f_r\}$ at every time step, and $P\{f = f_i\} = p_i > 0$, where $f_i : \Delta_k^{m+n} \mapsto \Delta_k^n, i = 1,2,\cdots,r$ are given logical functions, and $\sum_{i=1}^r p_i = 1$. $h : \Delta_k^n \mapsto \Delta_k^p$ is a given logical function.

Given a constant reference signal $Y^* = (y_1^*,\cdots,y_p^*) \in \Delta_k^p$. For the probabilistic logical control network (6.27), *the state feedback based output tracking control problem* is to design a state feedback control in the form of:

$$U(t) = g(X(t)), \tag{6.28}$$

under which there exists an integer $T > 0$ such that:

$$P\{Y(t) = Y^* \mid X(0) = X_0, U(t) = g(X(t))\} = 1$$

holds for $\forall\ X_0 \in \Delta_k^n$ and $\forall\ t \geq T$, where $g : \Delta_k^n \mapsto \Delta_k^m$ is a logical function to be designed.

In order to convert the system (6.27) and the state feedback control (6.28) into equivalent algebraic forms, respectively, we recall the definition and some useful properties of the semi-tensor product of matrices. For details, please refer to [1].

Using the vector form of logical variables, we set $x(t) = \ltimes_{i=1}^{n} x_i(t) \in \Delta_{k^n}$, $u(t) = \ltimes_{i=1}^{m} u_i(t) \in \Delta_{k^m}$ and $y(t) = \ltimes_{i=1}^{p} y_i(t) \in \Delta_{k^p}$. By Lemma 4.5, one can obtain the structural matrices of f_i $(i = 1, 2, \cdots, r)$, h and g as $L_i \in \mathscr{L}_{k^n \times k^{m+n}}$ $(i = 1, 2, \cdots, r)$, $H \in \mathscr{L}_{k^p \times k^n}$ and $G \in \mathscr{L}_{k^m \times k^n}$, respectively. Then, one can convert (6.27) and (6.28) into:

$$\begin{cases} x(t+1) = Lu(t)x(t), \\ y(t) = Hx(t), \end{cases} \tag{6.29}$$

and,

$$u(t) = Gx(t), \tag{6.30}$$

respectively, where $L \in \mathscr{L}_{k^n \times k^{m+n}}$ is chosen from the set $\{L_1, L_2, \cdots, L_r\}$ at every time step with $P\{L = L_i\} = p_i$. Moreover, the reference signal becomes $y^* = \ltimes_{i=1}^{p} y_i^* = \delta_{k^p}^{q}$, where q is uniquely determined by $y_i^*, i = 1, \cdots, p$. As was proved in [1], the system (6.29) is equivalent to the probabilistic logical control network (6.27), and the control (6.30) is equivalent to (6.28). Hence, the state feedback based output tracking control problem becomes how to design the state feedback gain matrix $G \in \mathscr{L}_{k^m \times k^n}$ for the system (6.29).

In this section, we study the existence and design of the state feedback gain matrix $G \in \mathscr{L}_{k^m \times k^n}$.

Consider the system (6.29) with the reference signal $y^* = \delta_{k^p}^{q}$. Define,

$$\mathscr{O}(y^*) = \{\delta_{k^n}^{i} : Col_i(H) = y^*, i \in \mathbb{N}, 1 \le i \le k^n\}. \tag{6.31}$$

It is easy to see that the set $\mathscr{O}(y^*)$ contains all the states of the system (6.29) whose outputs form the vector y^*. In the following, we presuppose $\mathscr{O}(y^*) \neq \emptyset$. Otherwise, if $\mathscr{O}(y^*) = \emptyset$, the state feedback based output tracking control problem is not solvable. We also assume that $\mathscr{O}(y^*) \neq \Delta_{k^n}$. Otherwise, the state feedback based output tracking control problem is solvable by any control input, which makes the problem trivial. The following lemma is straightforward.

Lemma 6.2

For the system (6.29), let $x_0 \in \Delta_{k^n}$ and $G \in \mathscr{L}_{k^m \times k^n}$. Then,

$$\begin{aligned} & P\{y(t) = y^* \mid x(0) = x_0, u(t) = Gx(t)\} \\ = & \sum_{a \in \mathscr{O}(y^*)} P\{x(t) = a \mid x(0) = x_0, u(t) = Gx(t)\} \end{aligned}$$

holds for $\forall\, t \in \mathbb{N}$. ■

Given a nonempty set $W \subseteq \Delta_{k^n}$ and $k \in \mathbb{Z}_+$, define a series of sets inductively as follows:

$$\Lambda_1(W) = \Big\{\delta_{k^n}^{i} : \text{ there exists } u \in \Delta_{k^m} \text{ such that: } \sum_{a \in W} P\{x(t+1) = a \mid x(t) = \delta_{k^n}^{i}, u(t) = u\} = 1\Big\}, \tag{6.32}$$

$$\Lambda_{k+1}(W) = \Big\{ \delta_{k^n}^i : \text{ there exists } u \in \Delta_{k^m} \text{ such that:}$$
$$\sum_{a \in \Lambda_k(W)} P\{x(t+1) = a \mid x(t) = \delta_{k^n}^i, u(t) = u\} = 1 \Big\}. \tag{6.33}$$

Then, we have the following lemma.

Lemma 6.3

(i) If $W \subseteq \Lambda_1(W)$, then $\Lambda_k(W) \subseteq \Lambda_{k+1}(W)$ holds for any integer $k \geq 1$.
(ii) If $\Lambda_1(W) = W$, then $\Lambda_k(W) = W$ holds for any integer $k \geq 1$.
(iii) If $\Lambda_j(W) = \Lambda_{j+1}(W)$ holds for some integer $j \geq 1$, then $\Lambda_k(W) = \Lambda_j(W)$ holds for any integer $k \geq j$. ■

Proof. First, we prove Conclusion (i) by induction.

Suppose that $\delta_{k^n}^i \in \Lambda_1(W)$. Then, there exists $u \in \Delta_{k^m}$ such that:

$$\sum_{a \in W} P\{x(t+1) = a \mid x(t) = \delta_{k^n}^i, u(t) = u\} = 1,$$

which together with $W \subseteq \Lambda_1(W)$ implies that:

$$\sum_{a \in \Lambda_1(W)} P\{x(t+1) = a \mid x(t) = \delta_{k^n}^i, u(t) = u\} = 1.$$

Hence, $\delta_{k^n}^i \in \Lambda_2(W)$, and thus $\Lambda_1(W) \subseteq \Lambda_2(W)$.

Assume that $\Lambda_{k-1}(W) \subseteq \Lambda_k(W)$ holds for the integer $k > 1$. Then, for $\delta_{k^n}^i \in \Lambda_k(W)$, there exists $u \in \Delta_{k^m}$ such that:

$$\sum_{a \in \Lambda_{k-1}(W)} P\{x(t+1) = a \mid x(t) = \delta_{k^n}^i, u(t) = u\} = 1.$$

Since $\Lambda_{k-1}(W) \subseteq \Lambda_k(W)$, we have:

$$\sum_{a \in \Lambda_k(W)} P\{x(t+1) = a \mid x(t) = \delta_{k^n}^i, u(t) = u\} = 1.$$

Hence, $\delta_{k^n}^i \in \Lambda_{k+1}(W)$, which shows that $\Lambda_k(W) \subseteq \Lambda_{k+1}(W)$.

By induction, $\Lambda_k(W) \subseteq \Lambda_{k+1}(W)$ holds for any integer $k \geq 1$.

Now, we prove Conclusion (ii) by induction.

It is obvious that $\Lambda_k(W) = W$ holds for $k = 1$. Assume that $\Lambda_k(W) = W$ holds for the integer $k \geq 1$. On one hand, from $\Lambda_1(W) = W$ and Conclusion (i), one can see that $W = \Lambda_k(W) \subseteq \Lambda_{k+1}(W)$. On the other hand, for $\delta_{k^n}^i \in \Lambda_{k+1}(W)$, there exists $u \in \Delta_{k^m}$ such that:

$$\sum_{a \in \Lambda_k(W)} P\{x(t+1) = a \mid x(t) = \delta_{k^n}^i, u(t) = u\} = 1.$$

Thus,

$$\sum_{a \in W} P\{x(t+1) = a \mid x(t) = \delta_{k^n}^i, u(t) = u\} = 1,$$

which implies that $\delta_{k^n}^i \in \Lambda_1(W) = W$. Hence, $\Lambda_{k+1}(W) \subseteq W$. This together with $W \subseteq \Lambda_{k+1}(W)$ shows that $\Lambda_{k+1}(W) = W$. By induction, $\Lambda_k(W) = W$ holds for any integer $k \geq 1$.

Finally, we prove Conclusion (iii).

Obviously, $\Lambda_k(W) = \Lambda_j(W)$ holds for $k = j$. Suppose that $\Lambda_l(W) = \Lambda_j(W)$ holds for any integer $j \leq l \leq k$ $(k > j)$. On one hand, for $\delta_{k^n}^i \in \Lambda_k(W)$, there exists $u \in \Delta_{k^m}$ such that:

$$\sum_{a \in \Lambda_{k-1}(W) = \Lambda_j(W) = \Lambda_k(W)} P\{x(t+1) = a \mid x(t) = \delta_{k^n}^i, u(t) = u\} = 1.$$

Thus, $\delta_{k^n}^i \in \Lambda_{k+1}(W)$, which implies that $\Lambda_j(W) = \Lambda_k(W) \subseteq \Lambda_{k+1}(W)$.

On the other hand, for $\delta_{k^n}^i \in \Lambda_{k+1}(W)$, there exists $u \in \Delta_{k^m}$ such that:

$$\sum_{a \in \Lambda_k(W) = \Lambda_j(W)} P\{x(t+1) = a \mid x(t) = \delta_{k^n}^i, u(t) = u\} = 1.$$

Hence, $\delta_{k^n}^i \in \Lambda_{j+1}(W) = \Lambda_j(W)$ and thus $\Lambda_{k+1}(W) \subseteq \Lambda_j(W)$, which together with $\Lambda_j(W) \subseteq \Lambda_{k+1}(W)$ shows that $\Lambda_{k+1}(W) = \Lambda_j(W)$. By induction, $\Lambda_k(W) = \Lambda_j(W)$ holds for any integer $k \geq j$. ∎

In the following, based on Lemmas 6.2 and 6.3, we present the main result of this paper.

Theorem 6.7

Consider the probabilistic logical control network (6.27) with the reference signal $y^* = \delta_{k^p}^q$. The state feedback based output tracking control problem is solvable, if and only if there exists an integer $1 \leq T \leq k^n - |\mathscr{O}(y^*)|$ such that $\mathscr{O}(y^*) \subseteq \Lambda_1(\mathscr{O}(y^*))$ and $\Lambda_T(\mathscr{O}(y^*)) = \Delta_{k^n}$. ∎

Proof. (Necessity) Suppose that the state feedback based output tracking control problem is solvable. Then, there exist a state feedback gain matrix $G \in \mathscr{L}_{k^m \times k^n}$ and a positive integer T such that for any $x_0 \in \Delta_{k^n}$ and any integer $t \geq T$, we have:

$$\begin{aligned}
1 &= P\{y(t) = y^* \mid x(0) = x_0, u(t) = Gx(t)\} \\
&= \sum_{a \in \mathscr{O}(y^*)} P\{x(t) = a \mid x(0) = x_0, u(t) = Gx(t)\} \\
&= \sum_{a \in \mathscr{O}(y^*)} \sum_{a_1, \cdots, a_{t-1} \in \Delta_{k^n}} P\{x(1) = a_1 \mid x(0) = x_0, u(0) = Gx_0\} \\
&\quad \times P\{x(2) = a_2 \mid x(1) = a_1, u(1) = Ga_1\} \times \cdots \\
&\quad \times P\{x(t) = a \mid x(t-1) = a_{t-1}, u(t-1) = Ga_{t-1}\} \qquad (6.34)
\end{aligned}$$

and

$$\begin{aligned}
1 &= P\{y(t+1)=y^* \mid x(0)=x_0, u(t)=Gx(t)\} \\
&= \sum_{a\in\mathscr{O}(y^*)} P\{x(t+1)=a \mid x(0)=x_0, u(t)=Gx(t)\} \\
&= \sum_{a\in\mathscr{O}(y^*)} \sum_{a_1,\cdots,a_t\in\Delta_{k^n}} P\{x(1)=a_1 \mid x(0)=x_0, u(0)=Gx_0\} \\
&\quad \times P\{x(2)=a_2 \mid x(1)=a_1, u(1)=Ga_1\} \times \cdots \\
&\quad \times P\{x(t+1)=a \mid x(t)=a_t, u(t)=Ga_t\} \\
&= \sum_{a\in\mathscr{O}(y^*)} \sum_{a_t\in\Delta_{k^n}} P\{x(t+1)=a \mid x(t)=a_t, u(t)=Ga_t\} \\
&\quad \times \Big[\sum_{a_1,\cdots,a_{t-1}\in\Delta_{k^n}} P\{x(1)=a_1 \mid x(0)=x_0, u(0)=Gx_0\} \\
&\quad \times \cdots \times P\{x(t)=a_t \mid x(t-1)=a_{t-1}, u(t-1)=Ga_{t-1}\}\Big].
\end{aligned} \tag{6.35}$$

One can see from (6.34) that:

$$\begin{aligned}
&\sum_{a_1,\cdots,a_{t-1}\in\Delta_{k^n}} P\{x(1)=a_1 \mid x(0)=x_0, u(0)=Gx_0\} \times \cdots \\
&\times P\{x(t)=a_t \mid x(t-1)=a_{t-1}, u(t-1)=Ga_{t-1}\} = 0
\end{aligned}$$

holds for any $a_t \in \Delta_{k^n} \setminus \mathscr{O}(y^*)$. This together with (6.35) shows that:

$$\begin{aligned}
1 &= \sum_{a\in\mathscr{O}(y^*)} \sum_{a_t\in\mathscr{O}(y^*)} P\{x(t+1)=a \mid x(t)=a_t, u(t)=Ga_t\} \\
&\quad \times \Big[\sum_{a_1,\cdots,a_{t-1}\in\Delta_{k^n}} P\{x(1)=a_1 \mid x(0)=x_0, u(0)=Gx_0\} \\
&\quad \times \cdots \times P\{x(t)=a_t \mid x(t-1)=a_{t-1}, u(t-1)=Ga_{t-1}\}\Big] \\
&= \sum_{a_t\in\mathscr{O}(y^*)} \sum_{a\in\mathscr{O}(y^*)} P\{x(t+1)=a \mid x(t)=a_t, u(t)=Ga_t\} \\
&\quad \times \Big[\sum_{a_1,\cdots,a_{t-1}\in\Delta_{k^n}} P\{x(1)=a_1 \mid x(0)=x_0, u(0)=Gx_0\} \\
&\quad \times \cdots \times P\{x(t)=a_t \mid x(t-1)=a_{t-1}, u(t-1)=Ga_{t-1}\}\Big].
\end{aligned} \tag{6.36}$$

Now, we prove that:

$$\sum_{a\in\mathscr{O}(y^*)} P\{x(t+1)=a \mid x(t)=a_t, u(t)=Ga_t\} = 1 \tag{6.37}$$

holds for any $a_t \in \mathscr{O}(y^*)$.

In fact, if there exists some $\widehat{a}_t \in \mathscr{O}(y^*)$ such that:

$$0 \le \sum_{a\in\mathscr{O}(y^*)} P\{x(t+1)=a \mid x(t)=\widehat{a}_t, u(t)=Ga_t\} < 1,$$

then from (6.36) we have:

$$\begin{aligned}
1 &= \sum_{a_t \in \mathcal{O}(y^*)} \sum_{a \in \mathcal{O}(y^*)} P\{x(t+1)=a \mid x(t)=a_t, u(t)=Ga_t\} \\
&\quad \times \Big[\sum_{a_1,\cdots,a_{t-1} \in \Delta_{k^n}} P\{x(1)=a_1 \mid x(0)=x_0, u(0)=Gx_0\} \\
&\quad \times \cdots \times P\{x(t)=a_t \mid x(t-1)=a_{t-1}, u(t-1)=Ga_{t-1}\} \Big] \\
&< \sum_{a_t \in \mathcal{O}(y^*)} \sum_{a_1,\cdots,a_{t-1} \in \Delta_{k^n}} P\{x(1)=a_1 \mid x(0)=x_0, u(0)=Gx_0\} \\
&\quad \times \cdots \times P\{x(t)=a_t \mid x(t-1)=a_{t-1}, u(t-1)=Ga_{t-1}\} = 1,
\end{aligned}$$

which is a contradiction. Thus, (6.37) holds, which together with (6.32) implies that $\mathcal{O}(y^*) \subseteq \Lambda_1(\mathcal{O}(y^*))$.

In the following, in order to show that $\Lambda_T(\mathcal{O}(y^*)) = \Delta_{k^n}$, we prove that if (6.34) holds for any $x_0 \in \Delta_{k^n}$ and $t \geq T$, then $x_0 \in \Lambda_t(\mathcal{O}(y^*))$.

Suppose that (6.34) holds for $t=2$ and any $x_0 \in \Delta_{k^n}$. Then, one can see from:

$$\sum_{a_1 \in \Delta_{k^n}} P\{x(1)=a_1 \mid x(0)=x_0, u(0)=Gx_0\} = 1$$

that $P\{x(1)=a_1 \mid x(0)=x_0, u(0)=Gx_0\} > 0$ implies that:

$$\sum_{a \in \mathcal{O}(y^*)} P\{x(2)=a \mid x(1)=a_1, u(1)=Ga_1\} = 1.$$

Thus, $a_1 \in \Lambda_1(\mathcal{O}(y^*))$ and:

$$\sum_{a_1 \in \Lambda_1(\mathcal{O}(y^*))} P\{x(1)=a_1 \mid x(0)=x_0, u(0)=Gx_0\} = 1,$$

which together with (6.33) shows that $x_0 \in \Lambda_2(\mathcal{O}(y^*))$.

Now, assume that the conclusion holds for the integer $t \geq 2$. For the case of $t+1$, we have:

$$\begin{aligned}
& \sum_{a \in \mathcal{O}(y^*)} \sum_{a_1,\cdots,a_t \in \Delta_{k^n}} P\{x(1)=a_1 \mid x(0)=x_0, u(0)=Gx_0\} \\
& \times P\{x(2)=a_2 \mid x(1)=a_1, u(1)=Ga_1\} \times \cdots \\
& \times P\{x(t+1)=a \mid x(t)=a_t, u(t)=Ga_t\} \\
= & \sum_{a_1 \in \Delta_{k^n}} P\{x(1)=a_1 \mid x(0)=x_0, u(0)=Gx_0\} \\
& \times \sum_{a \in \mathcal{O}(y^*)} \sum_{a_2,\cdots,a_t \in \Delta_{k^n}} P\{x(2)=a_2 \mid x(1)=a_1, u(1)=Ga_1\} \\
& \times \cdots \times P\{x(t+1)=a \mid x(t)=a_t, u(t)=Ga_t\} = 1.
\end{aligned}$$

Since $\sum_{a_1 \in \Delta_{k^n}} P\{x(1) = a_1 \mid x(0) = x_0, u(0) = Gx_0\} = 1$, one can see from $P\{x(1) = a_1 \mid x(0) = x_0, u(0) = Gx_0\} > 0$ that:

$$\begin{aligned} &\sum_{a \in \mathscr{O}(y^*)} \sum_{a_2, \cdots, a_t \in \Delta_{k^n}} P\{x(2) = a_2 \mid x(1) = a_1, u(1) = Ga_1\} \\ &\times \cdots \times P\{x(t+1) = a \mid x(t) = a_t, u(t) = Ga_t\} = 1. \end{aligned}$$

Thus, $a_1 \in \Lambda_t(\mathscr{O}(y^*))$ and:

$$\sum_{a_1 \in \Lambda_t(\mathscr{O}(y^*))} P\{x(1) = a_1 \mid x(0) = x_0, u(0) = Gx_0\} = 1,$$

which together with (6.33) shows that $x_0 \in \Lambda_{t+1}(\mathscr{O}(y^*))$.

By induction, if (6.34) holds for any $x_0 \in \Delta_{k^n}$ and $t \geq T$, then $x_0 \in \Lambda_t(\mathscr{O}(y^*))$. Therefore, $\Lambda_t(\mathscr{O}(y^*)) = \Delta_{k^n}$, $\forall\, t \geq T$, and thus $\Lambda_T(\mathscr{O}(y^*)) = \Delta_{k^n}$.

Finally, let T be the smallest positive integer such that $\Lambda_T(\mathscr{O}(y^*)) = \Delta_{k^n}$. We prove that $T \leq k^n - |\mathscr{O}(y^*)|$. It is enough to show that:

$$|\Lambda_t(\mathscr{O}(y^*))| \geq t + |\mathscr{O}(y^*)|, \ \forall\, 1 \leq t \leq T. \tag{6.38}$$

For the case of $t = 1$, if $|\Lambda_1(\mathscr{O}(y^*))| < 1 + |\mathscr{O}(y^*)|$, then $\Lambda_1(\mathscr{O}(y^*)) = \mathscr{O}(y^*)$. From Lemma 6.3, $\Lambda_T(\mathscr{O}(y^*)) = \mathscr{O}(y^*) \neq \Delta_{k^n}$, which is a contradiction to $\Lambda_T(\mathscr{O}(y^*)) = \Delta_{k^n}$. Thus, (6.38) holds for $t = 1$.

Assume that (6.38) holds for some $1 \leq t < T$. Since $\mathscr{O}(y^*) \subseteq \Lambda_1(\mathscr{O}(y^*))$, by Lemma 6.3, $\Lambda_t(\mathscr{O}(y^*)) \subseteq \Lambda_{t+1}(\mathscr{O}(y^*))$. Thus, $|\Lambda_{t+1}(\mathscr{O}(y^*))| \geq |\Lambda_t(\mathscr{O}(y^*))| \geq t + |\mathscr{O}(y^*)|$. If $|\Lambda_{t+1}(\mathscr{O}(y^*))| < t + 1 + |\mathscr{O}(y^*)|$, then $\Lambda_{t+1}(\mathscr{O}(y^*)) = \Lambda_t(\mathscr{O}(y^*))$. By Lemma 6.3, $\Lambda_t(\mathscr{O}(y^*)) = \Lambda_T(\mathscr{O}(y^*)) = \Delta_{k^n}$, which is a contradiction to the minimality of T. Hence, (6.38) holds for some $t+1$.

By induction, (6.38) holds. Letting $t = T$, we have:

$$k^n = |\Lambda_T(\mathscr{O}(y^*))| \geq T + |\mathscr{O}(y^*)|,$$

which shows that $T \leq k^n - |\mathscr{O}(y^*)|$.

(Sufficiency) Assuming that there exists an integer $1 \leq T \leq k^n - |\mathscr{O}(y^*)|$ such that $\mathscr{O}(y^*) \subseteq \Lambda_1(\mathscr{O}(y^*))$ and $\Lambda_T(\mathscr{O}(y^*)) = \Delta_{k^n}$. We, first construct a state feedback gain matrix $G \in \mathscr{L}_{k^m \times k^n}$.

Since $\mathscr{O}(y^*) \subseteq \Lambda_1(\mathscr{O}(y^*))$, by Lemma 6.3, it is easy to obtain that $\Lambda_1(\mathscr{O}(y^*)) \subseteq \cdots \subseteq \Lambda_T(\mathscr{O}(y^*)) = \Delta_{k^n}$. Set:

$$\widetilde{\Lambda}_i(\mathscr{O}(y^*)) = \Lambda_i(\mathscr{O}(y^*)) \setminus \Lambda_{i-1}(\mathscr{O}(y^*)), i = 1, \cdots, T, \tag{6.39}$$

where $\Lambda_0(\mathscr{O}(y^*)) := \emptyset$. It is obvious that $\widetilde{\Lambda}_i(\mathscr{O}(y^*)) \bigcap \widetilde{\Lambda}_j(\mathscr{O}(y^*)) = \emptyset$, $\forall\ i, j \in \{1, \cdots, T\}, i \neq j$, and $\bigcup_{i=1}^T \widetilde{\Lambda}_i(\mathscr{O}(y^*)) = \Delta_{k^n}$. Hence, for any integer $1 \leq i \leq k^n$, there exists a unique integer $1 \leq t_i \leq T$ such that $\delta_{k^n}^i \in \widetilde{\Lambda}_{t_i}(\mathscr{O}(y^*))$.

For $t_i = 1$, from (6.32), one can find an integer $1 \leq v_i \leq k^m$ such that:

$$\sum_{a \in \mathscr{O}(y^*)} P\{x(t+1) = a \mid x(t) = \delta_{k^n}^i, u(t) = \delta_{k^m}^{v_i}\} = 1. \tag{6.40}$$

For $2 \leq t_i \leq T$, from (6.33), one can find an integer $1 \leq v_i \leq k^m$ such that:

$$\sum_{a \in \Lambda_{t_i-1}(\mathscr{O}(y^*))} P\{x(t+1) = a \mid x(t) = \delta_{k^n}^i, u(t) = \delta_{k^m}^{v_i}\} = 1. \quad (6.41)$$

Set $G = \delta_{k^m}[v_1\ v_2\ \cdots\ v_{k^n}] \in \mathscr{L}_{k^m \times k^n}$.

In the following section, we prove that for the system (6.27) with the control $u(t) = Gx(t)$, $P\{y(t) = y^* \mid x(0) = \delta_{k^n}^i, u(t) = Gx(t)\} = 1$ holds for any $t \geq t_i$ and any $i = 1, 2, \cdots, k^n$.

When $t_i = 1$, from Lemma 6.2 and (6.40), we have:

$$\begin{aligned} & P\{y(1) = y^* \mid x(0) = \delta_{k^n}^i, u(t) = Gx(t)\} \\ = & \sum_{a \in \mathscr{O}(y^*)} P\{x(1) = a \mid x(0) = \delta_{k^n}^i, u(0) = \delta_{k^m}^{v_i}\} = 1. \end{aligned}$$

Assume that $P\{y(t) = y^* \mid x(0) = \delta_{k^n}^i, u(t) = Gx(t)\} = 1$ holds for some integer $t \geq 1$. Then, for the case of $t+1$, one can see from (6.40) that:

$$\begin{aligned} & P\{y(t+1) = y^* \mid x(0) = \delta_{k^n}^i, u(t) = Gx(t)\} \\ = & \sum_{a \in \mathscr{O}(y^*)} P\{x(t+1) = a \mid x(0) = \delta_{k^n}^i, u(t) = Gx(t)\} \\ = & \sum_{a \in \mathscr{O}(y^*)} \sum_{a_t \in \Delta_{k^n}} P\{x(t+1) = a \mid x(t) = a_t, u(t) = Ga_t\} \\ & \times \Big[\sum_{a_1, \cdots, a_{t-1} \in \Delta_{k^n}} P\{x(1) = a_1 \mid x(0) = \delta_{k^n}^i, u(0) = \delta_{k^m}^{v_i}\} \\ & \times \cdots \times P\{x(t) = a_t \mid x(t-1) = a_{t-1}, u(t-1) = Ga_{t-1}\} \Big] \\ = & \sum_{a_t \in \mathscr{O}(y^*) \subseteq \Lambda_1(\mathscr{O}(y^*))} \sum_{a \in \mathscr{O}(y^*)} P\{x(t+1) = a \mid x(0) = \delta_{k^n}^i, u(t) = Gx(t)\} \\ & \times \Big[\sum_{a_1, \cdots, a_{t-1} \in \Delta_{k^n}} P\{x(1) = a_1 \mid x(0) = \delta_{k^n}^i, u(0) = \delta_{k^m}^{v_i}\} \\ & \times \cdots \times P\{x(t) = a_t \mid x(t-1) = a_{t-1}, u(t-1) = Ga_{t-1}\} \Big] = 1. \end{aligned}$$

By induction, when $t_i = 1$, $P\{y(t) = y^* \mid x(0) = \delta_{k^n}^i, u(t) = Gx(t)\} = 1$ holds for any $t \geq t_i$.

When $2 \le t_i \le T$, it is easy to obtain from (6.40) and (6.41) that:

$$\begin{aligned}
& P\{y(t_i) = y^* \mid x(0) = \delta_{k^n}^i, u(t) = Gx(t)\} \\
= & \sum_{a \in \mathscr{O}(y^*)} P\{x(t_i) = a \mid x(0) = \delta_{k^n}^i, u(t) = Gx(t)\} \\
= & \sum_{a_1 \in \Lambda_{t_i - 1}(\mathscr{O}(y^*))} P\{x(1) = a_1 \mid x(0) = \delta_{k^n}^i, u(0) = \delta_{k^m}^{v_i}\} \\
& \times \sum_{a_2 \in \Lambda_{t_i - 2}(\mathscr{O}(y^*))} P\{x(2) = a_2 \mid x(1) = a_1, u(1) = Ga_1\} \times \cdots \times \\
& \sum_{a_{t_i - 1} \in \Lambda_1(\mathscr{O}(y^*))} P\{x(t_i - 1) = a_{t_i - 1} \mid x(t_i - 2) = a_{t_i - 2}, u(t_i - 2) = Ga_{t_i - 2}\} \\
& \times \sum_{a \in \mathscr{O}(y^*)} P\{x(t_i) = a \mid x(t_i - 1) = a_{t_i - 1}, u(t_i - 1) = Ga_{t_i - 1}\} = 1.
\end{aligned}$$

Suppose that $P\{y(t) = y^* \mid x(0) = \delta_{k^n}^i, u(t) = Gx(t)\} = 1$ holds for some integer $t \ge t_i$. Then, for the case of $t+1$, one can see from (6.40) that:

$$\begin{aligned}
(*) \;:= & \; P\{y(t+1) = y^* \mid x(0) = \delta_{k^n}^i, u(t) = Gx(t)\} \\
= & \sum_{a \in \mathscr{O}(y^*)} P\{x(t+1) = a \mid x(0) = \delta_{k^n}^i, u(t) = Gx(t)\} \\
= & \sum_{a_t \in \Delta_{k^n}} \sum_{a \in \mathscr{O}(y^*)} P\{x(t+1) = a \mid x(t) = a_t, u(t) = Ga_t\} \\
& \times \Big[\sum_{a_1, \cdots, a_{t-1} \in \Delta_{k^n}} P\{x(1) = a_1 \mid x(0) = \delta_{k^n}^i, u(0) = \delta_{k^m}^{v_i}\} \\
& \times P\{x(2) = a_2 \mid x(1) = a_1, u(1) = Ga_1\} \times \cdots \times \\
& P\{x(t) = a_t \mid x(t-1) = a_{t-1}, u(t-1) = Ga_{t-1}\} \Big].
\end{aligned}$$

Since,

$$\begin{aligned}
1 = & \; P\{y(t) = y^* \mid x(0) = \delta_{k^n}^i, u(t) = Gx(t)\} \\
= & \sum_{a_t \in \mathscr{O}(y^*)} P\{x(t) = a_t \mid x(0) = \delta_{k^n}^i, u(t) = Gx(t)\} \\
= & \sum_{a_t \in \mathscr{O}(y^*)} \sum_{a_1, \cdots, a_{t-1} \in \Delta_{k^n}} P\{x(1) = a_1 \mid x(0) = \delta_{k^n}^i, u(0) = \delta_{k^m}^{v_i}\} \\
& \times P\{x(2) = a_2 \mid x(1) = a_1, u(1) = Ga_1\} \times \cdots \times \\
& P\{x(t) = a_t \mid x(t-1) = a_{t-1}, u(t-1) = Ga_{t-1}\},
\end{aligned}$$

it is easy to obtain from (6.32) that:

$$\begin{aligned}
(*) &= \sum_{a_t \in \mathscr{O}(y^*) \subseteq \Lambda_1(\mathscr{O}(y^*))} \sum_{a \in \mathscr{O}(y^*)} P\{x(t+1) = a \mid x(t) = a_t, u(t) = Ga_t\} \\
&\quad \times \Big[\sum_{a_1, \cdots, a_{t-1} \in \Delta_{k^n}} P\{x(1) = a_1 \mid x(0) = \delta_{k^n}^i, u(0) = \delta_{k^m}^{v_i}\} \\
&\quad \times P\{x(2) = a_2 \mid x(1) = a_1, u(1) = Ga_1\} \times \cdots \times \\
&\quad P\{x(t) = a_t \mid x(t-1) = a_{t-1}, u(t-1) = Ga_{t-1}\} \Big] \\
&= \sum_{a_t \in \mathscr{O}(y^*)} \sum_{a_1, \cdots, a_{t-1} \in \Delta_{k^n}} P\{x(1) = a_1 \mid x(0) = \delta_{k^n}^i, u(0) = \delta_{k^m}^{v_i}\} \\
&\quad \times P\{x(2) = a_2 \mid x(1) = a_1, u(1) = Ga_1\} \times \cdots \times \\
&\quad P\{x(t) = a_t \mid x(t-1) = a_{t-1}, u(t-1) = Ga_{t-1}\} = 1.
\end{aligned}$$

By induction, when $2 \leq t_i \leq T$, $P\{y(t) = y^* \mid x(0) = \delta_{k^n}^i, u(t) = Gx(t)\} = 1$ holds for any $t \geq t_i$.

Therefore, for the system (6.27) with the control $u(t) = Gx(t)$, $P\{y(t) = y^* \mid x(0) = x_0, u(t) = Gx(t)\} = 1$ holds for any $t \geq T$ and any $x_0 \in \Delta_{k^n}$, which implies that the state feedback based output tracking control problem is solvable. ∎

Suppose that the conditions in Theorem 6.7 hold. The proof of Theorem 6.7 provides a constructive procedure to design state feedback based output tracking controllers for the probabilistic logical control network (6.27). This procedure contains the following steps:

1) Calculate $\mathscr{O}(y^*)$, $\Lambda_1(\mathscr{O}(y^*))$ and $\Lambda_i(\mathscr{O}(y^*))$, $i = 2, \cdots, k^n - |\mathscr{O}(y^*)|$ according to (6.31), (6.32) and (6.33), respectively.
2) Find a positive integer $T \leq k^n - |\mathscr{O}(y^*)|$ such that $\Lambda_T(\mathscr{O}(y^*)) = \Delta_{k^n}$. Calculate $\widetilde{\Lambda}_i(\mathscr{O}(y^*))$, $i = 1, \cdots, T$ according to (6.39).
3) Find $v_i, i = 1, 2, \cdots, k^n$ such that (6.40) and (6.41) hold for $t_i = 1$ and $2 \leq t_i \leq T$, respectively.
4) The state feedback gain matrix can be designed as:

$$G = \delta_{k^m}[v_1 \ v_2 \ \cdots \ v_{k^n}].$$

The technique proposed in the proof of Theorem 6.7 can be applied to the following partial stabilization problem of the probabilistic logical control network (6.27). Given a positive integer $\gamma \leq n$ and $x_i^* \in \Delta_k, i = 1, \cdots, \gamma$. Design a state feedback control in the form of (6.28), under which there exists an integer $T > 0$ such that:

$$P\{x_i(t) = x_i^*, i = 1, \cdots, \gamma \mid X(0) = X_0, U(t) = g(X(t))\} = 1$$

holds for $\forall\, X_0 \in \Delta_k^n$ and $\forall\, t \geq T$. In fact, consider the probabilistic logical control network (6.27) with $p = \gamma$ and $h(X(t)) = (x_1(t), \cdots, x_\gamma(t))$. Set $Y^* = (x_1^*, \cdots, x_\gamma^*)$. Then, one can convert the partial stabilization problem to the state feedback based output tracking control problem.

The differences between Theorem 1 of [6] and Theorem 6.7 are summarized as follows:

(i) The sets $\Lambda_k(W), k \in \mathbb{Z}_+$ defined in (6.32) and (6.33) are quite different from the sets $R_k(S), k \in \mathbb{Z}_+$ defined in [6].

(ii) The upper bound of T in Theorem 1 of [6] is k^n, while that in Theorem 6.7 is $k^n - |\mathscr{O}(y^*)|$. Obviously, $k^n - |\mathscr{O}(y^*)| \ll k^n$ when $|\mathscr{O}(y^*)|$ is not small.

(iii) In [6], we need to find a non-empty set $S \subseteq \mathscr{O}(y^*)$ which satisfies the conditions of Theorem 1. However, in this paper, $\mathscr{O}(y^*)$ itself should satisfy the conditions of Theorem 6.7. This may be caused by the ergodicity of the Markov chain produced by the dynamics of probabilistic logical control networks.

Finally, we discuss how to calculate the set $\Lambda_t(\mathscr{O}(y^*)), t \in \mathbb{Z}_+$ based on the algebraic form (6.29).

For $i = 1,2,\cdots,k^n$, $j = 1,2,\cdots,k^m$, $k = 1,\cdots,r$ and $t \geq 2$, define:

$$\Psi_1(i,j,k) = \begin{cases} 1, \text{ if } Col_i(Blk_j(L_k)) \in \mathscr{O}(y^*), \\ 0, \text{ otherwise} \end{cases}$$

and,

$$\Psi_t(i,j,k) = \begin{cases} 1, \text{ if } Col_i(Blk_j(L_k)) \in \Lambda_{t-1}(\mathscr{O}(y^*)), \\ 0, \text{ otherwise.} \end{cases}$$

Then, it is easy to obtain that:

$$\sum_{a \in \mathscr{O}(y^*)} P\{x(t+1) = a \mid x(t) = \delta_{k^n}^i, u(t) = \delta_{k^m}^j\} = \sum_{k=1}^{r} p_k \Psi_1(i,j,k)$$

and,

$$sum_{a \in \Lambda_{t-1}(\mathscr{O}(y^*))} P\{x(t+1) = a \mid x(t) = \delta_{k^n}^i, u(t) = \delta_{k^m}^j\} = \sum_{k=1}^{r} p_k \Psi_t(i,j,k).$$

Hence, $\sum\limits_{a \in \mathscr{O}(y^*)} P\{x(t+1) = a \mid x(t) = \delta_{k^n}^i, u(t) = \delta_{k^m}^j\} = 1$ if and only if $\Psi_1(i,j,k) = 1$, $\forall\, k = 1,\cdots,r$, that is, $Col_i(Blk_j(L_k)) \in \mathscr{O}(y^*)$, $\forall\, k = 1,\cdots,r$. For the integer $t \geq 2$, $\sum\limits_{a \in \Lambda_{t-1}(\mathscr{O}(y^*))} P\{x(t+1) = a \mid x(t) = \delta_{k^n}^i, u(t) = \delta_{k^m}^j\} = 1$ if and only if $\Psi_{t-1}(i,j,k) = 1$, $\forall\, k = 1,\cdots,r$, that is, $Col_i(Blk_j(L_k)) \in \Lambda_{t-1}(\mathscr{O}(y^*))$, $\forall\, k = 1,\cdots,r$.

Based on the above analysis, we have the following result.

Theorem 6.8

(i) $\delta_{k^n}^i \in \Lambda_1(\mathscr{O}(y^*))$ if and only if there exists a positive integer $v_i \leq k^m$ such that:

$$Col_i(Blk_{v_i}(L_k)) \in \mathscr{O}(y^*)$$

holds for any $k = 1,\cdots,r$.

(ii) Given an integer $t \geq 2$. $\delta_{k^n}^i \in \Lambda_t(\mathscr{O}(y^*))$ if and only if there exists a positive integer $v_i \leq k^m$ such that:

$$Col_i(Blk_{v_i}(L_k)) \in \Lambda_{t-1}(\mathscr{O}(y^*))$$

holds for any $k = 1, \cdots, r$. ∎

The rest of this subsection gives two illustrative examples to show how to use the results obtained in this paper to design state feedback based output tracking controllers for probabilistic logical control networks.

Example 6.3: The probabilistic logical control network model of an apoptosis network

Consider the probabilistic logical control network model of an apoptosis network [7] containing four nodes "IAP", "C3a", "C8a" and "TNF", where "IAP" (denoted by x_1) stands for the concentration level (high or low) of the inhibitor of apoptosis proteins, "C3a" (denoted by x_2) the concentration level of the active caspase 3, "C8a" (denoted by x_3) the concentration level of the active caspase 8, and "TNF" (a stimulus, denoted by u) the concentration level of the tumor necrosis factor. The model consists of eight sub-networks: $f_1 = (\neg x_2 \wedge u, \neg x_1 \wedge x_3, x_2 \vee u)$, $f_2 = (x_1, \neg x_1 \wedge x_3, x_2 \vee u)$, $f_3 = (\neg x_2 \wedge u, x_2, x_2 \vee u)$, $f_4 = (x_1, x_2, x_2 \vee u)$, $f_5 = (\neg x_2 \wedge u, \neg x_1 \wedge x_3, x_3)$, $f_6 = (x_1, \neg x_1 \wedge x_3, x_3)$, $f_7 = (\neg x_2 \wedge u, x_2, x_3)$, $f_8 = (x_1, x_2, x_3)$ with the network selection probabilities $p_1 = 0.336$, $p_2 = 0.224$, $p_3 = 0.144$, $p_4 = 0.096$, $p_5 = 0.084$, $p_6 = 0.056$, $p_7 = 0.036$, $p_8 = 0.024$.

In the apoptosis network, $\{x_1 = 1,\ x_2 = 0\}$ stands for the cell survival [8]. Thus, we are interested in observing x_1 and x_2 at each time instance, and have the following output equation:

$$\begin{cases} y_1(t) = x_1(t), \\ y_2(t) = x_2(t). \end{cases}$$

Our objective is to design a state feedback control (if possible) such that the output of the apoptosis network tracks the reference signal $Y^* = (1,0)$ (the cell survival).

Using the vector form of logical variables and setting $x(t) = \ltimes_{i=1}^3 x_i(t)$ and $y(t) = \ltimes_{i=1}^2 y_i(t)$, by the semi-tensor product of matrices, we have the following algebraic form of the considered apoptosis network:

$$\begin{cases} x(t+1) = Lu(t)x(t), \\ y(t) = Hx(t), \end{cases}$$

where $L \in \{L_i : i = 1, \cdots, 8\}$,

$$L_1 = \delta_8[7\ 7\ 3\ 3\ 5\ 7\ 1\ 3\ 7\ 7\ 8\ 8\ 5\ 7\ 6\ 8],\ L_2 = \delta_8[3\ 3\ 3\ 3\ 5\ 7\ 5\ 7\ 3\ 3\ 4\ 4\ 5\ 7\ 6\ 8],$$
$$L_3 = \delta_8[5\ 5\ 3\ 3\ 5\ 5\ 3\ 3\ 5\ 5\ 8\ 8\ 5\ 5\ 8\ 8],\ L_4 = \delta_8[1\ 1\ 3\ 3\ 5\ 5\ 7\ 7\ 1\ 1\ 4\ 4\ 5\ 5\ 8\ 8],$$
$$L_5 = \delta_8[7\ 8\ 3\ 4\ 5\ 8\ 1\ 4\ 7\ 8\ 7\ 8\ 5\ 8\ 5\ 8],\ L_6 = \delta_8[3\ 4\ 3\ 4\ 5\ 6\ 7\ 8\ 3\ 4\ 3\ 4\ 5\ 6\ 7\ 8],$$
$$L_7 = \delta_8[5\ 6\ 3\ 4\ 5\ 6\ 3\ 4\ 5\ 6\ 7\ 8\ 5\ 6\ 7\ 8],\ L_8 = \delta_8[1\ 2\ 3\ 4\ 5\ 6\ 7\ 8\ 1\ 2\ 3\ 4\ 5\ 6\ 7\ 8],$$

and $H = \delta_4[1\ 1\ 2\ 2\ 3\ 3\ 4\ 4]$. Moreover, the reference signal becomes $y^* = \delta_2^1 \ltimes \delta_2^2 = \delta_4^2$.

A simple calculation gives $\mathcal{O}(y^*) = \{\delta_8^3, \delta_8^4\}$. Moreover, one can obtain from Theorem 6.8 that $\Lambda_1(\mathcal{O}(y^*)) = \{\delta_8^3, \delta_8^4\} = \mathcal{O}(y^*)$. By Lemma 6.3, $\Lambda_t(\mathcal{O}(y^*)) = \mathcal{O}(y^*), \forall t \in \mathbb{Z}_+$. Thus, we cannot find a positive integer T such that $\Lambda_T(\mathcal{O}(y^*)) = \Delta_8$. By Theorem 6.7, one cannot design a state feedback control such that the output of the apoptosis network tracks the reference signal $Y^* = (1,0)$ (the cell survival). This fact is consistent to the bistability of apoptosis networks [8].

Example 6.4:

Consider the probabilistic logical control network (6.29) with $n = 4$ and $m = p = 2$, where L is chosen from the set $\{L_1, L_2\}$ at every time step,

$$\begin{aligned} L_1 &= \delta_{16}[1\ 10\ 8\ 7\ 9\ 2\ 6\ 8\ 2\ 10\ 3\ 7\ 3\ 8\ 13\ 15 \\ &\quad 2\ 6\ 12\ 9\ 16\ 11\ 12\ 10\ 11\ 12\ 4\ 11\ 3\ 11\ 15\ 15 \\ &\quad 15\ 6\ 1\ 13\ 10\ 12\ 10\ 2\ 12\ 2\ 15\ 7\ 13\ 11\ 13\ 16 \\ &\quad 14\ 8\ 9\ 14\ 15\ 13\ 8\ 13\ 13\ 16\ 16\ 14\ 13\ 12\ 15\ 11], \\ L_2 &= \delta_{16}[9\ 2\ 8\ 5\ 1\ 9\ 7\ 12\ 9\ 1\ 5\ 8\ 7\ 3\ 11\ 16 \\ &\quad 10\ 9\ 12\ 10\ 12\ 7\ 13\ 1\ 6\ 15\ 6\ 12\ 7\ 12\ 16\ 16 \\ &\quad 12\ 8\ 1\ 7\ 2\ 7\ 2\ 2\ 7\ 9\ 12\ 8\ 11\ 12\ 11\ 16 \\ &\quad 11\ 8\ 10\ 13\ 13\ 8\ 3\ 7\ 8\ 14\ 13\ 12\ 11\ 12\ 16\ 12], \end{aligned}$$

$H = \delta_4[1\ 1\ 2\ 2\ 3\ 3\ 4\ 4\ 1\ 1\ 2\ 2\ 3\ 3\ 4\ 4]$, $p_1 = 0.4$, and $p_2 = 0.6$.

We aim to design a state feedback gain matrix $G = \delta_4[v_1\ v_2\ \cdots\ v_{16}]$ (if possible) such that the output of the considered probabilistic logical control network tracks the reference signal $y^* = \delta_4^1$.

By a simple calculation, we have $\mathcal{O}(y^*) = \{\delta_{16}^1, \delta_{16}^2, \delta_{16}^9, \delta_{16}^{10}\}$. One can obtain that $\Lambda_1(\mathcal{O}(y^*)) = \{\delta_{16}^i : i = 1, \cdots, 10\}$, $\Lambda_2(\mathcal{O}(y^*)) = \{\delta_{16}^i : i = 1, \cdots, 14\}$ and $\Lambda_3(\mathcal{O}(y^*)) = \Delta_{16}$. Thus, $\mathcal{O}(y^*) \subseteq \Lambda_1(\mathcal{O}(y^*))$ and $\Lambda_3(\mathcal{O}(y^*)) = \Delta_{16}$. By Theorem 6.7, we can obtain 512 state feedback controls such that the output of the considered probabilistic logical control network tracks the reference signal $y^* = \delta_4^1$, one of which is given as follows:

$$u(t) = \delta_4[1\ 2\ 3\ 2\ 1\ 1\ 3\ 3\ 1\ 1\ 1\ 1\ 1\ 1\ 1\ 4]x_1(t)x_2(t).$$

REFERENCES

1. Cheng, D., Qi, H. and Li, Z. (2011). Analysis and Control of Boolean Networks: A Semi-Tensor Product Approach. London, Springer.
2. Shmulevich, I., Dougherty, E. R., Kim, S. and Zhang, W. (2002). Probabilistic Boolean networks: a rule-based uncertainty model for gene regulatory networks. Bioinformatics, 18(2): 261–274.
3. Li, F. and Sun, J. (2011). Controllability of probabilistic Boolean control networks. Automatica, 47(12): 2765–2771.
4. Cheng, D. and Qi, H. (2009). Controllability and observability of Boolean control networks. Automatica, 45(7): 1659–1667.

5. Faryabi, B., Vahedi, G. and Chamberland, J. (2009). Intervention in context-sensitive probabilistic Boolean networks revisited. EURASIP J. Bioinformatics and Systems Biology, Article ID: 360864, 13 pages.
6. Li, H., Wang, Y. and Xie, L. (2015). Output tracking control of Boolean control networks via state feedback: Constant reference signal case. Automatica, 59: 54–59.
7. Li, R., Yang, M. and Chu, T. (2014). State feedback stabilization for probabilistic Boolean networks. Automatica, 50: 1272–1278.
8. Chaves, M. (2009). Methods for qualitative analysis of genetic networks. Proc. 10th European Control Conference, 671–676.

Part II

Applications

7 Networked Evolutionary Games

7.1 INTRODUCTION TO NETWORKED EVOLUTIONARY GAMES

In the last two decades, there is a great deal of interest in the research of evolutionary games on graphs, namely, networked evolutionary games [1]. It can be applied to investigate some practical problems, where each individual just plays game with some specific players (for example, trading partners) other than all players on a network. The network, whose nodes and edges represent, respectively, players and interaction relationship among players, depicts the topological structure of the corresponding game. Hence, in a networked evolutionary games, a player's reward, or payoff, depends on the strategies taken by both his neighbors and himself [2–4, 8–12].

This section introduces some necessary preliminaries of networked evolutionary games.

DEFINITION 7.1: see [5]

A normal form finite game (N, S, C) consists of three factors:

(i) n players $N = \{1, 2, \cdots, n\}$;
(ii) Player i has its strategy set S_i, and $S := \prod_{i=1}^{n} S_i$ is the set of strategy profiles, $i \in N$;
(iii) Player i has its payoff function $c_i : S \to \mathbb{R}$, and $C := \{c_1, c_2, \cdots, c_n\}$ is the set of payoff functions, $i \in N$.

DEFINITION 7.2: see [5]

In a n-player normal form finite game $G = \{S_1, \cdots, S_n; c_1, \cdots, c_n\}$, the strategy profile $(s_1^*, s_2^*, \cdots, s_n^*)$ is a pure strategy Nash Equilibrium (NE) if, for each player i, s_i^* is (at least tied for) player i's best response to the strategies specified for the $n-1$ other players, $(s_1^*, \cdots, s_{i-1}^*, s_{i+1}^*, \cdots, s_n^*)$, that is,

$$c_i\left(s_1^*, \cdots, s_{i-1}^*, s_i^*, s_{i+1}^*, \cdots, s_n^*\right) \geq c_i\left(s_1^*, \cdots, s_{i-1}^*, s_i, s_{i+1}^*, \cdots, s_n^*\right),$$

for every feasible strategy $s_i \in S_i$, where S_i is the set of strategies of player i and c_i is the corresponding payoff function.

DEFINITION 7.3: see [4]

A networked evolutionary game (NEG), denoted by $((N,E),G,F)$, consists of three factors:

1. (N,E) is a network graph, $(i,j) \in E$ means player i and player j are neighbors, $U(i)$ is the set of neighbors of player i;
2. G is the fundamental network game (FNG) which is played by each pair of neighbors over the network (N,E);
3. $F = (f_1, f_2, \cdots, f_n)$ is the strategy updating rule (SUR) based on the local neighborhood information.

Let $c_{ij}(t)$ be the payoff of player i when playing with player j. The overall payoff of player i at time t could be the total payoff:

$$c_i(t) = \sum_{j \in U(i)} c_{ij}(t),$$

or the average payoff:

$$c_i(t) = \frac{1}{|U(i) - 1|} \sum_{j \in U(i)} c_{ij}(t).$$

The following example is a simple NEG model, which is based on the classical prisoners' dilemma game.

Example 7.1: Networked evolutionary prisoners' dilemma game

Assume that there are four players in the NEG and the network topology is a cycle (see Figure 7.1). The FNG between two neighbors is prisoners' dilemma game, and the payoff bi-matrix is shown in Table 7.1. From Definition 7.2, the NE of this FNG is that both players choose to defect.

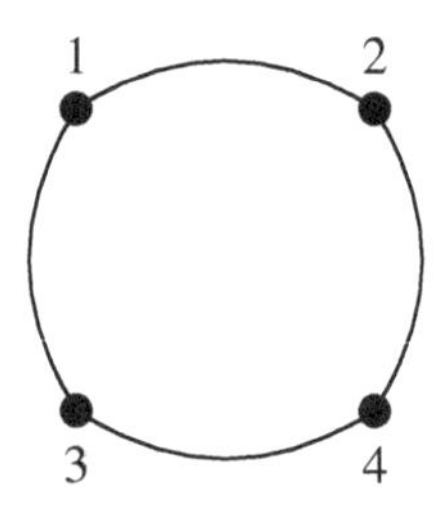

Figure 7.1 The network topology of the game in Example 7.1

Table 7.1: Payoff bi-matrix of the classical prisoners' dilemma model

Prisoner A \ Prisoner B	B stays silent (cooperates)	B betrays (defects)
A stays silent (cooperates)	(–1, –1)	(–5, 0)
A betrays (defects)	(0, –5)	(–2, –2)

All players' SURs are best response strategy updating rule (the detailed explanation about this SUR will be given later). Thus, the game dynamics can be expressed as the following form,

$$x_i(t+1) = f_i(x_j(t) \mid j \in U(i)),\ t \geq 0, \tag{7.1}$$

where $x_i(t)$ is player i's strategy at time t, and f_i is the best response SUR, $i = 1, 2, 3, 4$.

Because the player set and the strategy set are finite sets, the game dynamics (7.1) is a kind of finite-value system. We can make use of the obtained theory results in Part I to investigate the finite networked evolutionary game.

Next, we describe three ingredients of NEG in Definition 7.3 one by one.

I Network Topology

DEFINITION 7.4

1. *Undirected graph:* If $(i, j) \in E$, then i is in the neighborhood of j, denoted by $i \in U(j)$. Simultaneously, $j \in U(i)$.

   Directed graph: If the FNG is not symmetric, the directed edge is used to distinguish different roles of two players. Assume $(i, j) \in E$, i.e., there is an edge from i to j, then in the game i is player 1 and j is player 2.
2. *Homogeneous graph:* If the graph is undirected and each node has same degree, or the graph is directed and each node has same in-degree and same out-degree.

   Heterogeneous graph: If a graph is not homogeneous, it is said to be heterogeneous.

We consider two kinds of network graphs in this chapter.

(a) Time invariant graph: The network topology is fixed, that is, it does not change with time.

(b) Time varying graph: The network topology changes with time based on some special varying law, such as network topology with random entrance.

II Fundamental Network Game

DEFINITION 7.5

(i) A normal game with two players is called a fundamental network game, if $S_1 = S_2 := S_0 = \{1,2,\cdots,k\}$ and player i's payoff function is $c_i = c_i(x,y)$, where x is player 1's strategy, y is player 2's strategy, and $i = 1,2$. Namely, $N = \{1,2\}$, $S = S_0 \times S_0$, and $C = \{c_1,c_2\}$; (ii) A FNG is symmetric, if $c_1(x,y) = c_2(y,x)$, $\forall x,y \in S_0$.

In the existing research, there are many kinds of common games which can be served as the FNG, such as Prisoner's Dilemma, Traveler's Dilemma, Public goods game, Snowdrift game, Hawk-Dove game, Boxed pig game and so on.

III Strategy Updating Rule

Note that there are many commonly used strategy updating rules F, such as myopic best response adjustment rule, unconditional imitation, Fermi rule, which may be deterministic or probabilistic. Here we list several ones.

Myopic best response adjustment rule: Each player forecasts that its rivals will repeat their last step decisions, and the strategy choice at present time is the best response against its neighbors' strategies of the last step. Based on this, it holds that:

$$x_i(t+1) \in Q_i := arg \max_{x_i \in S_0} c_i\Big(x_i, x_j(t) \mid j \in U(i)\Big). \tag{7.2}$$

When player i has more than one best responses, define a priority for the strategy choice as follows: for $s_i,\ s_j \in S_0$, $s_i > s_j$ if and only if $i > j$. Then player i chooses its strategy according to $x_i(t+1) = \max\{x \mid x \in Q_i\}$, $i \in N$.

Unconditional imitation: For each player i, find the set of neighbors including itself who have the highest payoff, and denote the set by:

$$\widehat{Q}_i := arg \max_{j \in U(i)} c_j\Big(x(t) \mid j \in U(i)\Big). \tag{7.3}$$

If $i \notin \widehat{Q}_i$, then the i-th player chooses its strategy at time $t+1$ according to the following rule:

$$x_i(t+1) = x_{j^*}(t), \tag{7.4}$$

where $j^* = \max\{i \mid i \in \widehat{Q}_i\}$. Otherwise, considering the cost of strategy transformation, the strategy of player i remains unchanged. The above two rules lead to a k-valued logical dynamics.

Fermi rule: Randomly choose a neighbor $j \in U(i)$. Compare $c_j(t)$ with $c_i(t)$ to determine $x_i(t+1)$ as:

$$x_i(t+1) = \begin{cases} x_j(t), & \text{with probability } p_t; \\ x_i(t), & \text{with probability } 1-p_t. \end{cases} \tag{7.5}$$

where p_t is decided by the Fermi function:

$$p_t = \frac{1}{1+exp(-\zeta(c_j(t)-c_i(t)))}.$$

This rule leads to a probabilistic k-valued logical dynamics.

Thus, as one kind of finite-value systems, the algebraic formulation, convergence and optimization problems of finite NEGs will be explored in the following sections.

7.2 ALGEBRAIC FORMULATION OF NETWORKED EVOLUTIONARY GAMES

This section investigates the algebraic formulation of NEGs. For time invariant network and network with random entrance, based on myopic best response adjustment rule, we convert the game dynamics into equivalent algebraic forms, respectively.

First, we consider time invariant network.

To obtain the algebraic formulation of the NEG, one can take the following two key steps: (i) convert the payoff function of each player into an algebraic form by constructing its structural matrix, and (ii) identify the best strategy for each player by comparing the components of the obtained structural matrix.

For Step (i), using the vector form of finite variables, identify $S \sim \Delta_k$, where "$\sim$" denotes that the strategy $s_j \in S$ is equivalent to $\delta_k^j \in \Delta_k$, $j=1,\cdots,k$. Then, the payoff function of player i can be expressed as:

$$\begin{aligned}
& c_i\Big(x_i(t),x_j(t) \mid j \in U(i)\Big) \\
= \ & V_r^T(A) \sum_{j\in U(i)} x_i(t)x_j(t) = V_r^T(A) \sum_{j\in U(i)} W_{[k]}x_j(t)x_i(t) \\
= \ & V_r^T(A^T)\Big(\sum_{j<i,\ j\in U(i)} x_j(t)x_i(t) + \sum_{j>i,\ j\in U(i)} x_j(t)x_i(t)\Big) \\
= \ & V_r^T(A^T)(D_f^{k,k})^{n-2}\Big(\sum_{j<i,\ j\in U(i)} W_{[k^j,k^{n-j-1}]} + \\
& \sum_{j>i,\ j\in U(i)} W_{[k^{j-1},k^{n-j}]}\Big)x_{-i}(t)x_i(t) \\
:= \ & M_{c_i}x_{-i}(t)x_i(t),
\end{aligned}$$

where A is the payoff matrix, $M_{c_i} \in \mathbb{R}^{1\times k^n}$ is the structural matrix of c_i, $x_i(t) \in \Delta_k$ is the strategy of player i at time t and $x_{-i}(t) := x_1(t) \ltimes x_2(t) \ltimes \cdots \ltimes x_{i-1}(t) \ltimes x_{i+1}(t) \ltimes \cdots \ltimes x_n(t) \in \Delta_{k^{n-1}}$.

For Step (ii), divide M_{c_i} into k^{n-1} equal blocks as:

$$M_{c_i} = \Big[Blk_1(M_{c_i}),\cdots,Blk_{k^{n-1}}(M_{c_i})\Big],$$

where the elements in the l-th block of M_{c_i} correspond to all possible benefits of player i with other players' strategy profile $x_{-i}(t) = \delta_{k^{n-1}}^l$, $l=1,2,\cdots,k^{n-1}$.

Next, find the column index of the largest element for each block of M_{c_i}. For all $l = 1, 2, \cdots, k^{n-1}$, let ξ_l be the column index such that:

$$Col_{\xi_l}(Blk_l(M_{c_i})) \geq Col_{\xi}(Blk_l(M_{c_i})), \ \forall \ \xi = 1, \cdots, k.$$

If there are more than one maximum columns, one can pick out the largest column index as ξ_l according to the priority of strategy choice.

Letting $\widetilde{L}_i = \delta_k[\xi_1, \cdots, \xi_{k^{n-1}}]$, we can obtain the algebraic form for player i as:

$$x_i(t+1) = \widetilde{L}_i x_{-i}(t) = \widetilde{L}_i D_f^{k,k} W_{[k^{i-1},k]} x(t),$$

where $x(t) = \ltimes_{i=1}^{n} x_i(t)$.

Based on the above analysis, we have the following algorithm to construct the algebraic form.

ALGORITHM 7.1

1). Calculate the structural matrix, M_{c_i}, of the payoff function of player $i \in N$. $M_{c_i} = V_r^T(A^T)(D_f^{k,k})^{n-2}\Big(\sum_{j<i,\, j\in U(i)} W_{[k^j,k^{n-j-1}]} + \sum_{j>i,\, j\in U(i)} W_{[k^{j-1},k^{n-j}]}\Big)$.

2). Divide the matrix M_{c_i} into k^{n-1} equal blocks as:

$$M_{c_i} = \Big[Blk_1(M_{c_i}), \cdots, Blk_{k^{n-1}}(M_{c_i})\Big]. \tag{7.6}$$

For all $l = 1, 2, \cdots, k^{n-1}$, find the column index ξ_l, such that $\xi_l = \max\Big\{\xi_l \mid Col_{\xi_l}(Blk_l(M_{c_i})) = \max_{1\leqslant\xi\leqslant k} Col_{\xi}(Blk_l(M_{c_i}))\Big\}$.

3). Construct the algebraic form of the game as:

$$x(t+1) = Lx(t), \tag{7.7}$$

where $Col_i(L) = Col_i(L_1) \ltimes \cdots \ltimes Col_i(L_n)$, $L_i = \widetilde{L}_i D_f^{k,k} W_{[k^{i-1},k]}$, $\widetilde{L}_i = \delta_k[\xi_1, \cdots, \xi_{k^{n-1}}]$, $i \in N$, and $L \in \mathscr{L}_{k^n \times k^n}$.

Example 7.2: Algebraic form of an NEG with time invariant network

Consider a networked evolutionary game with the following conditions:

- four players. Denote by $N = \{1, 2, 3, 4\}$ the player set, and each player has the same strategy set $S = \{s_1, s_2\}$;
- network over the four players shown in Figure 7.1;
- payoff matrix $A = \begin{bmatrix} 2 & 4 \\ 0 & 10 \end{bmatrix}$ for any pair of players on the network;
- the evolutionary rule is the myopic best response adjustment rule.

Using the vector form of finite-value variables and identifying $s_i \sim \delta_2^i$, $i = 1,2$, according to Algorithm 7.1, we can establish the algebraic form of the game as follows. Calculate the structural matrix of the payoff function for each player:

$$M_{c_1} = [2\ 0\ 4\ 10](D_f^{2,2})^2(W_{[2,4]} + W_{[4,2]}) = [\underline{4}\ 0 \vdots \underline{4}\ 0 \vdots 6\ \underline{10} \vdots 6\ \underline{10} \vdots 6\ \underline{10} \vdots 6\ \underline{10} \vdots 8\ \underline{20} \vdots 8\ \underline{20}],$$

$$M_{c_2} = [2\ 0\ 4\ 10](D_f^{2,2})^2(W_{[2,4]} + I_8) = [\underline{4}\ 0 \vdots 6\ \underline{10} \vdots \underline{4}\ 0 \vdots 6\ \underline{10} \vdots 6\ \underline{10} \vdots 8\ \underline{20} \vdots 6\ \underline{10} \vdots 8\ \underline{20}],$$

$$M_{c_3} = [2\ 0\ 4\ 10](D_f^{2,2})^2(W_{[2,4]} + I_8) = [\underline{4}\ 0 \vdots 6\ \underline{10} \vdots \underline{4}\ 0 \vdots 6\ \underline{10} \vdots 6\ \underline{10} \vdots 8\ \underline{20} \vdots 6\ \underline{10} \vdots 8\ \underline{20}],$$

$$M_{c_4} = [2\ 0\ 4\ 10](D_f^{2,2})^2(W_{[4,2]} + I_8) = [\underline{4}\ 0 \vdots 6\ \underline{10} \vdots 6\ \underline{10} \vdots 8\ \underline{20} \vdots \underline{4}\ 0 \vdots 6\ \underline{10} \vdots 6\ \underline{10} \vdots 8\ \underline{20}].$$

Divide M_{c_i} into 8 equal blocks and underline the largest column for each block. Then, we obtain:

$$\widetilde{L}_1 = \delta_2[1\ \ 1\ \ 2\ \ 2\ \ 2\ \ 2\ \ 2\ \ 2],\ \widetilde{L}_2 = \delta_2[1\ \ 2\ \ 1\ \ 2\ \ 2\ \ 2\ \ 2\ \ 2],$$

$$\widetilde{L}_3 = \delta_2[1\ \ 2\ \ 1\ \ 2\ \ 2\ \ 2\ \ 2\ \ 2],\ \widetilde{L}_4 = \delta_2[1\ \ 2\ \ 2\ \ 2\ \ 1\ \ 2\ \ 2\ \ 2].$$

Thus, the algebraic form of the game is given as:

$$x(t+1) = Lx(t), \tag{7.8}$$

where $L = \delta_{16}[1\ \ 7\ \ 10\ \ 16\ \ 10\ \ 16\ \ 10\ \ 16\ \ 7\ \ 7\ \ 16\ \ 16\ \ \cdots\ \ 16]$.

Second, we consider network with random entrance.

First of all, we give the description of the NEG with random entrance, a major player of infinite horizon and minor players of finite horizon T. It means that the minor player only stays in the game for T times. Define n_m which is the maximum possible number of active players. Consider a countably infinite set of minor players $\Omega = \{1,2,\cdots\}$. At the time t, the number of the minor players that participate in the game may be described by the vector:

$$y(t) = (n_0(t), n_1(t), \cdots, n_{T-1}(t)), \tag{7.9}$$

where $n_l(t) = |I_l(t)|$, $I_l(t)$ is the set of players with entrance time $t-l$, $l = 0,1,\cdots,T-1$.

The number of new minor players that enter the game at the time $t+1$ is a random variable with a distribution depending on $y(t)$. Thus, the random entrance is modeled by the Markov chain $y(t)$ having a finite state space. Let $1,2,\cdots,S$ be an enumeration of the state space and $\Pi = (p_{i,j})_{S\times S}$ be the transition matrix of the Markov chain.

Then, the description of a NEG is provided in the following section.

Consider a NEG, which consists of the following three ingredients:

(a) A set of finite networks $\mathscr{M} := \{1,2,\cdots,m\}$: each network's topological structure is a connected undirected graph $(N_z, \mathscr{E}_z)$ (see Figure 7.2), where $\emptyset \neq N_z \subset \{1,2,\cdots,(T+1)n_m\} := N$ is the set of nodes available for the minor players and the interaction set $\mathscr{E}_z := \{(0,i) \mid i \in N_z\}$;

(b) A FNG: in network $z \in \mathcal{M}$, if $(0,j) \in \mathcal{E}_z$, i.e., $j \in N_z$ holds, then the major player in node 0 and the minor player in node j play the FNG with strategies $x_0(t)$ and $x_j(t)$ at time t, respectively.
(c) Players' strategy updating rules: these rules can be expressed as

$$\begin{cases} x_0(t+1) = f_{0,z}\left(x_0(t), x_i(t)\right), \ i \in N_z, \\ x_i(t+1) = f_{i,z}\left(x_i(t), x_0(t)\right), \ i \in N_z, \end{cases} \tag{7.10}$$

where $x_0(t), x_i(t) \in S_0$ are the strategies of the players in the node 0 and i, separately, at time t, $i \in N_z$, and $z \in \mathcal{M}$.

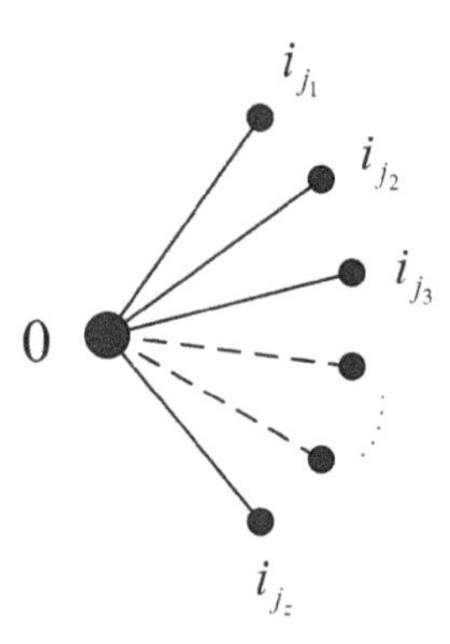

Figure 7.2 The network $(N_z, \mathcal{E}_z)$, where $N_z = \{i_{j_1}, \cdots, i_{j_z}\}$, $\mathcal{E}_z = \{(0, i_{j_1}), \cdots, (0, i_{j_z})\}$, and $z \in \mathcal{M}$

Divide the node set $\{1, 2, \cdots, (T+1)n_m\}$ into $T+1$ subset and label them as follows:

$$\underbrace{\{1, 2, \cdots, n_m\}}_{N^1}, \underbrace{\{n_m+1, n_m+2, \cdots, 2n_m\}}_{N^2}, \cdots, \underbrace{\{Tn_m+1, Tn_m+2, \cdots, (T+1)n_m\}}_{N^{T+1}}.$$

Define node set $N_j^i = \{(i-1)n_m+1, (i-1)n_m+2, \cdots, (i-1)n_m+j\}$, where $1 \leq i \leq T+1$ and $0 \leq j \leq n_m$.

From the description of random entrance, at time t, $I_0(t)$ enters the NEG and $I_{T-1}(t-1)$ leaves the NEG. Then, new participators $I_0(t)$ activates the new nodes, which do not exist in the network where the NEG proceeds at the time $t-1$, and the nodes, which are activated by $I_{T-1}(t-1)$, their interactions with node 0 are deleted. Thus, a new network is constructed and the NEG processes in this network at time t. The network $(N_{z(t)}, \mathcal{E}_{z(t)})$ changes as the following procedure:

- Define $d(t) := t(mod(T+1)) + 1$;
- Let $N_{z(t)} = \{N^1_{n_{d(t)-1}(t)}, \cdots, N^{d(t)-1}_{n_1(t)}, N^{d(t)}_{n_0(t)}, N^{d(t)+1}_0, N^{d(t)+2}_{n_{T-1}(t)}, \cdots, N^{T+1}_{n_{d(t)}(t)}\}$;
- Let $\mathcal{E}_{z(t)} = \{(0,i) \mid i \in N_{z(t)}\}$.

From this procedure, $d(t)$ and $y(t)$ determine a unique network $z(t) \in \mathcal{M}$. That is,

$$z(t) = f_z(d(t), y(t)), \tag{7.11}$$

where $z(t) \in \mathcal{M}$. Furthermore, $(T+1)S$ different networks are enough for the NEG, i.e., $|\mathcal{M}| \le (T+1)S$.

Now, we are ready to construct the algebraic form of this game. Consider *the myopic best response adjustment rule*, that is,

$$x_i(t) \in Q_{i,z} := arg\max_{x_i \in S} p_{i,z}(x_i, x_{-i}(t-1)), \ i \in N_z \cup \{0\}, \ z \in \mathcal{M}.$$

When $|Q_{i,z}| > 1$, randomly choose one with equal probability:

$$P\{x_i(t) = s_{j^*} \in Q_{i,z}\} = \frac{1}{|Q_{i,z}|}. \tag{7.12}$$

For the rest of the nodes $j \notin N_z$, which are not activated in the network z, we assign the initial state $\bar{x}_0$ for them, that is, $x_j(t) = \bar{x}_0$ holds.

To obtain the algebraic form of the NEG with random entrance, one follow the following two steps: (i) Construct the structural matrices of the updating laws for every node in each network; (ii) Construct the structural matrix of the evolving laws for the networks.

In Step (i), using the vector form of finite-value variables, we identify $S_0 \sim \Delta_k$, where "$\sim$" denotes that the strategy $j \in S_0$ is equivalent to $\delta_k^j \in \Delta_k, \ j = 1, 2, \cdots, k$. The payoff function of major player can be expressed as:

$$\begin{aligned} c_{0,z}(x_0(t), x(t)) &= \sum_{i \in N_z} m_c x_0(t) x_i(t) = \sum_{i \in N_z} m_c W_{[k]} x_i(t) x_0(t) \\ &= \sum_{i \in N_z} m_c W_{[k]} D_r^{k^{i-1},k} D_f^{k^i,k^{n-i}} x_1(t) x_2(t) \cdots x_n(t) x_0(t) \\ &:= M_{0,z} x_1(t) x_2(t) \cdots x_n(t) x_0(t), \end{aligned}$$

where $m_c \in \mathbb{R}_{1 \times k^2}$ is the structural matrix of the FNG's payoff function, $M_{0,z} \in \mathbb{R}_{1 \times k^{n+1}}$ is the structural matrix of $c_{0,z}$, and $x_i(t) \in \Delta_k$ is the strategy of the minor player in node i at time t. The payoff function of the minor player who activates the node i is:

$$c_{i,z}(x_i(t), x_{-i}(t)) = m_c W_{[k]} x_0(t) x_i(t) := M_{i,z} x_0(t) x_i(t), \tag{7.13}$$

where $m_c \in \mathbb{R}_{1 \times k^2}$ is the structural matrix of the FNG's payoff function, $M_{i,z} \in \mathbb{R}_{1 \times k^2}$ is the structural matrix of $c_{i,z}$, and $i \in N_z$.

For the major player, divide $M_{0,z}$ into k^n equal blocks as:

$$M_{0,z} = \left[Blk_1(M_{0,z}), Blk_2(M_{0,z}), \cdots, Blk_{k^n}(M_{0,z}) \right],$$

where the elements in the l-th block of $M_{0,z}$ correspond to all possible benefits of the major player with other players' strategy profile $x(t) = \ltimes_{i=1}^n x_i(t) = \delta_{k^n}^l, \ l = 1, 2, \cdots, k^n$.

Then, find the column index set of the largest element for each block of $M_{0,z}$. For all $l = 1,2,\cdots,k^n$, let the column index set $\Xi_{0,l,z}$, such that:

$$\Xi_{0,l,z} = \left\{ \xi_l \mid Col_{\xi_l}(Blk_l(M_{0,z})) = \max_{1 \leqslant \xi \leqslant k} Col_{\xi}(Blk_l(M_{0,z})) \right\}.$$

Define $r_{0,l,z} = |\Xi_{0,l,z}| \geq 1$ and construct the algebraic form of the updating law for the major player in network z as:

$$\begin{aligned} x_0(t+1) &= \tilde{L}_{0,z}x_1(t)x_2(t)\cdots x_n(t) = \tilde{L}_{0,z}D_r^{k,k}x_0(t)x_1(t)x_2(t)\cdots x_n(t) \\ &:= L_{0,z}x_0(t)x_1(t)\cdots x_n(t), \end{aligned} \tag{7.14}$$

where $Row_q(Col_l(\tilde{L}_{0,z})) = \frac{1}{r_{0,l,z}}$, $Row_p(Col_l(\tilde{L}_{0,z})) = 0$, $q \in \Xi_{0,l,z}$, $p \notin \Xi_{0,l,z}$, $l = 1,2\cdots,k^n$, and $z \in \mathscr{M}$.

For the minor player in node i of network z, $i \in N_z$, divide $M_{i,z}$ into k equal blocks as:

$$M_{i,z} = [Blk_1(M_{i,z}), Blk_2(M_{i,z}), \cdots, Blk_k(M_{i,z})],$$

where the elements in the l-th block of $M_{i,z}$ correspond to all possible benefits of player i with the major player's strategy $x_0(t) = \delta_k^l$, $l = 1,2,\cdots,k$.

Then, find the column index set of the largest element for each block of $M_{i,z}$. For all $l = 1,2,\cdots,k$, let the column index set $\Xi_{i,l,z}$, such that:

$$\Xi_{i,l,z} = \left\{ \xi_l \mid Col_{\xi_l}(Blk_l(M_{i,z})) = \max_{1 \leqslant \xi \leqslant k} Col_{\xi}(Blk_l(M_{i,z})) \right\}.$$

Define $r_{i,l,z} = |\Xi_{i,l,z}| \geq 1$ and construct the algebraic form of the updating law for the minor players in network z as:

$$x_i(t+1) = \tilde{L}_{i,z}x_0(t) = \tilde{L}_{i,z}D_f^{k,k^n}x_0(t)x_1(t)\cdots x_n(t) := L_{i,z}x_0(t)x_1(t)\cdots x_n(t), \tag{7.15}$$

where $Row_q(Col_l(\tilde{L}_{i,z})) = \frac{1}{r_{i,l,z}}$, $Row_p(Col_l(\tilde{L}_{i,z})) = 0$, $q \in \Xi_{i,l,z}$, $p \notin \Xi_{i,l,z}$, $l = 1,2\cdots,k$, and $z \in \mathscr{M}$.

For the node $j \notin N_z$, we assign them the initial state $\bar{x}_0$. Then:

$$x_j(t+1) = \bar{x}_0 = D_f^{k,k^{n+1}}\bar{x}_0x_0(t)x_1(t)\cdots x_n(t) := L_{j,z}x_0(t)x_1(t)\cdots x_n(t). \tag{7.16}$$

Thus, every new minor player entering the NEG will have the initial state $\bar{x}_0$.

Thus, by (7.14), (7.15), and (7.16), one gets the NEGs evolving on a fixed network z as follows:

$$X(t+1) = L_zX(t), \tag{7.17}$$

where $X(t) = x_0(t)x_1(t)\cdots x_n(t)$ and $L_z = L_{0,z} * L_{1,z} * \cdots * L_{n,z}$.

In Step (ii), using the vector form of logical variables, we identify $\{0,1,\cdots,T\} \sim \Delta_{T+1}$, $\{1,2,\cdots,S\} \sim \Delta_S$, and $\mathscr{M} \sim \Delta_M$, where $i \in \{0,1,\cdots,T\}$ is equivalent to $\delta_{T+1}^{i+1} \in \Delta_{T+1}$, $j \in \{1,2,\cdots,S\} \sim \delta_S^j \in \Delta_S$, and $z \in \mathscr{M} \sim \delta_M^z \in \Delta_M$.

It is easy to know that:

$$d(t+1) = \delta_{T+1}[2\ 3\ \cdots\ T+1\ 1]D_f^{T+1,S}d(t)y(t) := L_d d(t)y(t), \tag{7.18}$$

which guarantees that the logical sequence $\{d(t)\}$ is periodic. Then, we can rewrite (7.11) as:

$$z(t) = \tilde{L}_z d(t)y(t). \tag{7.19}$$

By the description of random entrance, there exists a probability matrix Π such that:

$$y(t+1) = \Pi y(t) = \Pi D_r^{T+1,S}d(t)y(t) := L_y d(t)y(t) \tag{7.20}$$

holds. By (7.18) and (7.20), one has:

$$d(t+1)y(t+1) = L_{d,y}d(t)y(t), \tag{7.21}$$

where $L_{d,y} = L_d * L_y$.

Define $L = [L_1\ L_2\ \cdots\ L_M]$, then we have:

$$X(t+1) = Lz(t)X(t) = L\tilde{L}_z d(t)y(t)X(t),$$

where $X(t) = \ltimes_{i=0}^{n} x_i(t)$. Furthermore, letting $Y(t) = d(t)y(t)X(t)$, one gets:

$$Y(t+1) = TY(t),$$

where $T = (L_{d,y}D_f^{(T+1)S,k^{n+1}}) * (L\tilde{L}_z)$.

Based on the above analysis, we have the following algorithm to construct the algebraic form of the NEG under study.

ALGORITHM 7.2

1). Calculate the structural matrix, M_i, of the payoff functions of the player in node $i \in N_z \cup \{0\}$ for each network z by:

$$M_{i,z} = \begin{cases} \sum_{i\in N_z} m_c W_{[k]} D_r^{k^{i-1},k} D_f^{k^i,k^{n-i}}, & i=0, \\ m_c W_{[k]}, & i \neq 0. \end{cases}$$

2). For each network z, respectively, divide the matrices $M_{0,z}$ and matrices $M_{i,z}$ into k^n and k equal blocks as:

$$\begin{aligned} M_{0,z} &= [Blk_1(M_{0,z}), Blk_2(M_{0,z}), \cdots, Blk_{k^n}(M_{0,z})], \\ M_{i,z} &= [Blk_1(M_{0,z}), Blk_2(M_{0,z}), \cdots, Blk_k(M_{0,z})], \end{aligned}$$

and for all $l_1 = 1,2,\cdots,k^n$ and $l_2 = 1,2,\cdots,k$, find the column index sets $\Xi_{0,l_1,z}$ and $\Xi_{i,l_2,z}$, such that:

$$\begin{aligned}\Xi_{0,l_1,z} &= \max\left\{\xi_{l_1} \mid Col_{\xi_{l_1}}(Blk_{l_1}(M_{0,z})) = \max_{1\leqslant\xi\leqslant k} Col_{\xi}(Blk_{l_1}(M_{0,z}))\right\},\\ \Xi_{i,l_2,z} &= \max\left\{\xi_{l_2} \mid Col_{\xi_{l_2}}(Blk_{l_2}(M_{i,z})) = \max_{1\leqslant\xi\leqslant k} Col_{\xi}(Blk_{l_2}(M_{0,z}))\right\},\end{aligned}$$

where $i \in N_z$.

3). Construct the algebraic form of the NEG evolving on network z under study as:

$$x(t+1) = L_z x(t),$$

where $L = L_{0,z} * L_{1,z} * \cdots * L_{n,z}$, $L_{0,z} = \tilde{L}_{0,z} D_r^{k,k}$, $Row_{q_1}(Col_{l_1}(\tilde{L}_{0,z})) = \frac{1}{r_{0,l_1,z}}$, $Row_{p_1}(Col_{l_1}(\tilde{L}_{0,z})) = 0$, $q_1 \in \Xi_{0,l_1,z}$, $p_1 \notin \Xi_{0,l_1,z}$, $l_1 = 1,2\cdots,k^n$, $L_{i,z} = \tilde{L}_{i,z} D_f^{k,k^n}$, $Row_{q_2}(Col_{l_2}(\tilde{L}_{i,z})) = \frac{1}{r_{i,l_2,z}}$, $Row_{p_2}(Col_{l_2}(\tilde{L}_{i,z})) = 0$, $q_2 \in \Xi_{i,l_2,z}$, $p_2 \notin \Xi_{i,l_2,z}$, $l_2 = 1,2\cdots,k$, $r_{0,l,z} = |\Xi_{0,l,z}| \geq 1$, $r_{i,l,z} = |\Xi_{i,l,z}| \geq 1$, $i \in N_z$, $L_{j,z} = D_f^{k,k^{n+1}} \bar{x}_0$, and $j \notin N_z$.

4). Define $L = [L_1\ L_2\ \cdots\ L_m]$, then we have the algebraic formulation as follows:

$$Y(t+1) = TY(t) \tag{7.22}$$

where $T = (L_{d,y} D_f^{(T+1)S,k^{n+1}}) * (L\tilde{L}_z)$, $Y(t) = d(t)y(t)X(t)$, and $X(t) = \ltimes_{i=0}^{n} x_i(t)$.

7.3 CONVERGENCE ANALYSIS OF NETWORKED EVOLUTIONARY GAMES

There are two cases for the final dynamical behavior of the NEGs with fixed network. One is that all players' strategies remain stationary at one profile, which is called a fixed point, and the other is that several strategy profiles are chosen periodically with period $s \geq 2$, which is called a cycle with length s. For the time varying network, more than one networks make the analysis of the NEG much more complicated. All those matrices in Algorithm 7.2 are probabilistic matrices. Thus, consider (7.22) as a finite homogenous Markov chain, which is a classical mathematical tool to investigate probabilistic logical networks. In this section, we only talk about the case of time invariant network. From the algebraic form, we can analyze the dynamical process of the game by investigating the properties of L. Using the lemma below, one can obtain all the final equilibrium states of the game, including the fixed points and cycles.

Lemma 7.1: see [6]

1. The number of cycles of length s for the dynamics of the evolutionary networked game, denoted by N_s, are inductively determined by:

$$\begin{cases} N_1 = tr(L), \\ N_s = \frac{tr(L^s) - \sum_{k \in \mathscr{P}(s)} kN_k}{s}, \ 2 \leq s \leq k^n, \end{cases}$$

   where $\mathscr{P}(s)$ denotes the set of proper factors of s, the proper factor of s is a positive integer $k < s$ satisfying $s/k \in \mathbb{Z}_+$, and $\mathbb{Z}_+$ is the set of positive integers.
2. The set of elements on cycles of length s, denoted by $\mathscr{C}_s$, is:

$$\mathscr{C}_s = \mathscr{D}_a(L^s) \setminus \bigcup_{t \in \mathscr{P}(s)} \mathscr{D}_a(L^t),$$

   where $\mathscr{D}_a(L)$ is the set of diagonal nonzero columns of L. ■

Next, we consider the stable degree of strategy profiles, that is, local convergence.

DEFINITION 7.6

In an evolutionary networked game, a strategy profile $s^* \in S^n := S \times \cdots \times S$ is called l-degree stable, if the number of mutant strategies in s^* is no more than l, the iterative sequence of strategy profiles will still converge to the eventual equilibrium s^* through evolution. Otherwise, the eventual equilibrium s^* may not be played.

Actually, l-degree stability we consider refers to the sequence of iterates $\{x(t)\}$ converging to a fixed point x^* of F when it starts "near" x^*, where F is the strategy updating rule. Thus, to describe the localized perturbations of the strategy profile, we need to define a "neighborhood" in the space of strategy profiles S^n.

DEFINITION 7.7

Given $s^* = (s^*_{(1)}, \cdots, s^*_{(n)}) \in S^n$, the l-step neighborhood of s^* is defined as:

$$V^l_{s^*} = \{s \in S^n \mid \|s - s^*\| \leq l, \ s \neq s^*\}, \tag{7.23}$$

where $\|\cdot\|$ is the Hamming distance, which denotes the numbers of different strategies between two strategy profiles.

Denote by $x(t;s)$ the strategy profile of all players at time t starting from an initial strategy profile s. Then, by Definition 7.7, we have the following proposition.

PROPOSITION 7.1

Strategy profile s^* is l-degree stable if and only if:

(C1). $\lim_{t\to+\infty} x(t;s) = s^*$, $\forall\, s \in V_{s^*}^l \bigcup \{s^*\}$, and

(C2). there exists an $s \in V_{s^*}^{l+1}$ such that $\lim_{t\to+\infty} x(t;s) \neq s^*$.

The number l determines the stable degree of strategy profile. The larger the l is, the more resistent the strategy profile is against the invasion of mutants. If $l = n$, then any mutant strategy profile will globally converge to s^*. Otherwise, only a part of strategy profiles converge to s^*, which can be called "local" or "neighbor" convergence. *Our aim is* to investigate the stable degree of strategy profile.

Using the vector form of logical variables, we have $s \sim \delta_{k^n}^j$, where $s \in S^n$ and $j \in \{1,2,\cdots,k^n\}$. Then, given a strategy profile $s^* = \delta_{k^n}^{j_0}$, the l-step neighborhood of s^* can be rewritten as:

$$V_{s^*}^l = \left\{ s = \delta_{k^n}^j \mid n - \sum_{i=1}^n V_r^T(I_k) S_{i,k}^n (I_{k^n} \otimes S_{i,k}^n) \delta_{k^n}^{j_0} \delta_{k^n}^j \leq l, j \neq j_0 \right\}, \tag{7.24}$$

where $V_r(I_k)$ is the row stacking form of the identity matrix I_k. To make it simple, we define a set as $\widehat{V}_{s^*}^l = \left\{ j \mid \delta_{k^n}^j \in V_{s^*}^l \right\}$.

Since the calculation of $V_{s^*}^l$ is fundamental for our main result, we would like to find a simple formula to get to the l-step neighborhood of a strategy profile. First, we consider a special strategy profile $s^* = \delta_{k^n}^1$, that is, all players choose the strategy s_1.

Construct a sequence of row vectors $\{\Gamma_i | i = 1,2,\cdots\}$ as:

$$\Gamma_1 = [\underbrace{0\ 1\ 1\ \cdots\ 1}_{k}] \in \mathbb{R}^{k\times k},$$

$$\Gamma_i = [\underbrace{\Gamma_{i-1} \vdots \Gamma_{i-1} + \mathbf{1}_{k^{i-1}} \vdots \Gamma_{i-1} + \mathbf{1}_{k^{i-1}} \vdots \cdots \vdots \Gamma_{i-1} + \mathbf{1}_{k^{i-1}}}_{k}] \in \mathbb{R}^{k\times k^i},\ i = 2,3,\cdots$$

where $\mathbf{1}_n := [\underbrace{1\ 1\ \cdots\ 1}_{n}]$.

By this construction, it is easy to obtain the following lemma.

Lemma 7.2

Consider the special strategy profile $s^* = \delta_{k^n}^1$. Then, the l-step neighborhood of s^* can be calculated by:

$$V_{s^*}^l = \{\delta_{k^n}^j \mid Col_j(\Gamma_n) \leq l\}.$$

■

For a general strategy profile $s^* = \delta_{k^n}^{j_0} = \delta_k^{j_1}\delta_k^{j_2}\cdots\delta_k^{j_n}$, we can use a coordinate transformation $y_i = T_i x_i$, where $T_i = \delta_k[\,k-j_i+2 \;\; k-j_i+3 \;\cdots\; k \;1\; 2 \;\cdots\; k-j_i+1]$, under which the strategy profile $s^* = \delta_{k^n}^{j_0}$ becomes $\widehat{s^*} = \delta_{k^n}^1$. Meanwhile, the algebraic form of the game dynamics is transformed into:

$$y(t+1) = TLT^{-1}y(t),$$

where $T = T_1(I_k \otimes T_2)(I_{k^2} \otimes T_3)\cdots(I_{k^{n-1}} \otimes T_n)$ and $y(t) = \ltimes_{i=1}^n y_i(t) \in \Delta_{k^n}$. Thus, Lemma 7.2 can also be applied to a general case.

Example 7.3:

Suppose $n=4$, $k=2$, $s^* = \delta_{16}^1$. First, construct Γ_i, $i=1,2,3,4$ as follows:

$$\Gamma_1 = [0\ 1],\ \Gamma_2 = [0\ 1\ 1\ 2],\ \Gamma_3 = [0\ 1\ 1\ 2\ 1\ 2\ 2\ 3],$$

$$\Gamma_4 = [0\ 1\ 1\ 2\ 1\ 2\ 2\ 3\ 1\ 2\ 2\ 3\ 2\ 3\ 3\ 4].$$

By Lemma 7.2, it is easy to obtain that $\widehat{V}_{s^*}^1 = \{2,3,5,9\}$, $\widehat{V}_{s^*}^2 = \{2,3,4,5,6,7,9,10,11,13\}$.

Next, we analyze the stable degree of strategy profile s^*. For $t \in \mathbb{Z}_+$ and the given strategy profile $s^* = \delta_{k^n}^{j_0}$, let $R_t(j_0)$ be a series of sets generated inductively by:

$$R_1(j_0) = \{j \mid Col_j(L) = \delta_{k^n}^{j_0}\},$$
$$R_2(j_0) = \{j \mid Col_j(L) = \delta_{k^n}^{j_1},\ \forall\ j_1 \in R_1(j_0)\backslash\{j_0\}\},$$
$$R_t(j_0) = \{j \mid Col_j(L) = \delta_{k^n}^{j_{t-1}},\ \forall\ j_{t-1} \in R_{t-1}(j_0)\},\quad t \geq 3,$$

where $j \in R_t(j_0)$ indicates that the strategy profile $\delta_{k^n}^j$ can reach s^* at the t-th step under the given strategy updating rule of the game. Then, based on the algebraic formulation of game dynamics, we have the following result.

Theorem 7.1

Consider the evolutionary networked game $((N,E),\ G,\ F)$ with its algebraic form: $x(t+1) = Lx(t)$. Then, strategy profile $s^* = \delta_{k^n}^{j_0}$ is l-degree stable if and only if,

$$\begin{cases} j_0 \in R_1(j_0), \\ \widehat{V}_{s^*}^{l} \subseteq \bigcup_{t=1}^{k^n} R_t(j_0), \\ \widehat{V}_{s^*}^{l+1} \nsubseteq \bigcup_{t=1}^{k^n} R_t(j_0). \end{cases} \quad (7.25)$$

■

Proof. (Sufficiency) Assuming that (7.25) is satisfied, we only need to prove (C1) and (C2) hold (see Proposition 7.1).

Since $j_0 \in R_1(j_0)$, one can obtain $Ls^* = Col_{j_0}(L) = \delta_{k^n}^{j_0}$. Hence, $\delta_{k^n}^{j_0}$ is a fixed point of $x(t+1) = Lx(t)$, that is, once the strategy profile $s^* = \delta_{k^n}^{j_0}$ is chosen, it will remain unchanged forever under the strategy updating rule of the game. Moreover, for any $j \in \widehat{V}_{s^*}^{l}$, since $\widehat{V}_{s^*}^{l} \subseteq \bigcup_{t=1}^{k^n} R_t(j_0)$, we have $j \in \bigcup_{t=1}^{k^n} R_t(j_0)$. Then there exists a positive integer $\widetilde{t} \leq k^n$, such that $j \in R_{\widetilde{t}}(j_0)$, which means that strategy profile $\delta_{l^n}^{j}$ will converge to s^* at the $\widetilde{t}$-th step. Based on the above discussion, we conclude that (C1) holds.

Moreover, from the condition $\widehat{V}_{s^*}^{l+1} \nsubseteq \bigcup_{t=1}^{k^n} R_t(j_0)$, one can see that there exists a $j \in \widehat{V}_{s^*}^{l+1}$ such that $j \notin \bigcup_{t=1}^{k^n} R_t(j_0)$. Now, we show that $\lim_{t \to +\infty} x(t; \delta_{k^n}^{j}) \neq s^*$. In fact, if $\lim_{t \to +\infty} x(t; \delta_{k^n}^{j}) = s^*$, then, there exists a $\tau \geq k^n + 1$, such that $x(\tau; \delta_{k^n}^{j}) = s^*$. However, since there are only k^n different strategy profiles in the game, there must exist two strategy profiles satisfying $x(\tau_1; \delta_{k^n}^{j}) = x(\tau_2; \delta_{k^n}^{j})$, $\tau_1, \tau_2 \leq \tau$. In this case, a cycle of strategy profiles or another fixed point appears, which is a contradiction to the fact that s^* is an only local fixed point of $x(t+1) = Lx(t)$. Therefore, (C2) is satisfied and the sufficiency is completed.

(Necessity) Suppose that strategy profile $s^* = \delta_{k^n}^{j_0}$ is l-degree stable. From Definition 7.6, it is easy to see that $j_0 \in R_1(j_0)$.

On the other hand, similar to the proof of sufficiency, we know that for an arbitrary strategy profile s, if $\lim_{t \to +\infty} x(t; s) = s^*$, then the strategy profile s will converge to s^* within k^n steps. Therefore, the set $\left\{ \delta_{k^n}^{j} \mid j \in \bigcup_{t=1}^{k^n} R_t(j_0) \right\}$ contains all the strategy profiles that can converge to s^* through evolution. Thus, the conditions (C1) and (C2) guarantee that:

$$\widehat{V}_{s^*}^{l} \subseteq \bigcup_{t=1}^{k^n} R_t(j_0) \text{ and } \widehat{V}_{s^*}^{l+1} \nsubseteq \bigcup_{t=1}^{k^n} R_t(j_0),$$

and the proof is completed. ■

From Theorem 7.1, to verify the stable degree, we need to calculate k^n sets. However, for a given strategy profile, some computations are redundant since the mutant strategy profile may be restored in less than k^n steps. To reduce the computation, we now consider the transient time, i.e., the longest time, needed to restore the mutant strategy profile $\delta_{k^n}^{j_0}$ through evolution and then remain unchanged.

First, construct a sequence of row vectors $\xi_1, \cdots, \xi_t, \cdots$ by the following equation:

$$\xi_t = \xi_{t-1} L = \xi_0 L^t, \ t \geq 1, \quad (7.26)$$

where $\xi_0 = (\delta_{k^n}^{j_0})^T$, and $\xi_t = [\xi_{t,1}, \cdots, \xi_{t,k^n}] \in \mathscr{B}_{1\times k^n}$.

Second, define an order, denoted by "$\succeq$", on the m-dimensional vector space $\mathbb{R}^m$ as follows:

$$(x_1, x_2, \cdots, x_m) \succeq (y_1, y_2, \cdots, y_m)$$

if and only if $x_i \geq y_i$, $i = 1, 2, \cdots, m$. Similarly, orders "$\preceq$" and "$\prec$" denote $x_i \leq y_i$ and $x_i < y_i$, respectively. Based on the above, we have the following proposition.

PROPOSITION 7.2

Suppose that $\xi_{1,j_0} = 1$, then for any $t \geq 1$, we have $\xi_{t+1} \succeq \xi_t$, that is, $\xi_{t+1,j} \geq \xi_{t,j}$, $j = 1, 2, \cdots, k^n$.

Proof. We prove it by mathematical induction. First, we show that $\xi_2 \succeq \xi_1$. Note that $\xi_{1,j_0} = 1$, then it follows that:

$$\begin{aligned}
\xi_2 &= \xi_1 L = \sum_{i=1}^{k^n} \xi_{1,i} Row_i(L) \\
&= Row_{j_0}(L) + \sum_{i=1, i\neq j_0}^{k^n} \xi_{1,i} Row_i(L) \\
&= \xi_1 + \sum_{i=1, i\neq j_0}^{k^n} \xi_{1,i} Row_i(L) \succeq \xi_1,
\end{aligned}$$

which implies that the conclusion is true for $t = 1$.

Assuming that $\xi_t \succeq \xi_{t-1}$, $t \geq 2$, we consider the case of $t+1$. In this case, one can obtain:

$$\begin{aligned}
\xi_{t+1} &= \xi_t L = \sum_{i=1}^{k^n} \xi_{t,i} Row_i(L) \\
&\succeq \sum_{i=1}^{k^n} \xi_{t-1,i} Row_i(L) = \xi_{t-1} L = \xi_t,
\end{aligned}$$

that is, $\xi_{t+1} \succeq \xi_t$. By induction, the conclusion holds. ∎

From Proposition 7.2 and the fact that $L \in \mathscr{L}_{k^n \times k^n}$, there must exist an integer τ $(0 \leq \tau \leq k^n)$ such that $\xi_\tau = \xi_{\tau+1}$. Let:

$$\tau_0 = \min\{\tau \mid \xi_\tau = \xi_{\tau+1}\}. \tag{7.27}$$

It is obvious that $\tau_0 \leq k^n$. Then, we have the following result.

Theorem 7.2

For the given strategy profile $s^* = \delta_{k^n}^{j_0}$, τ_0 defined in (7.27) is just the transient time. ∎

Proof. First, we define a series of sets as follows:

$$\Omega_t = \{i \mid \xi_{t,i} \neq 0,\ i = 1,2,\cdots,k^n\},\ t = 1,2,\cdots.$$

Considering that strategy profile s^* remains unchanged under the strategy updating rule, one can see that $L\delta_{k^n}^{j_0} = \delta_{k^n}^{j_0}$, which yields $\xi_{1,j_0} = 1$. Moreover, since $\xi_0 = (\delta_{k^n}^{j_0})^T$ and $\xi_1 = \xi_0 L = Row_{j_0}(L)$, it is easy to obtain:

$$L\delta_{k^n}^{i_1} = \delta_{k^n}^{j_0},\ \forall\, i_1 \in \Omega_1,$$

which implies that $\left\{\delta_{k^n}^{i_1} | i_1 \in \Omega_1\right\}$ contains all the strategy profiles, which reach s^* in one step.

Similarly, from the equation $\xi_2 = \xi_1 L = \xi_0 L^2$, we can obtain:

$$L^2\delta_{k^n}^{i_2} = \delta_{k^n}^{j_0},\ \forall\, i_2 \in \Omega_2.$$

Therefore, $\left\{\delta_{k^n}^{i_2} | i_2 \in \Omega_2\right\}$ contains all the strategy profiles, which reach s^* at the second step.

Generally speaking, the column indices of nonzero elements in ξ_t, $t \geq 3$, correspond to strategy profiles that can reach s^* at the t-th step, that is, $L^t\delta_{k^n}^{i_t} = \delta_{k^n}^{j_0}$, $\forall\, i_t \in \Omega_t$. Combining this and the definition of τ_0, when $t = \tau_0$, all the strategy profiles that can reach s^* under the strategy updating rule are obtained. Hence, τ_0 defined in (7.27) is the transition time. ∎

From Theorem 7.2, the following result is obvious.

COROLLARY 7.1

$s^* = \delta_{k^n}^{j_0}$ is l-degree stable if and only if:

$$\eta_l \preceq \xi_{\tau_0} \prec \eta_{l+1},$$

where τ_0 is the transient time, and $\eta_l = \sum_{i \in \widehat{V}_{s*}^l} (\delta_{k^n}^i)^T$, $1 \leq l \leq k^n$.

It should be pointed out that when the number of mutant strategies in s^* is no more than k, s^* will be restored within τ_0^* steps, where $\tau_0^* = \min\{\tau \mid \xi_\tau \succeq \eta_k\}$. With this, we have the following corollary.

COROLLARY 7.2

If strategy profile $s^* = \delta_{k^n}^{j_0}$ is l-degree stable, then $Col_j(L^{\tau_0^*}) = \delta_{k^n}^{j_0}$, $\forall\ j \in \widehat{V}_{s^*}^{l}$.

Based on Corollary 7.1, we establish an algorithm to calculate the stable degree of strategy profile $s^* = \delta_{k^n}^{j_0}$.

ALGORITHM 7.3

1). Setting $\xi_0 = (\delta_{k^n}^{j_0})^T$, calculate the vector ξ_1 and judge whether $\xi_{1,j_0} = 1$. If $\xi_{1,j_0} \neq 1$, then stop the computation. Otherwise, go to next step;
2). Compute ξ_t, $2 \leq t \leq \tau_0$, by the equation (7.26), where τ_0 is defined in (7.27);
3). Calculate the stable degree of strategy profile s^* as follows:

$$k = \min\left\{\alpha \mid n - \sum_{i=1}^{n} V_r^T(I_k) S_{i,k}^n (I_{k^n} \otimes S_{i,k}^n) \delta_{k^n}^{j_0} \delta_{k^n}^{j} = \alpha,\ \xi_{\tau_0,j} = 0,\ 1 \leq j \leq k^n\right\} - 1.$$

Example 7.4:

Consider an evolutionary networked game, described as follows:

- its network topological structure is shown in Figure 7.3;
- its player set is $N = \{1,2,3,4\}$, and the basic game between two players is Stag-hunt Game, where the strategy set is {"stag-hunt", "hare-hunt"} and the corresponding payoff matrix is $A = \begin{bmatrix} 3 & 0 \\ 1 & 1 \end{bmatrix}$;
- the adopted strategy updating rule is "best imitate" strategy updating rule.

From the payoff matrix A of Stag-hunt Game, we can see that each player will gain the maximum benefit when all the players adopt "stag-hunt" strategy. In the following, we use the results obtained in this paper to analyze the stable degree of the strategy profile.

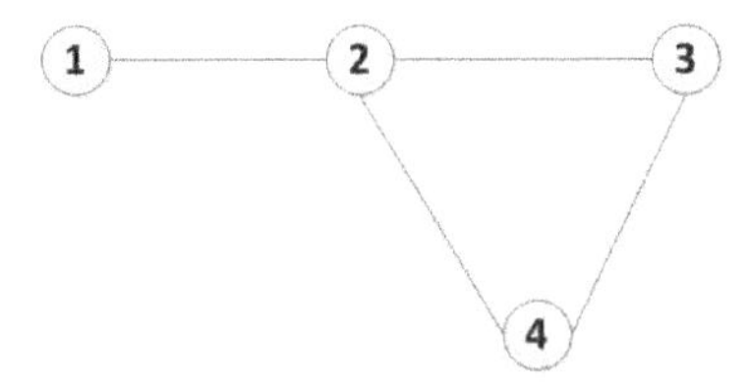

Figure 7.3 The network of Example 7.4

First, we convert the game dynamics into an algebraic form. Using the vector form of logical variables, by identifying the strategy "stag-hunt"$\sim \delta_2^1$, and "hare-hunt"$\sim \delta_2^2$, the payoff function of each player can be calculated in the following way:

$$\begin{aligned}
p_1(x(t)) &= V_r^T(A)x_1(t)x_2(t) = V_r^T(A)(E_d)^2 W_{[4]}x(t),\\
p_2(x(t)) &= V_r^T(A)[x_2(t)x_1(t) + x_2(t)x_3(t) + x_2(t)x_4(t)]\\
&= V_r^T(A)\left[W_{[2]}(E_d)^2 W_{[4]} + (E_d)^2 W_{[8,2]} + (E_d)^2 W_{[4,2]}\right]x(t),\\
p_3(x(t)) &= V_r^T(A)[x_3(t)x_2(t) + x_3(t)x_4(t)] = V_r^T(A)\left[W_{[2]}(E_d)^2 W_{[8,2]} + (E_d)^2\right]x(t),\\
p_4(x(t)) &= V_r^T(A)[x_4(t)x_2(t) + x_4(t)x_3(t)] = V_r^T(A)\left[W_{[2]}(E_d)^2 W_{[4,2]} + W_{[2]}(E_d)^2\right]x(t),
\end{aligned}$$

where $x(t) = \ltimes_{i=1}^4 x(t) \in \Delta_{16}$. It is easy to obtain the average payoff matrix as:

$$M = \begin{bmatrix}
3 & 3 & 3 & 3 & 0 & 0 & 0 & 0 & 1 & 1 & 1 & 1 & 1 & 1 & 1 & 1\\
3 & 2 & 2 & 1 & 1 & 1 & 1 & 1 & 2 & 1 & 1 & 0 & 1 & 1 & 1 & 1\\
3 & \frac{3}{2} & 1 & 1 & \frac{3}{2} & 0 & 1 & 1 & 3 & \frac{3}{2} & 1 & 1 & \frac{3}{2} & 0 & 1 & 1\\
3 & 1 & \frac{3}{2} & 1 & \frac{3}{2} & 1 & 0 & 1 & 3 & 1 & \frac{3}{2} & 1 & \frac{3}{2} & 1 & 0 & 1
\end{bmatrix}.$$

Thus, the algebraic form of game dynamics is given as:

$$x(t+1) = Lx(t), \tag{7.28}$$

where $L = \delta_{16}[1\ 1\ 1\ 4 \vdots 9\ \ 16\ \ 16\ \ 16 \vdots 1\ \ 9\ \ 9\ \ 16 \vdots 9\ \ 16\ \ 16\ \ 16]$.

Next, we investigate the stable degree of strategy profile $s^* = \delta_{16}^1$, which stands for that all players choose the "stag-hunt" strategy. Since $Col_1(L) = \delta_{16}^1$, s^* is a fixed point of (7.28), with which we obtain $R_1(1) = \{1,2,3,9\}, R_2(1) = \{5,10,11,13\}$, and $R_t(1) = \emptyset,\ 3 \le t \le 16$.

It is noted that the 1-step neighborhood and 2-step neighborhood of s^* are $\widehat{V}_{s^*}^1 = \{2,3,5,9\}$ and $\widehat{V}_{s^*}^2 = \{4,6,7,10,11,13\} \bigcup \widehat{V}_{s^*}^1$, respectively. Thus, we have:

$$\widehat{V}_{s^*}^1 \subseteq \bigcup_{t=1}^{16} R_t(1) \text{ and } \widehat{V}_{s^*}^2 \nsubseteq \bigcup_{t=1}^{16} R_t(1).$$

Hence, the stable degree of s^* is $k = 1$, which implies that strategy profile s^* will be chosen again through evolution and remain unchanged when the number of mutant strategies in s^* is no more than one.

Finally, we calculate the transient time. Letting $\xi_0 = (\delta_{16}^1)^T$ and by (7.26), we have $\xi_1 = [0\ 1\ 1\ 0\ 0\ 0\ 0\ 0\ 1\ 0\ 0\ 0\ 0\ 0\ 0\ 0]$, $\xi_2 = [0\ 1\ 1\ 0\ 1\ 0\ 0\ 0\ 1\ 1\ 1\ 0\ 1\ 0\ 0\ 0]$, and $\xi_3 = \xi_2$, which implies that the transient time of s^* is $\tau_0 = 2$. Thus, it can be concluded that when a mutation arises in s^*, it will never be chosen again by players if s^* is not restored within two steps.

7.4 OPTIMIZATION OF NETWORKED EVOLUTIONARY GAMES

This section investigates strategy optimization problems of NEGs.

First, for time invariant network, we add a pseudo-player to the game, which can be perceived as an external control input. Without loss of generality, assume that the first player is the pseudo-player who can take strategies freely. Let $u(t) \in \Delta_k$ be the strategy of the pseudo-player at time t, and $x_i(t) \in \Delta_k$ be the strategy of player $i \in \{2,3,\cdots,n\}$ at time t. The aim is to present a necessary and sufficient condition under which the dynamic process of the game starting from any strategy initial profile can globally converge to the best strategy profile and design a free-type control sequence to maximize the following average payoff of the pseudo-player in the long run:

$$J(u) = \limsup_{T\to\infty} \frac{1}{T}\sum_{t=1}^{T} c_1\Big(u(t), x_j(t) \mid j \in U(1)\Big), \tag{7.29}$$

where $c_1\Big(u(t), x_j(t) \mid j \in U(1)\Big) = V_r^T(A) \sum_{j\in U(1)} u(t)x_j(t)$.

Assume that the algebraic form of the game under the pseudo-player's control is:

$$x(t+1) = Lu(t)x(t), \tag{7.30}$$

where $x(t) := \ltimes_{i=2}^{n} x_i(t) \in \Delta_{k^{n-1}}$, and $L \in \mathscr{L}_{k^{n-1}\times k^n}$.

Let the strategy profile $\delta_{k^{n-1}}^{k^{n-1}}$ be the best strategy profile, where $\delta_{k^{n-1}}^{k^{n-1}} = \delta_k^k \ltimes \cdots \ltimes \delta_k^k$ denotes that each player $i \in \{2,3,\cdots,n\}$ chooses the strategy s_k. Obviously, if all players, including the pseudo-player, adopt the strategy δ_k^k at time t, the strategy remains the same for the next players $i \in \{2,3,\cdots,n\}$ according to the myopic best reply adjustment rule. Therefore, we have the following proposition.

PROPOSITION 7.3

Consider the algebraic form (7.30) for the evolutionary networked game. Then,

$$Col_{k^{n-1}}(Blk_k(L)) = \delta_{k^{n-1}}^{k^{n-1}}, \tag{7.31}$$

that is to say, once each player $i \in \{2,3,\cdots,n\}$ takes the strategy s_k at time t, we can let the pseudo-player also choose the strategy s_k so that the strategy s_k is same for all the players $i \in \{2,3,\cdots,n\}$.

Next, to make the evolutionary process of the game globally converge to the best strategy profile, we only need to guarantee that the best strategy profile is reachable globally under a free-type control sequence. For any $t \in \mathbb{Z}_+$, we have:

$$\begin{aligned} x(t) &= Lu(t-1)x(t-1) = Lu(t-1)Lu(t-2)x(t-2) \\ &= L(I_k \otimes L)u(t-1)u(t-2)x(t-2) \\ &= L(I_k \otimes L)\cdots(I_{k^{t-1}} \otimes L)u(t-1)\cdots u(0)x(0) \\ &:= \widetilde{L}u(t-1)u(t-2)\cdots u(0)x(0), \end{aligned} \tag{7.32}$$

where $\widetilde{L} = L(I_k \otimes L)\cdots(I_{k^{t-1}} \otimes L) \in \mathscr{L}_{k^{n-1}\times k^{(n-1)+t}}$.

Based on (7.32) and Proposition 7.3, we have the following result.

Theorem 7.3

Consider the evolutionary networked game with the algebraic form (7.30). The evolutionary process of the game globally converges to the best strategy profile $\delta_{k^{n-1}}^{k^{n-1}}$ by a free-type control sequence, if and only if there exist integers $\tau > 0$ and $1 \leq \alpha \leq k^{\tau}$, such that:

$$Col_j(Blk_{\alpha}(\widetilde{L})) = \delta_{k^{n-1}}^{k^{n-1}}, \ \forall\ 1 \leq j \leq k^{n-1}, \tag{7.33}$$

where $\widetilde{L} = L(I_k \otimes L) \cdots (I_{k^{\tau-1}} \otimes L) \in \mathscr{L}_{k^{n-1} \times k^{(n-1)+\tau}}$. In addition, if (7.33) holds, then the free-type control sequence which maximizes the average payoff (7.29) of the pseudo-player can be designed as:

$$u(t) = \begin{cases} \widetilde{u}(t),\ 0 \leq t \leq \tau - 1, \\ \delta_k^k,\ t \geq \tau, \end{cases} \tag{7.34}$$

where $\widetilde{u}(\tau-1) \ltimes \widetilde{u}(\tau-2) \ltimes \cdots \ltimes \widetilde{u}(0) = \delta_{k^{\tau}}^{\alpha}$. ■

Proof. (Necessity) Suppose that the evolutionary networked game globally converges to the best strategy profile $\delta_{k^{n-1}}^{k^{n-1}} := x_e$. Since the iterative sequence starting from any initial strategy profile $x(0) \in \Delta_{k^{n-1}}$ converges to x_e, there exist a free-type control sequence $\{u(t)\}$ and a time $\tau > 0$ such that:

$$x(\tau) = \widetilde{L}u(\tau-1) \cdots u(0)x(0) = x_e,\ \forall\ x(0) \in \Delta_{k^{n-1}},$$

which implies that (7.33) is satisfied.

(Sufficiency) Assuming that the condition (7.33) holds, we prove that the evolutionary process globally converges to x_e under the free-type control sequence (7.34). By (7.33) and Proposition 7.3, one can see that for any $x(0) \in \Delta_{k^{n-1}}$,

$$x(\tau) = Blk_{\alpha}(\widetilde{L})x(0) = \delta_{k^{n-1}}[k^{n-1} \cdots k^{n-1}]x(0) = x_e,$$

and

$$\begin{aligned} x(t) &= Lu(t-1)x(t-1) = Blk_k(L)x(t-1) \\ &= (Blk_k(L))^{t-\tau}x(\tau) = x_e,\ \forall\ t > \tau, \end{aligned} \tag{7.35}$$

that is, the evolutionary process of the game globally converges to the best strategy profile x_e. ■

It is noted that the designed free-type control sequence (7.34) implies that the pseudo-player adopts the strategy s_{α_t} for time $0 \leq t \leq \tau - 1$ and s_k for times $t \geq \tau$, where α_t, $0 \leq t \leq \tau - 1$ satisfy $\delta_{k^{\tau}}^{\alpha} = \delta_k^{\alpha_{\tau-1}} \ltimes \delta_k^{\alpha_{\tau-2}} \ltimes \cdots \ltimes \delta_k^{\alpha_0}$ and $\alpha = (\alpha_{\tau-1} - 1)k^{\tau-1} + (\alpha_{\tau-2} - 1)k^{\tau-2} + \cdots + (\alpha_1 - 1)k + \alpha_0$.

Example 7.5: Strategy optimization of an NEG with time invariant network

Recall Example 7.2. Letting the first player to be a pseudo-player, we can obtain the algebraic form, in this, case as:

$$x(t+1) = Lu(t)x(t),$$

where $x(t) = \ltimes_{i=2}^{4} x_i$, $L = \delta_8[1\ 8\ 2\ 8\ 2\ 8\ 2\ 8\ 7\ 8\ 8\ 8\ 8\ 8\ 8\ 8]$. A simple calculation shows that:

$L(I_2 \otimes L) = \delta_8[\,1\,8\,8\,8\,8\,8\,8\,8 \vdots 2\,8\,8\,8\,8\,8\,8\,8 \vdots 7\,8\,8\,8\,8\,8\,8\,8 \vdots 8\,8\,8\,8\,8\,8\,8\,8\,]$.
It is obvious that $Blk_4(L(I_2 \otimes L)) = \delta_8[\,8\,8\,8\,8\,8\,8\,8\,8\,]$, that is, $u(1)u(0) = \delta_4^4 = \delta_2^2\delta_2^2$.

From Theorem 7.3, the evolutionary process converges to the best strategy profile $\delta_8^8 = \delta_2^2\delta_2^2\delta_2^2$ by the free-type control sequence $\{u(t) = \delta_2^2,\ t \geq 0\}$, that is to say, if the pseudo-player adopts the strategy s_2 and it remains unchanged, the evolutionary result of the game starting from any initial strategy profile will be that all players choose the strategy s_2 and keep it in the long run. Meanwhile, the pseudo-player can receive the maximal average payoff $J(u) = 20$ in the long run.

Second, for network with random entrance, we also investigate the strategy optimization problem. The aim is to optimize the major player's updating law when the other players take *the myopic best response adjustment rule*. We first convert the given NEG into a control-depending Markov chain. Then, we use dynamic programming algorithm and receding horizon control to get a state feedback controller.

Define $L_i = [L_{i,1}\ L_{i,2} \cdots L_{i,m}]$ and $x(t) = \ltimes_{i=1}^{n} x_i(t)$, one can get:

$$\begin{aligned} x_i(t+1) &= L_i z(t) x_0(t) x(t) \\ &= L_i \tilde{L}_z d(t) y(t) x_0(t) x(t) = L_i \tilde{L}_z W_{[k,(T+1)S]} x_0(t) d(t) y(t) x(t) \\ &:= L_i^u x_0(t) d(t) y(t) x(t), \end{aligned} \tag{7.36}$$

where $i \in N$. Moreover, it:

$$\begin{aligned} d(t+1)y(t+1) &= L_{d,y} d(t) y(t) \\ &= L_{d,y} D_r^{k,(T+1)S} D_f^{(T+1)Sk,k^n} x_0(t) d(t) y(t) x(t) \\ &= L_{d,y}^u x_0(t) d(t) y(t) x(t). \end{aligned} \tag{7.37}$$

Thus, letting $u(t) = x_0(t)$, by (7.36) and (7.37), one has:

$$h(t+1) = L_u u(t) h(t), \tag{7.38}$$

where $L_u = L_{d,y}^u * L_1^u * L_2^u * \cdots * L_n^u$ and $h(t) = d(t)y(t)x(t)$.

Note that $L_u u(t)$ is the control-depending (column) Markovian transition matrix. Defining $L_u u(t) := A(u(t)) = (a_{i,j}(u(t)))_{H \times H}$ and $H := (T+1)Sk^n$, we get:

$$P\{h(t+1) = i \mid h(t) = j, u(t) = u\} = a_{i,j}(u). \tag{7.39}$$

From (7.12), when the NEG evolves, we get the payoff $p(u(t),h(t))$ for the major player at time t:

$$\begin{aligned} p(u(t),h(t)) &= p_{0,z(t)}(u(t),x(t)) \\ &= M_0 z(t)x(t)u(t) = M_0 \tilde{L}_z d(t)y(t)x(t)u(t) \\ &:= \Phi h(t)u(t), \end{aligned} \tag{7.40}$$

where $M_0 = [M_{0,1}\ M_{0,2} \cdots M_{0,m}]$.

When $h(0) = h_0$, there is the infinite horizon performance criterion:

$$J(h(0)) := E\left[\sum_{t=0}^{\infty} \lambda^t p(u(t),h(t))\right], \tag{7.41}$$

where $0 < \lambda < 1$ is a discount factor. Our problem is: finding a proper updating law for $u(t)$ to maximize $J(h(0))$.

We, first consider the finite horizon performance criterion as follows:

$$J_N(h(0)) := E\left[\sum_{t=0}^{N} \lambda^t p(u(t),h(t))\right]. \tag{7.42}$$

And, the following useful proposition is proven in [7] (Sec. 1.3, p.23).

PROPOSITION 7.4

For system (7.38), let $J^*(h(0))$ be the optimal value of (7.42). Then,

$$J^*(h(0)) = J_0(h(0)), \tag{7.43}$$

where the function J_0 is given by the last step of dynamic programming algorithm. The algorithm proceeds backward in time from $t = N$ to $t = 0$ as follows:

$$J_N(h(N)) = \max_{u(N)\in\Delta_k} \lambda^N p(u(N),h(N)), \tag{7.44}$$

and,

$$J_t(h(t)) = \max_{u(t)\in\Delta_k} E\left[\left(\lambda^t p(u(t),h(t)) + J_{t+1}(h(t+1))\right)\right]. \tag{7.45}$$

In the following section, we give a method to calculate the above functions. By (7.38) and (7.39), we can calculate the expectation of $J_{t+1}(h(t+1))$ in (7.45) being conditional on $u(t),h(t)$, that is,

$$E\left[J_{t+1}(h(t+1))\right] = \sum_{i=1}^{k} a_{i,h(t)}(u(t))J_{t+1}(h(t+1)). \tag{7.46}$$

Then, the dynamic programming solution to the above optimization problem is converted into:

$$\begin{cases} J_N(h(N)) = \max_{u(N)\in\Delta_k} \lambda^N \Phi h(N)u(N), \\ J_t(h(t)) = \max_{u(t)\in\Delta_k} \left[\lambda^t \Phi h(t)u(t) + \sum_{i=1}^{k} a_{i,j(t)}(u(t))J_{t+1}(h(t+1))\right], \\ \qquad t = N-1, N-2, \cdots, 1, 0. \end{cases} \tag{7.47}$$

Divide Φ into H equal blocks as:

$$\Phi = [Blk_1(\Phi)\ Blk_2(\Phi)\cdots Blk_H(\Phi)]. \tag{7.48}$$

Then, the first equation of (7.47) is rewritten as

$$J_N = \lambda^N \begin{pmatrix} \max_{u(N)\in\Delta_k} Blk_1(\Phi)u(N) \\ \max_{u(N)\in\Delta_k} Blk_2(\Phi)u(N) \\ \vdots \\ \max_{u(N)\in\Delta_k} Blk_N(\Phi)u(N) \end{pmatrix}. \tag{7.49}$$

Because of $u(N) \in \Delta_k$, we get the optimal control $u^*(N)$ for $J_N(h(N)) \mid_{h(N)=\delta_H^i}$ is $u^*(N) = \delta_k^{j^*}$, where:

$$j^* = \arg\max_j Col_j(Blk_i(\Phi)). \tag{7.50}$$

And, it is easy to rewrite the second equation of (7.47) as

$$J_t = \begin{pmatrix} \max_{u(t)\in\Delta_k} \left\{\lambda^t Blk_1(\Phi)u(t) + J_{t+1}^T Col_1(L_u u(t))\right\} \\ \max_{u(t)\in\Delta_k} \left\{\lambda^t Blk_2(\Phi)u(t) + J_{t+1}^T Col_2(L_u u(t))\right\} \\ \vdots \\ \max_{u(t)\in\Delta_k} \left\{\lambda^t Blk_H(\Phi)u(t) + J_{t+1}^T Col_H(L_u u(t))\right\} \end{pmatrix}. \tag{7.51}$$

Define $B_i = L_u W_{[H,k]} \delta_H^i$, it is easy to know that $J_{t+1}^T B_i u(t) = J_{t+1}^T Col_i(L_u u(t))$ holds. Thus, we rewrite (7.51) as:

$$J_t = \begin{pmatrix} \max_{u(t)\in\Delta_k} \left\{\lambda^t Blk_1(\Phi) + J_{t+1}^T B_1\right\} u(t) \\ \max_{u(t)\in\Delta_k} \left\{\lambda^t Blk_2(\Phi) + J_{t+1}^T B_2\right\} u(t) \\ \vdots \\ \max_{u(t)\in\Delta_k} \left\{\lambda^t Blk_H(\Phi) + J_{t+1}^T B_H\right\} u(t) \end{pmatrix}. \tag{7.52}$$

Since $u(t) \in \Delta_k$, setting:

$$\xi_i(t) = \lambda^t Blk_i(\Phi) + J_{t+1}^T B_i \in \mathbb{R}^k, \tag{7.53}$$

then, the optimal control u_i^* for $J_t(h(t)) \mid_{h(t)=\delta_H^i}$ is $u_i^*(t) = \delta_k^{j^*}$, where:

$$j^* = \arg\max_j \xi_j^i(t). \tag{7.54}$$

(7.54) shows that the optimal solution can be obtained very easily. (7.49)–(7.54) provide a complete solution to maximize the finite horizon performance criterion (7.42). The following two lemmas help us to maximize the infinite horizon performance criterion (7.41).

Lemma 7.3: see [13]

Define $d := \min_{h\in\Delta_H}\min_{u_i\neq u_j\in\Delta_k}|p(u_i,h)-p(u_j,h)|$ and $M := \max_{u\in\Delta_k,h\in\Delta_H}|p(u,h)| < \infty$. Assume $d > 0$ and $\{u_0^{\ell-1}(0), u_0^{\ell-1}(1), \cdots, u_0^{\ell-1}(\ell-1)\}$ is the optimal control sequence to maximize $J_0^{\ell-1} = J(h(0))$ with the initial state $h(0)$, which is defined as follow:

$$J_s^{s+\ell-1} = J(h(s)) := E\left[\sum_{t=s}^{s+\ell-1} \lambda^t p(u(t),h(t))\right]. \tag{7.55}$$

Then, the optimal control sequence $u^*(0), u^*(1), \cdots$ obtained by receding horizon control is exactly the optimal control for the infinite horizon case, provided that the filter length ℓ satisfies:

$$\ell > \log_\lambda \frac{(1-\lambda)d}{2M}.$$

In addition, one has $u^*(t) = u_t^{t+\ell-1}(t)$. ∎

Lemma 7.4: see [13]

For the infinite horizon game, the optimal control can be expressed as:

$$u^*(t) = \mathscr{K} h(t), \tag{7.56}$$

where $\mathscr{K}$ is a logical matrix. ∎

The logical feedback matrix $\mathscr{K}$ can be calculated from the method of solving ℓ-horizon optimization problem proposed above as the filter length ℓ satisfies $\ell > \log_\lambda \frac{(1-\lambda)d}{2M}$. Because the feedback rule is time-invariant, we use receding-horizon based optimization to find $u^*(0)$ with respect to all $x(0)$ provides all optimal controls $u^*(t)$. In the following, we provide an illustrative example.

Example 7.6: Strategy optimization of a NEG with random entrance

The minor players have time horizon 2 and the maximum number of minor players attending the game is 1. The number of new minor players entering the game at each instant of time is either 0 or 1. Thus, the entrance dynamics is described by a

Markov chain with state space: $(0,1)$, $(1,0)$, and $(1,1)$. For the states of the Markov chain, we shall use the enumeration $1,2,3$ respectively. The entrance dynamics is described by the following transition matrix:

$$\Pi = \begin{pmatrix} 0 & 1 & 0 \\ 0.6 & 0 & 0.5 \\ 0.4 & 0 & 0.5 \end{pmatrix}. \tag{7.57}$$

Consider a NEG with the following basic:

- Three network topological structures, denote by $(N_z, \mathscr{E}_z)$, where $N_1 = \{1\}$, $N_2 = \{2\}$, $N_3 = \{3\}$, $N_4 = \{1,2\}$, $N_5 = \{2,3\}$, $N_6 = \{1,3\}$, $\mathscr{E}_z = \{(0,i) \mid i \in N_z\}$, and $z \in \mathscr{M} = \{1,2,\cdots,6\}$.
- The FNG's payoff bi-matrix shown in Table 7.2;
- The evolutionary rule is **the myopic best response adjustment rule**;

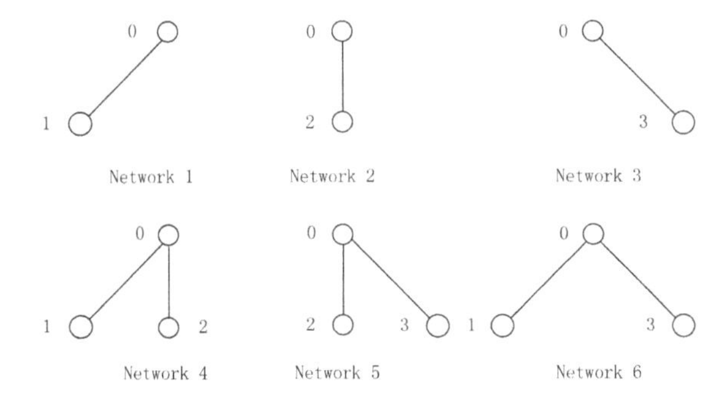

Figure 7.4 The network of Example 7.5

Table 7.2: Payoff bi-matrix of the FNG of Example 7.5

Player 1 \ Player 2	**M**	**F**
M	(2, 2)	(1, 0)
F	(0, 1)	(3, 3)

We, first convert the dynamics of the NEG into an algebraic form. Denote $M \sim \delta_2^1$, $F \sim \delta_2^2$, $(0,1) \sim \delta_3^1$, $(1,0) \sim \delta_3^2$, $(1,1) \sim \delta_3^3$, network $j \sim \delta_6^j$, $j = 1,2,\cdots,6$, $N^1 = \{1\}$, $N^2 = \{2\}$, and $N^3 = \{3\}$. Using the vector form of logical variables, it is easy to know from the network structure that the payoff function of each player has the form of $p_{0,z}(x_0(t),x(t)) = M_{0,z}x(t)x_0(t)$, $p_{i,z}(x_i(t),x_0(t)) = M_{i,z}x_0(t)x_i(t)$, $x(t) = (x_1(t),x_2(t),x_3(t))$, $x_i(t) \in \Delta_2$, and $i \in N_z$, and $z \in \mathscr{M}$.

With (7.12), one obtains

$$\begin{aligned} M_{0,1} &= [2\,0\,2\,0\,2\,0\,2\,0\,1\,3\,1\,3\,1\,3\,1\,3], & M_{0,2} &= [2\,0\,2\,0\,1\,3\,1\,3\,2\,0\,2\,0\,1\,3\,1\,3], \\ M_{0,3} &= [2\,0\,1\,3\,2\,0\,1\,3\,2\,0\,1\,3\,2\,0\,1\,3], & M_{0,4} &= [4\,0\,4\,0\,3\,3\,3\,3\,3\,3\,3\,3\,2\,6\,2\,6], \\ M_{0,5} &= [4\,0\,3\,3\,3\,3\,2\,6\,4\,0\,3\,3\,3\,3\,2\,6], & M_{0,6} &= [4\,0\,3\,3\,4\,0\,3\,3\,3\,3\,2\,6\,3\,3\,2\,6], \end{aligned}$$
$$M_{i,z} = [2\,0\,1\,3], i \in N_z, z \in \mathscr{M}. \tag{7.58}$$

Then, we have:

$$\begin{aligned} L_{i,z} &= \delta_2[1\ 1\ 1\ 1\ 1\ 1\ 1\ 1\ 2\ 2\ 2\ 2\ 2\ 2\ 2\ 2], \\ L_{j,z} &= \delta_2[1\ 1\ 1\ 1\ 1\ 1\ 1\ 1\ 1\ 1\ 1\ 1\ 1\ 1\ 1\ 1]. \end{aligned} \tag{7.59}$$

where $i \in N_z$, $j \notin N_z$, and $z \in \mathscr{M}$.

In order to reduce the calculation, by enumeration, we have:

$$z(t) = \delta_6[3\ 1\ 6\ 1\ 2\ 4\ 2\ 3\ 5]d(t)y(t), \tag{7.60}$$

and

$$h(t+1) = L_u u(t)h(t), \tag{7.61}$$

where $h(t) = d(t)y(t)x_1(t)x_2(t)x_3(t)$ and

$$L_u = \begin{pmatrix} 0 & 0 & \cdots & 1 & 1 & \cdots & 0 & 0 \\ \vdots & \vdots & \cdots & \vdots & \vdots & \cdots & 0.5 & 0.5 \\ 0 & 0 & \cdots & 0 & 0 & \cdots & \vdots & \vdots \\ \vdots & \vdots & \cdots & \vdots & \vdots & \cdots & 0.5 & 0.5 \\ 0 & 0 & \cdots & 0 & 0 & \cdots & \vdots & \vdots \\ \vdots & \vdots & \cdots & \vdots & \vdots & \cdots & 0 & 0 \\ 0.6 & 0.6 & \cdots & 0 & 0 & \cdots & \vdots & \vdots \\ \vdots & \vdots & \cdots & \vdots & \vdots & \cdots & 0 & 0 \\ 0.4 & 0.4 & \cdots & 0 & 0 & \cdots & \vdots & \vdots \\ \vdots & \vdots & \cdots & \vdots & \vdots & \cdots & 0 & 0 \\ 0 & 0 & \cdots & 0 & 0 & \cdots & 0 & 0 \end{pmatrix}.$$

By (7.40) and (7.58), we have:

$$\begin{aligned} \Phi &= [M_{0,1}\ M_{0,2}\ M_{0,3}\ M_{0,4}\ M_{0,5}\ M_{0,6}]\tilde{L}_z \\ &= [2\ 0\ 1\ 3\ 2\ 0\ 1\ 3\ 2\ 0\ 1\ 3\ 2\ 0\ 1\ 3\ 2\ 0\ 2\ 0\ 2\ 0\ 2\ 0\ 1\ 3\ 1\ 3\ 1\ 3\ 1\ 3\ 4\ 0\ 3\ 3 \\ &\quad\ 4\ 0\ 3\ 3\ 3\ 3\ 2\ 6\ 3\ 3\ 2\ 6\ 2\ 0\ 2\ 0\ 2\ 0\ 2\ 0\ 1\ 3\ 1\ 3\ 1\ 3\ 1\ 3\ 2\ 0\ 2\ 0\ 1\ 3\ 1\ 3 \\ &\quad\ 2\ 0\ 2\ 0\ 1\ 3\ 1\ 3\ 4\ 0\ 4\ 0\ 3\ 3\ 3\ 3\ 3\ 3\ 3\ 3\ 2\ 6\ 2\ 6\ 2\ 0\ 2\ 0\ 1\ 3\ 1\ 3\ 2\ 0\ 2\ 0 \\ &\quad\ 1\ 3\ 1\ 3\ 2\ 0\ 1\ 3\ 2\ 0\ 1\ 3\ 2\ 0\ 1\ 3\ 2\ 0\ 1\ 3\ 4\ 0\ 3\ 3\ 3\ 3\ 2\ 6\ 4\ 0\ 3\ 3\ 3\ 3\ 2\ 6]. \end{aligned} \tag{7.62}$$

It is obvious that $d = 1$ and $M = 6$. Letting $\lambda = 0.9$, we have the filter length $\ell = 45.4391$. Then, let $N = 46$, there is the structural matrix updating law for the major player as:

$$\begin{aligned} \mathscr{K} &= \delta_2[1\ 2\ 1\ 2\ 1\ 2\ 1\ 2\ 1\ 1\ 1\ 1\ 2\ 2\ 2\ 2\ 1\ 1 \\ &\quad\ 1\ 1\ 1\ 2\ 1\ 2\ 1\ 1\ 1\ 1\ 2\ 2\ 2\ 2\ 1\ 1\ 2\ 2 \\ &\quad\ 1\ 1\ 2\ 2\ 1\ 1\ 1\ 1\ 1\ 1\ 2\ 2\ 1\ 1\ 2\ 2\ 1\ 1 \\ &\quad\ 2\ 2\ 1\ 2\ 1\ 2\ 1\ 2\ 1\ 2\ 1\ 1\ 1\ 2\ 1\ 1\ 1\ 2], \end{aligned} \tag{7.63}$$

and the maximum expected payoff for the major player is:

$$
\begin{aligned}
J^* \;=\; & [25.08\ 26.08\ 25.08\ 26.08\ 25.08\ 26.08\ 25.08\ 26.08\ 24.55\ 24.55\ 24.55\ 24.55\\
& 26.45\ 26.45\ 26.45\ 26.45\ 27.33\ 26.33\ 27.33\ 26.33\ 26.33\ 28.88\ 26.33\ 28.88\\
& 25.08\ 25.08\ 25.08\ 25.08\ 26.08\ 26.08\ 26.08\ 26.08\ 24.55\ 24.55\ 26.45\ 26.45\\
& 24.55\ 24.55\ 26.45\ 26.45\ 27.33\ 27.33\ 26.33\ 26.33\ 26.33\ 26.33\ 28.88\ 28.88\\
& 25.08\ 25.08\ 26.08\ 26.08\ 25.08\ 25.08\ 26.08\ 26.08\ 24.55\ 26.45\ 24.55\ 26.45\\
& 24.55\ 26.45\ 24.55\ 26.45\ 27.33\ 26.33\ 26.33\ 28.88\ 27.33\ 26.33\ 26.33\ 28.88],
\end{aligned}
$$

where the l-th element in J^* is the maximum expected payoff for the major player under the updating law $u(t) = \mathscr{K} h(t)$ with the initial state $h(0) = \delta_h^l$, $l = 1, 2, \cdots, h$.

REFERENCES

1. Nowak, M. A. and May, R. M. (1992). Evolutionary games and spatial chaos. Nature, 359(6398): 826–829.
2. Hauert, C. and Doebeli, M. (2004). Spatial structure often inhibits the evolution of cooperation in the snowdrift game. Nature, 428(6983): 643.
3. Szabó, G. and Fath, G. (2007). Evolutionary games on graphs. Physics Reports, 446(4): 97–216.
4. Cheng, D., He, F., Qi, H. and Xu, T. (2015). Modeling, analysis and control of networked evolutionary games. IEEE Trans. Aut. Contr., 60(9): 2402–2415.
5. Gibbons, R. (1992). A primer in game theory. Harvester Wheatsheaf.
6. Cheng, D., Qi, H. and Li, Z. (2011). Analysis and Control of Boolean Networks: A Semi-Tensor Product Approach. London, Springer.
7. Bertsekas, D. P. (1995). Dynamic programming and optimal control (Vol. 1, No. 2). Belmont, MA: Athena Scientific.
8. Cheng, D., Qi, H., He, F., Xu, T. and Dong, H. (2014). Semi-tensor product approach to networked evolutionary games. Control Theory and Technology, 12(2): 198–214.
9. Guo, P., Wang, Y. and Li, H. (2013). Algebraic formulation and strategy optimization for a class of evolutionary networked games via semi-tensor product method. Automatica, 49(11): 3384–3389.
10. Guo, P., Wang, Y. and Li, H. (2016). Stable degree analysis for strategy profiles of evolutionary networked games. Science China Information Sciences, 59(5): 052204.
11. Zhao, G., Wang, Y. and Li, H. (2016). A matrix approach to modeling and optimization for dynamic games with random entrance. Applied Mathematics and Computation, 290: 9–20.
12. Zhao, G. and Wang, Y. (2016). Formulation and optimization control of a class of networked evolutionary games with switched topologies. Nonlinear Analysis: Hybrid Systems, 22: 98–107.
13. Cheng, D., Zhao, Y. and Xu, T. (2015). Receding horizon based feedback optimization for mix-valued logical networks. IEEE Trans. Aut. Contr., 60(12): 3362–3366.

8 Nonlinear Feedback Shift Registers

8.1 INTRODUCTION TO SHIFT REGISTERS

The feedback shift register (FSR) is used to create pseudo-random sequences in many fields, such as cell phones and digital cable, secure and privacy communications, error detecting and correcting codes [8], and cryptographic systems [1]. FSRs are divided into linear and nonlinear ones according to their feedback functions. Linear FSRs are widely studied, and up to now, many interesting results have been obtained [5, 9]. Since the linear FSR is relatively simple, the sequence generated by it can be easily forecasted and thus be cryptographically insecure. Therefore, the application of nonlinear feedback shift registers becomes more popular. The major research fields on nonlinear FSRs usually have the following three aspects [3]: (1) Analysis for a given nonlinear FSR, including the nonsingularity, the number of fixed points and cycles with different lengths of the sequence generated by nonlinear FSR, decomposition and synthesis of cycles, etc.; (2) Construction of the shortest nonlinear FSR which can generate a given sequence; (3) Construction of all full length nonlinear FSRs.

It is worth noting that the semi-tensor product of matrices has been successfully used in the study of FSRs [7, 10–12]. In [10], a generalization of the linear feedback shift register synthesis problem was presented for synthesizing minimum-length matrix feedback shift registers to generate prescribed matrix sequences and so a new complexity measure, that is, matrix complexity, was introduced. [12] studied the stability of nonlinear feedback shift registers (NFSRs) using a Boolean network approach. A Boolean network is an autonomous system that evolves as an automaton through Boolean functions. A NFSR can be viewed as a Boolean network. Based on its Boolean network representation, some sufficient and necessary conditions are provided for globally (locally) stable NFSRs.

This chapter introduces the main results of [7] and [11]. First, authors regarded NFSR as a special Boolean network in [11], and used semi-tensor product of matrices and matrix expression of logic to convert the dynamic equations of NFSR into an equivalent algebraic equation. Based on them, some novel and generalized techniques were proposed to study NFSR. Then, [7] investigated the multi-valued FSR and presented a new method to analyze its nonsingularity, number of cycles, and cycle decomposition and synthesis.

8.2 ANALYSIS OF FEEDBACK SHIFT REGISTERS VIA BOOLEAN NETWORKS

This section is devoted to studying synthesis and generation of NFSR for a given binary sequence. First, the definition of NFSR is introduced. Figure 8.1 shows a nNFSR which consists of n memory cells containing a 0 or 1. n is called the stage of the NFSR. The content of all the n cells is said to be a state of the nNFSR. f, called the feedback function, is a nonlinear polynomial function over $\mathbb{F}_2$, where $\mathbb{F}_2$ is a finite field with elements 0 and 1. The nNFSR works in the discrete time. The state of the nNFSR at a given moment $t+1(t>0)$ is determined by its state at the moment t and results from shifting the content of the rth cell to the $(r-1)$th cell $(2 \leq r \leq n)$, and putting the value $f(x_1, x_2, \cdots, x_n)$ of the function f for the state of this nNFSR at the moment t into the nth cell.

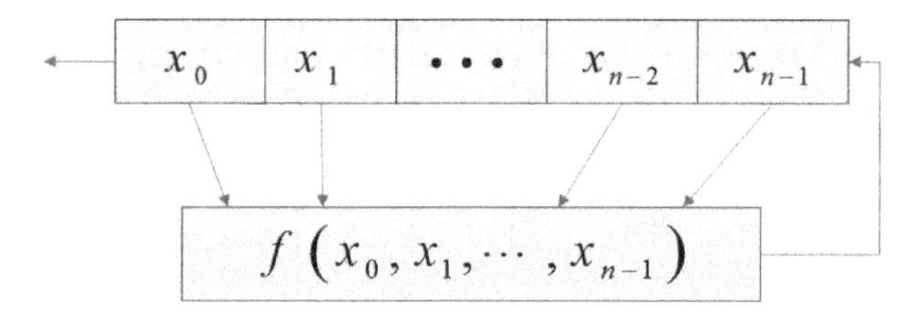

Figure 8.1 A n-stage NFSR with the feedback function f

Let $X_t = (x_1(t), x_2(t), \cdots, x_n(t))$ be the state of a nNFSR at moment t. The state X_0 from which the nNFSR starts its work is said to be the initial state. Therefore, the binary sequence generated by a nNFSR will be $x_1(0), x_1(1), x_1(2), \cdots$. Different initial states of nNFSR may lead to the production of different state sequences $\{X_i | i = 1, 2, \cdots\}$. Any state sequence produced by a nNFSR must be led into a cycle with period T, where T is the smallest positive integer such that for each integer $i(i = m, m+1, \cdots)$ the condition $X_{i+T} = X_i$ holds. Obviously, T satisfies $T \leq 2^n$. A nNFSR which can produce a cycle with period $T = 2^n$ is called full-length nNFSR, and the binary sequence generated by full-length nNFSR is called full-length sequence.

8.2.1 CONSTRUCTION OF THE SHORTEST NFSR FOR A GIVEN NONPERIODIC SEQUENCE

In order to construct the shortest nNFSR which can generate a given nonperiodic sequence, two steps are necessary. The first step is to determine the smallest stage n, and the second step is to calculate the feedback function. Let $S = s_1 s_2 \cdots s_l$ be a given binary nonperiodic sequence with length l. The concrete construction steps of the nNFSRs which can generate S are described below:

1) Use Blumer's algorithm to determine the shortest stage n of the FSR which can generate S.

2) Present the state transition process of the nFSR which can generate S, i.e.,

$$(s_1, s_2, \cdots, s_n) \to (s_2, s_3, \cdots, s_{n+1}) \to \cdots \to (s_{l-n+1}, s_{l-n+2}, \cdots, s_l).$$

Assume the algebraic form of this process is $\delta_{2^n}^{i_1} \to \delta_{2^n}^{i_2} \to \cdots \to \delta_{2^n}^{i_{l-n+1}}$.

3) For the dynamic system $x(t+1) = Lx(t)$ of the nFSR, we have:

$$Col_{i_1}(L) = \delta_{2^n}^{i_2}, Col_{i_2}(L) = \delta_{2^n}^{i_3}, \cdots, Col_{i_{l-n}}(L) = \delta_{2^n}^{i_{l-n+1}}.$$

Each state has two possible successors, which means that each column of L has two choices except the known $l-n$ columns. Hence, we can construct all $2^{2^n-(l-n)}$ Ls. Actually, for any L, we can calculate a linear or nonlinear feedback function; that is, one can construct all the $2^{2^n-(l-n)}$ linear or nonlinear nFSR which can generate S. However, in order to make sure the nFSR constructed is nonlinear, specify $Col_{2^n}(L) = \delta_{2^n}^{2^{n-1}}$ when $Col_{2^n}(L)$ is not yet determined even if $Col_{2^n}(L) = \delta_{2^n}^{2^n}$ does not mean that the nFSR is linear. Therefore, L has the following form,

$$L = \delta_{2^n}[* \cdots * i_2 * \cdots * i_3 * \cdots * i_{l-n+1} * \cdots * 2^{n-1}]$$

and the number of this type of L is $2^{2^n-(l-n)-1}$.

4) Calculate a nonlinear feedback function for each L. From the method mentioned above, we can easily construct all $2^{2^n-(l-n)}$ nFSRs and at least $2^{2^n-(l-n)-1}$ nNFSRs to output the given sequence S with length l.

Now we give an example to show how to construct shortest nNFSRs to generate a given sequence.

Example 8.1:

Assume $S = 1100010110100111$ is a sequence with length 16.

1) We know that a 4NFSR can generate S by using Blumer's algorithm.

2) The process of state transition is $(1,1,0,0) \to (1,0,0,0) \to (0,0,0,1) \to (0,0,1,0) \to (0,1,0,1) \to (1,0,1,1) \to (0,1,1,0) \to (1,1,0,1) \to (1,0,1,0) \to (0,1,0,0) \to (1,0,0,1) \to (0,0,1,1) \to (0,1,1,1)$. That is, $\delta_{16}^4 \to \delta_{16}^8 \to \delta_{16}^{15} \to \delta_{16}^{14} \to \delta_{16}^{11} \to \delta_{16}^5 \to \delta_{16}^{10} \to \delta_{16}^3 \to \delta_{16}^6 \to \delta_{16}^{12} \to \delta_{16}^7 \to \delta_{16}^{13} \to \delta_{16}^9$.

3) We have $L = \delta_{16}[*\ *\ 6\ 8\ 10\ 12\ 13\ 15\ *\ 3\ 5\ 7\ 9\ 11\ 14\ 15]$, and there are 3 columns undetermined. Thus, we can construct 2^3 different L as follows:

$$\begin{aligned}
L_1 &= \delta_{16}[1\ 3\ 6\ 8\ 10\ 12\ 13\ 15\ 1\ 3\ 5\ 7\ 9\ 11\ 14\ 15], \\
L_2 &= \delta_{16}[1\ 3\ 6\ 8\ 10\ 12\ 13\ 15\ 2\ 3\ 5\ 7\ 9\ 11\ 14\ 15], \\
L_3 &= \delta_{16}[1\ 4\ 6\ 8\ 10\ 12\ 13\ 15\ 1\ 3\ 5\ 7\ 9\ 11\ 14\ 15], \\
L_4 &= \delta_{16}[1\ 4\ 6\ 8\ 10\ 12\ 13\ 15\ 2\ 3\ 5\ 7\ 9\ 11\ 14\ 15], \\
L_5 &= \delta_{16}[2\ 3\ 6\ 8\ 10\ 12\ 13\ 15\ 1\ 3\ 5\ 7\ 9\ 11\ 14\ 15], \\
L_6 &= \delta_{16}[2\ 3\ 6\ 8\ 10\ 12\ 13\ 15\ 2\ 3\ 5\ 7\ 9\ 11\ 14\ 15], \\
L_7 &= \delta_{16}[2\ 4\ 6\ 8\ 10\ 12\ 13\ 15\ 1\ 3\ 5\ 7\ 9\ 11\ 14\ 15], \\
L_8 &= \delta_{16}[2\ 4\ 6\ 8\ 10\ 12\ 13\ 15\ 2\ 3\ 5\ 7\ 9\ 11\ 14\ 15].
\end{aligned}$$

4) For each $L_i, i = 1, 2, \cdots, 8$, we can calculate a nonlinear feedback function. Taking L_1 as an illustration, we construct a nonlinear feedback function which can generate the given S.

i) Let:

$$\begin{array}{l} S_1^4 = \delta_2[1\ 1\ 1\ 1\ 1\ 1\ 1\ 1\ 2\ 2\ 2\ 2\ 2\ 2\ 2\ 2], \\ S_2^4 = \delta_2[1\ 1\ 1\ 1\ 2\ 2\ 2\ 2\ 1\ 1\ 1\ 1\ 2\ 2\ 2\ 2], \\ S_3^4 = \delta_2[1\ 1\ 2\ 2\ 1\ 1\ 2\ 2\ 1\ 1\ 2\ 2\ 1\ 1\ 2\ 2], \\ S_4^4 = \delta_2[1\ 2\ 1\ 2\ 1\ 2\ 1\ 2\ 1\ 2\ 1\ 2\ 1\ 2\ 1\ 2]. \end{array}$$

ii) We have $M_i = S_i^4 L_1, i = 1, 2, 3, 4$, which are:

$$\begin{array}{l} M_1 = [1\ 1\ 1\ 1\ 2\ 2\ 2\ 2\ 1\ 1\ 1\ 1\ 2\ 2\ 2\ 2], \\ M_2 = [1\ 1\ 2\ 2\ 1\ 1\ 2\ 2\ 1\ 1\ 2\ 2\ 1\ 1\ 2\ 2], \\ M_3 = [1\ 2\ 1\ 2\ 1\ 2\ 1\ 2\ 1\ 2\ 1\ 2\ 1\ 2\ 1\ 2], \\ M_4 = [1\ 1\ 2\ 2\ 2\ 2\ 1\ 1\ 1\ 1\ 1\ 1\ 1\ 1\ 2\ 1]. \end{array}$$

iii) Considering M_1, it is easy to verify that:

$$\begin{cases} M_1(M_n - I_2) = 0, \\ M_1 W_{[2]}(M_n - I_2) \neq 0, \\ M_1 W_{[2,\ 4]}(M_n - I_2) = 0, \\ M_1 W_{[2,\ 8]}(M_n - I_2) = 0, \end{cases}$$

where $M_n := M_\neg = \delta_2[2,\ 1]$.

iv) Using the technique of semi-tensor product, the system can be obtained by the following equation:

$$\begin{cases} x_1(t+1) = & x_2(t), \\ x_2(t+1) = & x_3(t), \\ x_3(t+1) = & x_4(t), \\ x_4(t+1) = & [x_1(t) \wedge x_2(t) \wedge x_3(t)] \vee [x_1(t) \wedge (\neg x_2(t)) \wedge (\neg x_3(t))] \\ & \vee[(\neg x_1(t)) \wedge (\neg x_2(t)) \wedge ((x_3(t) \wedge x_4(t)) \bar{\vee} (\neg x_4(t)))] \\ & \vee[(\neg x_1(t)) \wedge x_2(t)]. \end{cases} \tag{8.1}$$

A 4NFSR with dynamic system (8.1) constructed according to L_1 can be used to generate the given sequence S when the initial state is $(1, 1, 0, 0)$. Similarly, the other 7 different 4NFSRs from L_i, $i = 2, 3, \cdots, 8$, can be obtained to generate S.

8.2.2 CONSTRUCTION OF THE SHORTEST NFSR FOR A GIVEN PERIODIC SEQUENCE

The method of the construction of the shortest NFSR for a given periodic sequence is similar to that for a nonperiodic sequence. There are only a few differences in the process of state transition. Assuming that $S = s_1 s_2 \cdots s_l$ is a given periodic sequence, the concrete construction steps of the nNFSRs which can generate S are as follows:

1) Use Blumer's algorithm to determine the shortest stage n of FSR which can generate S.

2) The state transition process is $(s_1,s_2,\cdots,s_n) \to (s_2,s_3,\cdots,s_{n+1}) \to \cdots \to (s_{l-n+1},s_{l-n+2},\cdots,s_l) \to (s_{l-n+2},s_{l-n+3},\cdots,s_l,s_1) \to \cdots \to (s_l,s_1,\cdots,s_{l-1}) \to (s_1,s_2,\cdots,s_n)$. The algebraic form of this process can be expressed by $\delta_{2^n}^{i_1} \to \delta_{2^n}^{i_2} \to \cdots \to \delta_{2^n}^{i_l} \to \delta_{2^n}^{i_1}$.

3) Construct L as follows:

$$L = \delta_{2^n}[*\cdots*\ i_2*\cdots*\ i_l*\cdots*\ i_1*\cdots*\ i_{2^n-1}].$$

The remaining steps are similar to those proposed in Subsection 8.2.1, so we omit them. In summary, we can construct all 2^{2^n-l} nFSRs and at least 2^{2^n-l-1} nNFSR to generate a given periodic sequence.

8.2.3 CONSTRUCTION OF FULL-LENGTH N-NFSR

Full-length nNFSR refers to the nNFSR that can generate the full-length binary sequence with period 2^n. Full-length nNFSR has major applications in a wide variety of systems such as cryptography and digital communication systems. Constructing the full-length nNFSR is a hard combination problem. In this subsection, simple cycles joining algorithm, called Positive Cycles Joining Algorithm, is proposed for the construction of full-length nNFSR.

Theorem 8.1

The state transition graph of a nNFSR contains cycles only if its transition matrix L is nonsingular.

Proof. If there is an appendage in the state graph, the appendage must be led into a cycle. This means $Col_i(L) = Col_{i+2^{n-1}}(L) = \delta_{2^n}^{2i}$. This conclusion contradicts the fact that L is nonsingular. ∎

Theorem 8.2

The successor state of $\delta_{2^n}^{i}$ is $\delta_{2^n}^{2i-1}$ which implies that the successor state of $\delta_{2^n}^{i+2^{n-1}}$ is $\delta_{2^n}^{2i}$ when L is nonsingular, and vice versa. In other words, $Col_i(L) = \delta_{2^n}^{2i-1}$ implies $Col_{i+2^{n-1}}(L) = \delta_{2^n}^{2i}$ when L is nonsingular, and vice versa.

Proof. If $\delta_{2^n}^{i}$ and $\delta_{2^n}^{i+2^{n-1}}$ have the same successor state, then we have $Col_i(L) = Col_{i+2^{n-1}}(L)$, i.e., L is singular. It is contradictory to the existing facts. ∎

When $\delta_{2^n}^{i} \to \delta_{2^n}^{2i-1}$ and $\delta_{2^n}^{i+2^{n-1}} \to \delta_{2^n}^{2i}$ belong to two different cycles, these two cycles can be joined together to one complete cycle by changing the successors. The algorithm presented below is based on the principle of cycles joining. Through cycles joining, all the cycles in the state transition graph of an nNFSR can be joined together to one complete cycle which is the state sequence with period 2^n. The nNFSR which can produce this complete cycle will be the full-length nNFSR.

ALGORITHM 8.1

1) Construct a nonsingular transition matrix $L \in \mathscr{L}_{2^n \times 2^n}$ randomly. According to Theorem 8.1, this ensures that the state transition graph that correspond to L contains cycles only.
2) List all the cycles of the state transition graph of L by a sorting rule, such that the superscript i of the first state $\delta_{2^n}^i$ of the tth cycle must be the minimum value of superscript which has not emerged in the first $t-1$ cycles.
3) From the second cycle on the list, update the predecessor of the first state and the successor of the last state of each cycle. According to the sorting rule in 2), the new predecessor of the first state and the successor of the last state of the tth cycle can be found in the first $t-1$ cycles. In this way, all the cycles can be joined together to one maximal cycle.
4) Construct a new L' according to the maximal cycle and calculate an unique feedback function $f(x)$ based on L'.
Finally, a nNFSR with feedback function $f(x)$ can output a full-length sequence whatever the initial state is.

Algorithm 8.1 is named by Positive Cycles Joining Algorithm.

8.2.4 CONSTRUCTION OF MORE FULL-LENGTH *N*NFSRS

It is well known that the number of full-length nNFSRs are $2^{2^{n-1}-n}$. Thus, it is natural to consider the problem of developing an algorithm for the construction of all the full-length nNFSRs. It is not only interesting from the mathematical viewpoint but also important in a practical sense. However, it is quite a difficult and challenging problem. In this subsection, two algorithms are presented for the construction of more full-length nNFSRs based on the Positive Cycles Joining Algorithm and its variation. Those two algorithms can construct $2^{2^{n-2}-1}$ full-length nNFSRs, respectively.

Theorem 8.3

For any $i \in \{2^{n-1}+1,\cdots,2^n-1\}$, $\delta_{2^n}^i$ must not be the first state of any cycles listed in Positive Cycles Joining Algorithm.

Theorem 8.4

Assume $L_1, L_2 \in \mathscr{L}_{2^n \times 2^n}$ are two nonsingular transition matrices which are randomly constructed in the first step of Positive Cycles Joining Algorithm, if there exists $j \in \{2^{n-2}+1, \cdots, 2^{n-1}-1\}$ such that $Col_j(L_1) \neq Col_j(L_2)$, then the two full-length nNFSRs obtained from Positive Cycles Joining Algorithm are different.

Proof of Theorems 8.3 and 8.4 can be seen in [11]. Next, we show how to construct $2^{2^{n-2}-1}$ different full-length nNFSRs.

ALGORITHM 8.2

1) Construct nonsingular transition matrices which satisfy two properties:

(1) $Col_{2^{n-1}}(L) = \delta_{2^n}^{2^n}$, $Col_{2^n}(L) = \delta_{2^n}^{2^n-1}$;

(2) For any L_1, L_2, there are $Col_j(L_1) = Col_j(L_2)$ for any $j \in \{1, 2, \cdots, 2^{n-2}\}$ and there exists $j \in \{2^{n-2}+1, \cdots, 2^{n-1}-1\}$ such that $Col_j(L_1) \neq Col_j(L_2)$. Obviously, the number of such nonsingular transition matrices are $2^{2^{n-2}-1}$.

2) Let each transition matrix obtained in 1) be the initial transition matrix of Positive Cycles Joining Algorithm and construct the corresponding full-length n-NFSRs. According to Theorems 8.3 and 8.4, the resulting $2^{2^{n-2}-1}$ full-length nNFSRs are different.

In order to construct more full-length nNFSRs, another algorithm, called Negative Cycles Joining Algorithm is proposed, which is similar with the Positive Cycles Joining Algorithm.

ALGORITHM 8.3

The only difference between Negative Cycles Joining Algorithm and Positive Cycles Joining Algorithm appears in step 2). We list step 2) below and omit the same parts.

2) List all the cycles of the state transition graph of L by a sorting rule such that the superscript i of the first state $\delta_{2^n}^i$ of the tth cycle must be the maximum value of superscript which has not emerged in the first $t-1$ cycles.

The following results are straightforward to verify that they are similar with Theorems 8.3 and 8.4.

Theorem 8.5

For any $i \in \{2, \cdots, 2^{n-1}\}$, $\delta_{2^n}^i$ must not be the first state of any cycles in Negative Cycles Joining Algorithm.

Theorem 8.6

Assume $L_1, L_2 \in \mathscr{L}_{2^n \times 2^n}$ are two nonsingular transition matrices which are randomly constructed in the first step of Negative Cycles Joining Algorithm. If there exists $j \in \{2, \cdots, 2^{n-2}\}$ such that $Col_j(L_1) \neq Col_j(L_2)$, then the two full-length nNFSRs obtained from Negative Cycles Joining Algorithm are different.

Algorithm 8.4 shows how to construct $2^{2^{n-2}-1}$ different full-length nNFSRs based on Theorems 8.5 and 8.6.

ALGORITHM 8.4

1) Construct nonsingular transition matrices which satisfy two properties:

(1) $Col_1(L) = \delta_{2^n}^2, Col_{2^{n-1}}(L) = \delta_{2^n}^{2^n}, Col_{2^{n-1}+1}(L) = \delta_{2^n}^1, Col_{2^n}(L) = \delta_{2^n}^{2^n-1}$;

(2) For any L_1, L_2, there are $Col_j(L_1) = Col_j(L_2)$ for any $j \in \{2^{n-2}+1, \cdots, 2^{n-1}\}$ and there exists $j \in \{2, \cdots, 2^{n-2}\}$ such that $Col_j(L_1) \neq Col_j(L_2)$. Obviously, the number of such nonsingular transition matrices are $2^{2^{n-2}-1}$.

2) Let each transition matrix obtained in 1) be the initial transition matrix of Negative Cycles Joining Algorithm and construct the corresponding full-length nNFSRs.

8.3 ANALYSIS OF FEEDBACK SHIFT REGISTERS VIA MULTI-VALUED LOGICAL NETWORKS

This section [7] investigates the multi-valued FSR and presents a new method to analyze its nonsingularity, number of cycles, and cycle decomposition and synthesis. First, the FSR is expressed as an algebraical form by using the semi-tensor product of matrices, based on which several necessary and sufficient conditions are given for the nonsingularity. Second, the structure of FSRs is defined, and a new method to determine the number of cycles with different lengths for arbitrary given FSR is introduced. Third, the problem on cycle decomposition and synthesis is considered, and some new results are obtained. Finally, an illustrative example is given to support the new results.

8.3.1 NONSINGULARITY OF FSRS

An FSR is called a nonsingular FSR if it only has cycles without branches in its transition diagram. For the nonsingularity of FSRs, there exist several results. It is well known that a given n-stage FSR is completely decided by a feedback function $f(x_1, x_2, \cdots, x_n)$. In [2], a necessary and sufficient condition was given for nonsingularity of binary FSRs (i.e., $x_i \in \{0,1\}, i = 1, 2, \cdots, n$). That is, a binary FSR is nonsingular iff its feedback function $f(x_1, x_2, \cdots, x_n)$ can be expressed as $x_1 + f_0(x_2, \cdots, x_n) \pmod 2$, where f_0 is independent of the variable x_1. But it is not generalized to FSRs over the general filed. Hence, it is necessary to establish some new methods to judge the nonsingularity of multi-valued FSRs. In [4], Lai gave a necessary and sufficient condition for nonsingularity of nonlinear FSRs over the q-valued filed. Since there are $q-1$ n-ary functions to be constructed, it is difficult to use this method in practice. Thus, nonsingularity of FSRs, is still an interesting and challenging topic to study.

Graph 8.1 denotes a n-stage k-valued FSR with a feedback logical function $f(x_0,x_1,\cdots,x_{n-1})$, where $x_i \in \mathscr{D}_k$, $i=0,1,\cdots,n-1$. Now, starting from the state $\alpha^0=(x_0,x_1,\cdots,x_{n-1})$, a state sequence $\alpha^0,\alpha^1,\alpha^2,\cdots$ is generated by the FSR, where $\alpha^1=(x_1,x_2,\cdots,x_n)$ is decided by the state α^0, $x_n=f(x_0,x_1,\cdots,x_{n-1})$, α^2 is decided by the state α^1, and so on. Thus, the algebraic expression of the FSR can be:

$$\alpha^{t+1}=(x_{t+1},x_{t+2},\cdots,x_{t+n})=T_f(\alpha^t),\ t=0,1,\cdots, \tag{8.2}$$

where $x_{t+n}=f(x_t,x_{t+1},\cdots,x_{t+n-1})$, and $T_f:\mathscr{D}_k^n \to \mathscr{D}_k^n$ is called the transition function of FSR.

DEFINITION 8.1

A feedback shift register is said to be nonsingular if its state transition diagram contains only cycles.

About the transition diagram of FSR, an example is provided below.

Let $f(x_0,x_1)=x_0\left(1+x_1^2\right) \pmod 3$, where $x_i \in \mathscr{D}_3=\{0,1,2\}$. Figure 8.2 depicts the state transition diagram of 3-valued FSR with feedback function f, denoted by G_f.

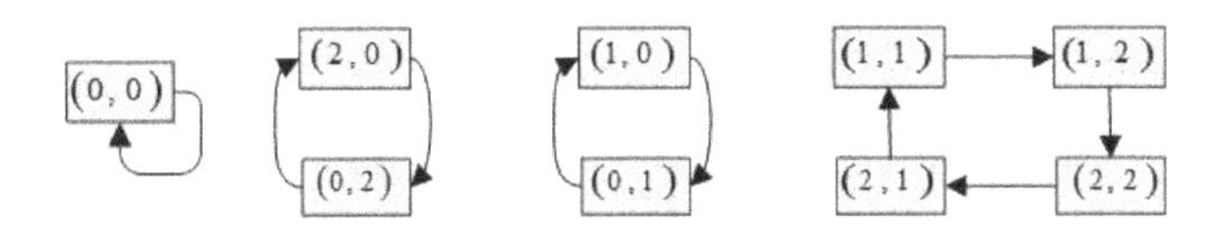

Figure 8.2 State transition diagram of the FSR with feedback function f

For the nonsingularity of a binary FSR, the result is as follows.

Lemma 8.1

A binary nonlinear FSR is nonsingular if and only if its feedback function $f(x_1,x_2,\cdots,x_n)$ can be represented as:

$$f(x_1,x_2,\cdots,x_n)=x_1+f_0(x_2,\cdots,x_n) \pmod 2. \tag{8.3}$$

where $x_i \in \mathscr{D}=\{0,1\}$.

Using the semi-tensor product, a new result is given to judge the nonsingularity of a binary FSR.

Theorem 8.7

A binary FSR is nonsingular if and only if:

$$M_1 = \delta_2[2\ 1]M_2,$$

where M_1 and $M_2 \in \mathscr{L}_{2\times 2^{n-1}}$ are two blocks of matrix $M_f = [M_1\ M_2]$, which is the structural matrix of feedback function $f(x_1,x_2,\cdots,x_n)$, and $x_i \in \mathscr{D}$, $i = 1,2,\cdots,n$.
Proof. (Necessity) Using the vector form of states, i.e., $1 \sim \delta_2^1$ and $0 \sim \delta_2^2$, we have by Lemma 8.1:

$$f(x_1,x_2,\cdots,x_n) = M_f x = M_f x_1 x_2 \cdots x_n,$$

where $x_i \in \Delta = \{\delta_2^1, \delta_2^2\}$, $M_f \in \mathscr{L}_{2\times 2^n}$ can be expressed as:

$$\begin{aligned} M_f &= M_+(I_2 \otimes M_{f_0}) \\ &= \delta_2[2\ 1\ 1\ 2]\begin{bmatrix} M_{f_0} & 0 \\ 0 & M_{f_0} \end{bmatrix} \\ &= \left[\delta_2[2\ 1]M_{f_0} \quad \delta_2[1\ 2]M_{f_0}\right], \end{aligned}$$

$M_+ = \delta_2[2\ 1\ 1\ 2]$ and M_{f_0} are structural matrices of mod 2 addition and function $f_0(x_2,\cdots,x_n)$, respectively. Therefore, $M_1 = \delta_2[2\ 1]M_{f_0}$ and $M_2 = \delta_2[1\ 2]M_{f_0} = M_{f_0}$.
(Sufficiency) From the proof of necessity, set $M_{f_0} = M_2$ and:

$$f_0(x_2,\cdots,x_n) = M_{f_0} x_2 \cdots x_n.$$

Then, $f(x_1,x_2,\cdots,x_n) = x_1 + f_0(x_2,\cdots,x_n)$ (mod 2) and the FSR is nonsingular by Lemma 8.1. The proof is completed. ∎

COROLLARY 8.1

If a binary FSR with feedback function f is nonsingular, then the number of column δ_2^1 is equal to that of column δ_2^2 in matrix M_f.
Proof. According to Theorem 8.7, $M_2 = \delta_2[2\ 1]M_1$. Then, $Col_i(M_1)$ and $Col_i(M_2)$ are different from each other, $i = 1,2,\cdots,2^{n-1}$. Moreover, each column only takes values from δ_2^1 and δ_2^2. Thus, the proof is completed. ∎

From Corollary 8.1, it is easy to see that a nonsingular binary FSR keeps the balance (the matrix M_f has the same number of δ_2^1 and δ_2^2), which is an important index of sequence cryptography.

It is noted that Theorem 8.1 cannot be generalized to the k-valued $(k > 2)$ FSR. Thus, we need to develop a new approach to judge the nonsingularity of k-valued FSRs.

DEFINITION 8.2

Consider FSR (8.2). Assume $\alpha, \beta \in \mathscr{D}_{k^n}$. A state α of FSR is called the successor of state β and β is called the predecessor of α, if $\alpha = T_f(\beta)$.

Lemma 8.2

An FSR is nonsingular if and only if each state has only one successor and one predecessor.

Then, we have the following results.

Lemma 8.3

Each state of an FSR has only one successor and one predecessor if and only if $T_f(\alpha) \neq T_f(\beta)$ for all α and β $(\alpha \neq \beta)$.

Proof. The proof is obvious. Thus, we omit it. ∎

By the semi-tensor product, there exists a structural matrix for the transition function T_f. We still denote it by T_f. That is,

$$x_{t+1} \cdots x_{n+t} = T_f x_t \cdots x_{n+t-1}.$$

In order to prove this lemma, we set $\alpha^t = x_{n+t-1} \cdots x_{t+1} x_t$. Then, using the semi-tensor product and the swap matrix, we can obtain another logical matrix $T_f' \in \mathscr{L}_{k^n \times k^n}$, such that:

$$\alpha^{t+1} = T_f' \alpha^t, \tag{8.4}$$

where $T_f' = W_{[k^{n-1}, k]} \cdots W_{[k,k]} T_f W_{[k,k]} \cdots W_{[k, k^{n-1}]}$.

COROLLARY 8.2

Each state of an FSR has only one successor and one predecessor if and only if T_f' is nonsingular, that is, $|T_f'| \neq 0$.

Proof. (Necessity) Assume that $T_f(\alpha) = T'_f\alpha$ and each state has only one predecessor. Then, for all $i \neq j$, $T'_f\delta^i_{k^n} \neq T'_f\delta^j_{k^n}$, i.e., $Col_i(T'_f) \neq Col_j(T'_f)$. On the other hand, T'_f is a logical matrix, that is to say, each column of T'_f has only one non-zero element. Thus, $|T'_f| \neq 0$.

(Sufficiency) For all $\alpha = \delta^i_{k^n}$ and $\beta = \delta^j_{k^n}$ $(i \neq j)$, if $|T_f| \neq 0$, then, $T'_f\alpha \neq T'_f\beta$. That is, $T'_f(\alpha) \neq T'_f(\beta)$. Thus, the proof is completed by Lemma 8.3. ∎

By the semi-tensor product, there exists a structural matrix $M_f \in \mathcal{L}_{k\times k^n}$ such that $f(x_0,x_1,\cdots,x_{n-1}) = M_f x_0 x_1 \cdots x_{n-1}$. Then, we obtain:

$$x_n = f(x_0,x_1,\cdots,x_{n-1}) = M'_f x_{n-1}\cdots x_1 x_0,$$

where $M'_f = M_f W_{[k,k]}\cdots W_{[k,k^{n-1}]}$. Split M'_f into k^{n-1} equal blocks as $M'_f = [M_1\ M_2\cdots M_{k^{n-1}}]$, where $M_i \in \mathcal{L}_{k\times k}$, $i = 1,2\cdots,k^{n-1}$.

Theorem 8.8

$|T'_f| \neq 0$ if and only if any two columns of M_i are different from each other, i.e.,

$$|M_i| \neq 0,\ i = 1,2\cdots,k^{n-1}.$$

Proof. Let $x_{n-1}\cdots x_1 = \delta^i_{k^{n-1}}$. Thus, $x_n = M'_f x_{n-1}\cdots x_1 x_0 = M_i x_0$. Then, the FSR has only one successor and one predecessor if and only if x_n and x_0 are one to one relationship, i.e., $|M_i| \neq 0$. Therefore, by Corollary 8.2, the proof is completed. ∎

Summarizing the above results, we have the following result.

Theorem 8.9

For an n-stage k-valued FSR with a feedback function $f(x_0,x_1,\cdots,x_{n-1})$, the following conditions are equivalent:

1) The FSR is nonsingular;
2) Its transition diagram has only cycles;
3) Each state has only one successor and one predecessor;
4) $|T'_f| \neq 0$, where $\alpha^{t+1} = T'_f\alpha^t$, $\alpha^t = x_{n+t-1}\cdots x_t$, $x_i \in \Delta_k$;
5) $|M_i| \neq 0$, $i = 1,2,\cdots,k^{n-1}$, which are blocks of the matrix $M'_f = [M_1\ M_2\ \cdots\ M_{k^{n-1}}]$, where $M_i \in \mathcal{L}_{k\times k}$, $i = 1,2,\cdots,k^{n-1}$.

Theorem 8.8 can be applied to all FSRs over the filed $GF(q)$ for arbitrary positive integer q, including linear and nonlinear ones. Moreover, it can solve the FSR of degree more than three, which could not be solved in [6].

8.3.2 NUMBER OF CYCLES FOR A FEEDBACK SHIFT REGISTER

This subsection first gives a definition of the structural matrix of FSR and the method how to compute the structural matrix. Then, we introduce a new method to determine the number of cycles with different lengths.

Consider an FSR with a feedback function $f(x_0, x_1, \cdots, x_{n-1})$, where $x_i \in \mathscr{D}_k$, $i = 0, 1, 2, \cdots, n-1$. Let $\alpha^t = x_{n+t-1} \cdots x_t$ and $\alpha^{t+1} = T'_f \alpha^t$, where $x_i \in \Delta_k$, $t = 0, 1, 2, \cdots$. In the following section, we call $T'_f \in \mathscr{L}_{k^n \times k^n}$ the structural matrix of FSR. Then, we have the following result.

Theorem 8.10

The structural matrix T'_f of FSR can be given as:

$$T'_f = [T_f^1 \; T_f^2 \; \cdots \; T_f^{k^{n-1}}], \tag{8.5}$$

where T_f^i is defined as:

$$Col_j(T_f^i) = Col_j(M_i)\delta_{k^{n-1}}^i, \; j = 1, 2, \cdots, k, \; i = 1, 2, \cdots, k^{n-1},$$

and $M_i \in \mathscr{L}_{k \times k}$, $i = 1, 2, \cdots, k^{n-1}$, are same as in Theorem 8.9.

Proof. Using the vector form of logical variables,

$$\begin{aligned} \alpha^1 &= x_n x_{n-1} \cdots x_1 = (M'_f x_{n-1} \cdots x_1 x_0) x_{n-1} \cdots x_1 \\ &= M'_f x_{n-1} \cdots x_1 W_{[k^{n-1}, k]} x_{n-1} \cdots x_1 x_0 \\ &= M'_f (I_{k^{n-1}} \otimes W_{[k^{n-1}, k]}) x_{n-1} \cdots x_1 x_{n-1} \cdots x_1 x_0 \\ &= M'_f (I_{k^{n-1}} \otimes W_{[k^{n-1}, k]}) M_{r, k^{n-1}} x_{n-1} \cdots x_1 x_0. \end{aligned} \tag{8.6}$$

Deducing by analogy, we have:

$$\alpha^{t+1} = M'_f (I_{k^{n-1}} \otimes W_{[k^{n-1}, k]}) M_{r, k^{n-1}} \alpha^t, \;\; t = 0, 1, 2, \cdots.$$

Since the logical function $T_f(\alpha)$ has only one structural matrix, from (8.4) and (8.6) we can prove:

$$T'_f = M'_f (I_{k^{n-1}} \otimes W_{[k^{n-1}, k]}) M_{r, k^{n-1}}.$$

By the semi-tensor product,

$$M_{r, k^{n-1}} = \left[\left(\delta_{k^{n-1}}^1\right)^2 \; \left(\delta_{k^{n-1}}^2\right)^2 \; \cdots \; \left(\delta_{k^{n-1}}^{k^{n-1}}\right)^2 \right] \in \mathscr{L}_{k^{2n-2} \times k^{n-1}},$$

and,

$$W_{[k^{n-1}, k]} = [B_1 \; B_2 \; \cdots \; B_{k^{n-1}}] \in \mathscr{L}_{k^n \times k^n}, \tag{8.7}$$

where $B_i = \left[\delta_k^1 \ltimes \delta_{k^{n-1}}^i \ \delta_k^2 \ltimes \delta_{k^{n-1}}^i \ \cdots \ \delta_k^k \ltimes \delta_{k^{n-1}}^i\right] \in \mathscr{L}_{k^n \times k}$, $i = 1,2,\cdots,k^{n-1}$. Set $M'_f = [M_1 \ M_2 \ \cdots \ M_{k^{n-1}}]$. Then, we have:

$$\begin{aligned}
T'_f &= M'_f(I_{k^{n-1}} \otimes W_{[k^{n-1},k]}M_{r,k^{n-1}} \\
&= [M_1 \ \cdots \ M_{k^{n-1}}] \begin{bmatrix} W_{[k^{n-1},k]} & & \\ & \ddots & \\ & & W_{[k^{n-1},k]} \end{bmatrix} M_{r,k^{n-1}} \qquad (8.8) \\
&= \left[M_1 W_{[k^{n-1},k]} \ \cdots \ M_{k^{n-1}} W_{[k^{n-1},k]}\right] \left[\left(\delta_{k^{n-1}}^1\right)^2 \cdots \left(\delta_{k^{n-1}}^{k^{n-1}}\right)^2\right] \\
&= \left[M_1 W_{[k^{n-1},k]} \delta_{k^{n-1}}^1 \ \cdots \ M_{k^{n-1}} W_{[k^{n-1},k]} \delta_{k^{n-1}}^{k^{n-1}}\right] \\
&= [M_1 B_1 \ \cdots \ M_{k^{n-1}} B_{k^{n-1}}] = \left[M_1 \otimes \delta_{k^{n-1}}^1 \ \cdots \ M_{k^{n-1}} \delta_{k^{n-1}}^{k^{n-1}}\right] \\
&:= \left[T_f^1 \ T_f^2 \ \cdots \ T_f^{k^{n-1}}\right] \in \mathscr{L}_{k^n \times k^n}.
\end{aligned}$$

Thus, $T_f^i = M_i \otimes \delta_{k^{n-1}}^i$, $i = 1,2,\cdots,k^{n-1}$, and the proof is completed. ∎

In order to obtain the number of cycles with different lengths, we need the following result.

Lemma 8.4

Consider a k-valued n-stage FSR with structural matrix T'_f. Then,

1) The number of fixed points for the FSRs, denoted by N_e, equals the number of i, for which $(T'_f)_{ii} = 1$. That is,

$$N_e = Trace(T'_f); \qquad (8.9)$$

2) The number of cycles, N_s, which have length s, are inductively determined by:

$$\begin{cases} N_1 &= N_e \\ N_s &= \dfrac{\text{Trace}\left((T'_f)^s\right) - \Sigma_{t \in \mathscr{P}(s)} t N_t}{s}, \quad 2 \le s \le k^n, \end{cases} \qquad (8.10)$$

where $\mathscr{P}(s)$ is a set of proper factors of s. For example, $\mathscr{P}(6) = \{1,2,3\}$.

From Lemma 8.4, in order to obtain the number of cycles it is only needed to compute the structural matrix of FSR.

8.3.3 CYCLE DECOMPOSITION AND SYNTHESIS FOR FSRS

Let $f(x_1,x_2,\cdots,x_n)$ be a feedback function of an FSR, $x_i \in \mathscr{D}_k$, $i = 1,2,\cdots,n$, and G_f be a graph of FSR. Suppose that $S_0 = (a_1,a_2,\cdots,a_n)$, $S_1 = (\oslash(a_1),a_2,\cdots,a_n)$, $\cdots$,

$S_{k-1} = (\oslash^{k-1}(a_1), a_2, \cdots, a_n)$ are k conjugate points of G_f, where $a_i \in \{0, 1, \cdots, k-1\}$, the operator $\oslash$ is defined as:

$$\oslash(a) = \begin{cases} a+1, & a = 0, 1 \cdots, k-2; \\ 0, & a = k-1. \end{cases} \tag{8.11}$$

Using the vector form of logical variables, i.e., $k - i \sim \delta_k^i$, $i = 0, 1, \cdots, n-1$, the structural matrix of operator $\oslash$ is $\delta_k[k\ 1\ 2 \cdots\ k-1]$.

Consider a new FSR with feedback function:

$$f_1(x_1, x_2, \cdots, x_n) = f(x_1, x_2, \cdots, x_n) + \nabla_{a_n,k}(x_n) \cdots \nabla_{a_2,k}(x_2) \pmod{k}, \tag{8.12}$$

where:

$$\nabla_{a,k}(x) = \begin{cases} k-1, & x = a, \\ 0, & x \neq a, \end{cases}$$

its structural matrix is $M_{\nabla_{a,k}} = \delta_k[\underbrace{k \cdots k}_{k-a-1}\ 1\ \underbrace{k \cdots k}_{a}]$, $a \in \{0, 1, \cdots, k-1\}$ and $x \in \mathscr{D}_k$.

Since only the k-valued FSR is considered in this section, $\nabla_{a,k}(x)$ is abbreviated as $\nabla_a(x)$.

Now, we have the following results.

Lemma 8.5

Assume $x, y \in \mathscr{D}_k$. Then,

$$x + y \pmod{k} = M_{+,k} \ltimes x \ltimes y, \tag{8.13}$$

where,

$$M_{+,k} = \delta_k[2\ \cdots\ k\ 1 \vdots 3\ \cdots\ k\ 1\ 2 \vdots \cdots \vdots 1\ 2\ \cdots\ k] = \begin{bmatrix} M_{o,k} & M_{o,k}^2 & \cdots & M_{o,k}^k \end{bmatrix}$$

$M_{o,k} = \delta_k[2\ \cdots\ k\ 1]$.

Proof. The proof can be directly obtained by the semi-tensor product and the truth table of mod k addition. Thus, we omit it. ∎

Lemma 8.6

Let $x \in \Delta_m$ and $y \in \Delta_n$. Then,

$$E_{d,m,n} x y = y, \tag{8.14}$$

and,

$$E_{d,m,n} W_{[n,m]} y x = y, \tag{8.15}$$

where $E_{d,m,n} = [\underbrace{I_n\, I_n\, \cdots I_n}_{m}]$.

Proof. For all $x = \delta_m^i \in \Delta_m$, $E_{d,m,n} \ltimes x \ltimes y = I_n y = y$. Moreover, (8.15) can be obtained from (8.14). Thus, the proof is completed. ∎

It is noted that $E_{d,m,n}$ is abbreviated as $E_{d,n}$ when $m = n$.

Lemma 8.7

Let $x \in \Delta_n$, $A \in \mathscr{L}_{m_1 \times n}$ and $B \in \mathscr{L}_{m_2 \times n}$. Then,

$$(Ax) \ltimes (Bx) = Cx, \tag{8.16}$$

where $C = A * B \in \mathscr{L}_{(m_1 m_2) \times n}$, i.e., $Col_i(C) = Col_i(A) \ltimes Col_i(B)$, $i = 1, 2, \cdots, n$.

Proof. For all $x = \delta_n^i$, $Cx = Col_i(C)$ and $(Ax) \ltimes (Bx) = Col_i(A) \ltimes Col_i(B)$. Thus, the proof is completed. ∎

Theorem 8.11

Let $f(x_1, x_2, \cdots, x_n)$ be a feedback function of a nonsingular FSR. Then,

1) The FSR with feedback function $f_1(x_1, x_2, \cdots, x_n)$ is still nonsingular;
2) If $S_0, S_1, \cdots, S_{k-1}$ are on k different cycles of G_f, then these k cycles will merge into one cycle in G_{f_1}.

Proof. 1) To prove the nonsingularity of new FSR, we need to obtain the matrix M'_{f_1} first. Suppose that:

$$f(x_1, x_2, \cdots, x_n) = M'_f x_n \cdots x_2 x_1,$$

where,

$$M'_f = \delta_k[b_1\, b_2 \cdots\, b_{k^n}] = [M_1\, M_2\, \cdots\, M_{k^{n-1}}], \tag{8.17}$$

$b_i \in \{1, 2, \cdots, k\}$, $i = 1, 2, \cdots, k^n$ and $M_j \in \mathscr{L}_{k \times k}$, $j = 1, 2, \cdots, k^{n-1}$. Then, by (8.12),

$$M'_{f_1} x_n \cdots x_2 x_1 = M_{+,k} M'_f x_n \cdots x_1 M_{\nabla_a} x_n \cdots x_2,$$

where $M_{\nabla_a} = \delta_k[\underbrace{k \cdots k}_{a-1}\ 1\ \underbrace{k \cdots k}_{k^{n-1}-a}]$ and $a = (k - a_n - 1)k^{n-2} + \cdots + (k - a_3 - 1)k + (k - a_2)$. From (8.14) and (8.15), we obtain:

$$M_{\nabla_a} x_n \cdots x_2 = M_{\nabla_a} E_{d,k,k^{n-1}} x_1 x_n \cdots x_2 = M_{\nabla_a} E_{d,k,k^{n-1}} W_{[k^{n-1},k]} x_n \cdots x_2 x_1,$$

and,

$$M_{\nabla_a} E_{d,k,k^{n-1}} W_{[k^{n-1},k]} = [\underbrace{M_{\nabla_a} M_{\nabla_a} \cdots M_{\nabla_a}}_{k}] W_{[k^{n-1},k]} = \delta_k[\underbrace{k\mathbf{1}_k\, k\mathbf{1}_k \cdots k\mathbf{1}_k}_{a-1}\ \mathbf{1}_k\ \underbrace{k\mathbf{1}_k \cdots k\mathbf{1}_k}_{k^{n-1}-a}], \tag{8.18}$$

where $\mathbf{1}_k = [\underbrace{1 \cdots 1}_{k}]$.

Then, according to Lemma 8.7,

$$M'_f x_n \cdots x_1 M_{\nabla_a} x_n \cdots x_2 = M x_n \cdots x_1,$$

where $M = M'_f * (M_{\nabla_a} E_{d,k,k^{n-1}} W_{[k^{n-1},k]})$ is given by:

$$Col_i(M) = Col_i(M'_f * (M_{\nabla_a} E_{d,k,k^{n-1}} W_{[k^{n-1},k]})) = \begin{cases} \delta_k^{b_i} \ltimes \delta_k^1 = \delta_{k^2}^{(b_i-1)k+1}, & (a-1)k < i \leq ak, \\ \delta_k^{b_i} \ltimes \delta_k^k = \delta_{k^2}^{b_i k}, & \text{otherwise.} \end{cases} \tag{8.19}$$

Therefore,

$$Col_i(M'_{f_1}) = M_{+,k} Col_i(M) = \begin{cases} M_{o,k}^{b_i} \ltimes \delta_k^1 = \delta_k^{b_i+1}, & (a-1)k < i \leq ak, \\ M_{o,k}^{b_i} \ltimes \delta_k^k = \delta_k^{b_i}, & \text{otherwise,} \end{cases} \tag{8.20}$$

where we define $k+1 = 1$ standing for the truth.

Now, when $i \leq (a-1)k$ or $i > ak$, $Col_i(M'_{f_1}) = Col_i(M'_f)$. Since the FSR with feedback function f is nonsingular, the columns of b-th $(b \neq a)$ block of M'_f, are different by Theorem 8.8. So are the columns of b-th block of M'_{f_1}.

When $(a-1)k < i \leq ak$, $\delta_k^{b_i} \in \text{Col}(M_a)$. Since columns of M_a are different, b_i, $i = (a-1)k+1, \cdots, ak$, are different. Then, $b_i + 1$, $i = (a-1)k+1, \cdots, ak$, are different, i.e., all columns of a-th equal block of $M_{f'_1}$ are different. Thus, according to Theorem 8.8, the FSR with feedback function $f_1(x_1, x_2, \cdots, x_n)$ is also nonsingular.

2) Assume that S_i, $i = 0, 1, \cdots, k-1$, are on different cycles of G_f. Let their transition diagrams be:

$$S_i \to T_f(S_i) \to \cdots \to S_i, \ i = 0, 1, 2, \cdots, k-1.$$

If $S = (x_1, x_2, \cdots, x_n) \notin \{S_0, S_1, \cdots, S_{k-1}\}$, then,

$$\nabla_{a_n}(x_n) \cdots \nabla_{a_2}(x_2) = 0, \ \ f_1(S) = f(S), \ \ T_{f_1}(S) = T_f(S).$$

Otherwise, according to the proof of 1),

$$f_1(S_j) = \delta_k^{b_{(a-1)k+c-j}+1},$$

where $j = 0,1,\cdots,k-1$, $a_n \ltimes \cdots \ltimes a_2 = \delta^a_{k^{n-1}}$, $a_1 = \delta^c_k$ and $k+1 = 1$. Since,

$$b_{(a-1)k+c-j} + 1 \in \left\{ b_{(a-1)k+i} \,|\, i = 0,1,\cdots,k-1 \right\},$$

for all $i_j \in \{0,1,\cdots,k-1\}$, there exists a unique i_{j+1} such that $f_1(S_{i_j}) = f(S_{i_{j+1}})$. Particularly, $f_1(S_{i_{k-1}}) = f(S_{i_0})$, $S_{i_0} := S_0$. Therefore, $T_{f_1}(S_{i_j}) = T_f(S_{i_{j+1}})$, $j = 0,1,\cdots,k-1$. The transition diagram containing $S_0, S_1, \cdots, S_{k-1}$ in G_{f_1} can be described as:

$$\begin{aligned} S_0 \to T_{f_1}(S_0) = T_f(S_{i_1}) \to \cdots \quad &\to S_{i_1} \to T_f(S_{i_2}) \to \cdots \\ &\to S_{i_{k-1}} \to T_{f_1}(S_{i_{k-1}}) = T_f(S_0) \to \cdots \to S_0, \end{aligned}$$

where $\{i_1,\cdots,i_{k-1}\} = \{1,\cdots,k-1\}$. Then, $S_0, S_1, \cdots, S_{k-1}$ are on one cycle of G_{f_1}, and the proof is completed. ∎

When S_i, $i = 0,1,\cdots,k-1$, are on the same cycle of G_f, this cycle might not split into some small cycles of G_{f_1}. This is different from the binary case.

For example, let $f(x_1,x_2) = \delta_3[1\ 2\ 3\ 3\ 2\ 1\ 2\ 3\ 1]x_2x_1$. Obviously, $S_0 = (0,0)$, $S_1 = (1,0)$ and $S_2 = (2,0)$ are three conjugate points of the FSR. Then, $f_1(x_1,x_2) = f(x_1,x_2) + \nabla_2(x_2) = \delta_3[1\ 2\ 3\ 3\ 2\ 1\ 3\ 1\ 2]x_2x_1$. The cycle of G_f containing the three points is:

$$S_0 \to T_f(S_0) = (0,2) \to S_2 \to T_f(S_2) = (0,1) \to (1,2) \to (2,1) \to S_1 \to T_f(S_1) = S_0.$$

For G_{f_1}, if it can not be split into some small cycles, it becomes the following cycle containing these three conjugate points:

$$\begin{aligned} S_0 \to T_{f_1}(S_0) = (0,1) = T_f(S_2) \quad &\to (1,2) \to (2,1) \to S_1 \\ &\to T_{f_1}(S_1) = (0,2) = T_f(S_0) \to S_2 \to S_0. \end{aligned}$$

8.3.4 AN ILLUSTRATIVE EXAMPLE

In this subsection, an example is given to show the effectiveness of the obtained results.

Example 8.2:

Consider a feedback shift register with feedback function:

$$f(x_1,x_2) = x_1\left(1 + x_2^2\right) \pmod 3. \tag{8.21}$$

We study the following three problems.

Nonsingularity. From (8.21), the truth value table of function f is as follows. Using the vector form, i.e., $2 \sim \delta_3^1$, $1 \sim \delta_3^2$ and $0 \sim \delta_3^3$, we have:

$$f(x_1,x_2) = \delta_3[2\ 1\ 3 \vdots 2\ 1\ 3 \vdots 1\ 2\ 3]x_2x_1 := M_f' x_2x_1.$$

It is obvious that $M_1 = \delta_3[2\ 1\ 3] = M_2$ and $M_3 = \delta_3[1\ 2\ 3]$. Since all columns of M_i are different, $i = 1,2,3$, the FSR is nonsingular by Theorem 8.8.

Table 8.1: Truth value table of function f

x_2	2	2	2	1	1	1	0	0	0
x_1	2	1	0	2	1	0	2	1	0
$f(x_1,x_2)$	1	2	0	1	2	0	2	1	0

Number of cycles. From (8.5), the structural matrix of FSR is:

$$\begin{aligned} T'_f &= [\delta_3^2\delta_3^1 \ \ \delta_3^1\delta_3^1 \ \ \delta_3^3\delta_3^1 \vdots \delta_3^2\delta_3^2 \ \ \delta_3^1\delta_3^2 \ \ \delta_3^3\delta_3^2 \vdots \delta_3^1\delta_3^3 \ \ \delta_3^2\delta_3^3 \ \ \delta_3^3\delta_3^3] \\ &= \delta_9[4\ 1\ 7\ 5\ 2\ 8\ 3\ 6\ 9]. \end{aligned} \tag{8.22}$$

Since $Trace(T'_f) = 1$, there is unique fixed point. Through computation, $(T'_f)^2 = \delta_9[5\ 4\ 3\ 2\ 1\ 6\ 7\ 8\ 9]$ and $Trace((T'_f)^2) = 5$. Then, the number of cycles with length 2 are $N_2 = \frac{5-1}{2} = 2$.

Now, $(T'_f)^3 = \delta_9[2\ 5\ 7\ 1\ 4\ 8\ 3\ 6\ 9]$. Since $Trace((T'_f)^3) - 1 = 0$, there is no cycle with length 3 by Lemma 8.4. $T'^4_f = \delta_9[1\ 2\ 3\ 4\ 5\ 6\ 7\ 8\ 9]$ and $Trace((T'_f)^4) = 9$. Thus, $N_4 = \frac{9-1-2\times 2}{4} = 1$ and there is one cycle with length 4. Therefore, the nonsingular FSR has four cycles.

Synthesis of cycles. It is obvious that $S_0 = (0,0)$, $S_1 = (1,0)$ and $S_2 = (2,0)$ are three conjugate points. The three points are on three different cycles in G_f. To obtain a new and large cycle, set:

$$f_1(x_1,x_2) = f(x_1,x_2) + \nabla_0(x_2) (\text{mod } 3).$$

By using the semi-tensor product and (8.20), we have $f_1(x_1,x_2) = M'_{f_1}x_2x_1 = \delta_3[2\ 1\ 3 \vdots 2\ 1\ 3 \vdots 2\ 3\ 1]x_2x_1$. From (8.5),

$$\begin{aligned} T'_{f_1} &= \left[\delta_3^2\delta_3^1 \ \ \delta_3^1\delta_3^1 \ \ \delta_3^3\delta_3^1 \vdots \delta_3^2\delta_3^2 \ \ \delta_3^1\delta_3^2 \ \ \delta_3^3\delta_3^2 \vdots \delta_3^2\delta_3^3 \ \ \delta_3^3\delta_3^3 \ \ \delta_3^1\delta_3^3\right] \\ &= \delta_9[4\ 1\ 7 \vdots 5\ 2\ 8 \vdots 6\ 9\ 3], \end{aligned} \tag{8.23}$$

and its transition diagram is as follows:

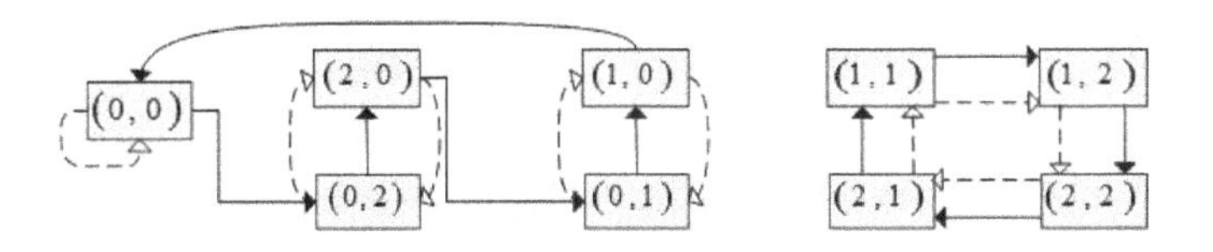

Figure 8.3 Transition diagram of the FSR with feedback function f_1

Note: the dotted lines denote original cycles in G_f (See Fig. 8.2). It is observed that the FSR with feedback function f_1 merges three cycles of G_f into one larger cycle of G_{f_1}.

REFERENCES

1. Bruen, A. A. and Mollin, R. A. (2009). Cryptography and shift registers. Open Math. J., 2: 16–21.
2. Golomb, S. W. (1967). Shift Register Sequence. CA, Holden.
3. Jansen, C. J. A. (1986). Investigation on nolinear stream ciper systems: construction and evolution methods. University of Delft.
4. Lai, X. (1987). Condition for the nonsingularity of a feedback shift-register over a general finite field. IEEE Trans. Information Theory, 33(5): 747–757.
5. Lewis, T. G. and Payne, W. H. (1973). Generalized feedback shift register pseudorandom number algorithm. Journal of the ACM, 20(3): 456–468.
6. Li, C. and Xie, R. (1995). Criterion for the nonsingularity of feedback shift registers. Journal of Electronics, 17(5): 500–505.
7. Liu, Z., Wang, Y. and Cheng, D. (2015). Nonsingularity of feedback shift registers. Automatica, 55: 247–253.
8. Rajpal, N., Kumar, A., Dudhani, S. and Jindal, P. R. (2004). Copyright protection using nonlinear forward feedback shift register and error-correction technique. The 7th Ann. Intl. Conf. Map India.
9. Wang, L. and Mccluskey, E. (1986). Condensed linear feedback shift register (LFSR) testing—a pseudoexhaustive test technique. IEEE Trans. Computers, 35(4): 367–370.
10. Wang, L. and Zeng, G. (2016). On the matrix feedback shift register synthesis for matrix sequences. Science China Information Sciences, 59: 032107.
11. Zhao, D., Peng, H. and Li, L. (2014). Novel way to research nonlinear feedback shift register. Science China Information Sciences, 57: 092114.
12. Zhong, J. and Lin, D. (2016). Stability of nonlinear feedback shift registers. Science China Information Sciences, 59: 012204.

9 Graph Theory

9.1 INTRODUCTION TO GRAPH THEORY

This section introduces some basic concepts of Graph Theory.

A graph $\mathscr{G}$ consists of a vertex (node) set $\mathscr{V} = \{v_1, v_2, \cdots, v_n\}$ and an edge set $\mathscr{E} \subset \mathscr{V} \times \mathscr{V}$, denoted by $\mathscr{G} = \{\mathscr{V}, \mathscr{E}\}$. If each edge of $\mathscr{G}$, denoted by $e_{ij} = (v_i, v_j) \in \mathscr{E}$, is an ordered pair of two vertices of $\mathscr{V}$, we call $\mathscr{G}$ directed graph (or digraph); if each edge $e_{ij} \in \mathscr{E}$ implies that $e_{ji} \in \mathscr{E}$, then we call $\mathscr{G}$ an undirected graph. A graph is called to be simple, if each edge $e \in \mathscr{E}$ is described by a pair of two distinct vertices. In a digraph $\mathscr{G}$, a directed path is a sequence of ordered edges of the form $(v_{i_1}, v_{i_2}), (v_{i_2}, v_{i_3}), \cdots$. For node i, its neighbor set, $\mathscr{N}_i$, is defined as:

$$\mathscr{N}_i := \left\{ j \mid e_{ji} = (v_j, v_i) \in \mathscr{E} \right\}.$$

DEFINITION 9.1: see [1]

Consider a graph $\mathscr{G} = \{\mathscr{V}, \mathscr{E}\}$. Given a vertex subset $S \subseteq \mathscr{V}$, if $i \notin \mathscr{N}_j$ and $j \notin \mathscr{N}_i$ hold for any $i, j \in S$ $(i \neq j)$, then S is called an internally stable set (an independent set, or a vertex packing) of $\mathscr{G}$. Furthermore, S is called a maximum internally stable set, if for any vertex subset strictly containing S is not an internally stable set. An internally stable S is called an absolute maximum internally stable set if $|S|$ is the largest among those of all the internally stable sets of $\mathscr{G}$, and the largest $|S|$ is called the internally stable number of $\mathscr{G}$, denoted by $\alpha(\mathscr{G}) = |S|$, where $|S|$ stands for the cardinality of S.

From Definition 9.1, it is easy to see that any subset of an internally stable set is also an internally stable one, and any internally stable set can be enlarged to a maximum internally stable one. Especially, the empty set $\emptyset$ is regarded as an internally stable of any graph.

DEFINITION 9.2: see [1]

Consider a graph $\mathscr{G} = \{\mathscr{V}, \mathscr{E}\}$. Given a weight function $w : \mathscr{V} \mapsto \mathbb{R}$, a vertex subset $S \subseteq \mathscr{V}$ is called a maximum weight stable set if S is an internally stable set and $\sum_{i \in S} w(v_i)$ is the largest among those of all the internally stable sets.

Lemma 9.1: see [2]

Every pseudo-Boolean function $f(x_1, x_2, \cdots, x_n)$ can be uniquely represented in the multi-linear polynomial form of:

$$f(x_1,x_2,\cdots,x_n) = c_0 + \sum_{k=1}^{m} c_k \prod_{i\in A_k} x_i,$$

where $c_0, c_1, \cdots, c_m$ are real coefficients, $A_1, A_2, \cdots, A_m$ are non-empty subsets of $\{1, 2, \cdots, n\}$, and the product is the conventional one. ■

9.2 GRAPH MAXIMUM STABLE SET AND COLORING PROBLEMS

In this section, we investigate the maximum (weight) stable set and vertex coloring problems by the semi-tensor product method. First, we consider the internally stable set problem.

Consider a graph $\mathscr{G}$ with n nodes $\mathscr{V} = \{v_1, v_2, \cdots, v_n\}$. Assume that the adjacency matrix, $A = [a_{ij}]$, of $\mathscr{G}$ is given as:

$$a_{ij} = \begin{cases} 1, & v_j \in \mathscr{N}_i; \\ 0, & v_j \notin \mathscr{N}_i. \end{cases} \tag{9.1}$$

If $\mathscr{G}$ is not a simple graph, say, there exists v_i such that $(v_i, v_i) \in \mathscr{E}$, then we just let $a_{ii} = 0$ in our study. Thus, without loss of generality, we can assume that $a_{ii} = 0$ holds for all i in the sequel.

Given a vertex subset $S \subseteq \mathscr{V}$, define a vector $V_S = [x_1, x_2, \cdots, x_n]$, called the characteristic logical vector of S, as follows:

$$x_i = \begin{cases} 1, & v_i \in S; \\ 0, & v_i \notin S. \end{cases} \tag{9.2}$$

Then, we have the following result to determine whether or not S is an internally stable set.

Theorem 9.1

S is an internally stable set of $\mathscr{G}$, if and only if the first row of matrix M_S has at least one component of zero, where,

$$M_S = \sum_{i=1}^{n} \sum_{j\neq i} a_{ij} M_{ij}, \quad M_{ij} = M_{ji} = M_c(E_d)^{n-2} W_{[2^j,2^{n-j}]} W_{[2^i,2^{j-i-1}]} \in \mathscr{L}^{2\times 2^n} \ (i<j). \tag{9.3}$$

■

Proof. ($\Rightarrow$) If S is an internally stable set of $\mathscr{G}$, it is easy to see from Definition 9.1 that for any nodes $v_i, v_j \in \mathscr{V}$, if $a_{ij} = 1$, then $v_i \notin S$ or $v_j \notin S$. Thus, from (9.2) we have $x_i x_j = 0$. Therefore, the characteristic vector $[x_1, x_2, \cdots, x_n]$ of S satisfies:

$$\sum_{i=1}^{n}\sum_{j=1}^{n} a_{ij}x_i x_j = \sum_{i=1}^{n}\sum_{j\neq i} a_{ij}x_i x_j = 0, \tag{9.4}$$

which is a pseudo-Boolean equation.

Let,

$$y_i = \begin{bmatrix} x_i \\ \bar{x}_i \end{bmatrix}, \quad \bar{x}_i = 1 - x_i, \ \ i = 1, 2, \cdots, n.$$

Since $x_i x_j = x_j x_i (= x_i \wedge x_j)$, without loss of generality, we assume that $i < j$. Then, we have:

$$\begin{aligned} y_i y_j &= (E_d)^{n-2} y_{j+1} \cdots y_n y_{i+1} \cdots y_{j-1} y_1 \cdots y_{i-1} y_i y_j \\ &= (E_d)^{n-2} W_{[2^j, 2^{n-j}]} y_{i+1} \cdots y_{j-1} y_1 \cdots y_i y_j y_{j+1} \cdots y_n \\ &= (E_d)^{n-2} W_{[2^j, 2^{n-j}]} W_{[2^i, 2^{j-i-1}]} y_1 \cdots y_i y_{i+1} \cdots y_{j-1} y_j \cdots y_n, \end{aligned}$$

where the product is "$\ltimes$". Thus, $x_i x_j = J_1 M_{ij} \ltimes_{i=1}^{n} y_i$, where $J_1 = [1, 0]$, and,

$$M_{ij} = M_c (E_d)^{n-2} W_{[2^j, 2^{n-j}]} W_{[2^i, 2^{j-i-1}]}.$$

Hence, the equation (9.4) can be expressed as:

$$0 = \sum_{i=1}^{n}\sum_{j\neq i} a_{ij}x_i x_j = J_1 \sum_{i=1}^{n}\sum_{j\neq i} a_{ij} M_{ij} \ltimes_{i=1}^{n} y_i = J_1 M_S Y, \tag{9.5}$$

where $Y := \ltimes_{i=1}^{n} y_i$.

Since $[x_1, x_2, \cdots, x_n]$ is a solution to (9.4), Y satisfies (9.5). Note that $Y \in \Delta_{2^n}$. The equation (9.5) has a solution Y is equivalent to that the first row of M_S, which has at least one component of zero. Thus, the necessity is proved.

($\Leftarrow$) If the first row of matrix M_S has at least one component of zero, then the equation $M_S Y = [0, *]^T$ has a solution $Y \in \Delta_{2^n}$, where "$*$" stands for some real number. Equivalently, (9.5) or (9.4) has a solution $[x_1, x_2, \cdots, x_n]$. Since, $x_i \in \mathscr{D}$ and $a_{ij} \geqslant 0$, from (9.4) we have $a_{ij}x_i x_j = 0$ holds for any i, j, which implies that $v_i \notin S$ or $v_j \notin S$ holds when $a_{ij} = 1$. From the definition of internally stable sets, S is an internally stable set. Thus, the proof is completed. ∎

From the above proof, we have the following corollary.

COROLLARY 9.1

Consider the graph $\mathscr{G}$ in Theorem 9.1. For each node $i \in \mathscr{V}$, we assign it a characteristic logical variable $x_i \in \mathscr{D}$ and let $y_i = [x_i, \bar{x}_i]^T$. Then, $\mathscr{G}$ has a non-empty internally stable set if and only if the equation:

$$J_1 M_S \ltimes_{i=1}^{n} y_i = 0$$

is solvable. Furthermore, the number of zero components in $J_1 M_S$ are just the number of internally stable sets of $\mathscr{G}$.

Based on the proof of Theorem 9.1 and Corollary 9.1, we now establish a new algorithm to find all the internally stable sets for any graph $\mathscr{G}$.

ALGORITHM 9.1

Given a graph $\mathscr{G}$ with nodes $\mathscr{V} = \{v_1, v_2, \cdots, v_n\}$, assume that its adjacency matrix is $A = [a_{ij}]$. For each node v_i, we assign it a characteristic logical variable $x_i \in \mathscr{D}$ and let $y_i = [x_i, \bar{x}_i]^T$. To find all the internally stable sets of $\mathscr{G}$, we follow the following steps:

1). Compute matrix $M_S = \sum_{i=1}^{n} \sum_{j \neq i} a_{ij} M_{ij}$ by (9.3).
2). Extract the first row of M_S, and denote it by $\beta = [b_1, b_2, \cdots, b_{2^n}]$. If for all i, $b_i \neq 0$, then $\mathscr{G}$ has no internally stable set and stop the calculation. Otherwise, find out all the zero components of β, and denote their positions in β by $i_1, i_2, \cdots, i_m$, that is, $b_{i_k} = 0$, $k = 1, 2, \cdots, m$. Then, each $b_{i_k} = 0$ will correspond to one internally stable set.
3). For each index i_k, $k = 1, 2, \cdots, m$, consider $\ltimes_{i=1}^{n} y_i = \delta_{2^n}^{i_k}$. Let [5],

$$\begin{cases} S_1^n = \delta_2[\underbrace{1 \cdots 1}_{2^{n-1}} \underbrace{2 \cdots 2}_{2^{n-1}}], \\ S_2^n = \delta_2[\underbrace{1 \cdots 1}_{2^{n-2}} \underbrace{2 \cdots 2}_{2^{n-2}} \underbrace{1 \cdots 1}_{2^{n-2}} \underbrace{2 \cdots 2}_{2^{n-2}}], \\ \vdots \\ S_n^n = \delta_2[\underbrace{\underbrace{1\,2}_{2} \cdots \underbrace{1\,2}_{2}}_{2^{n-1}}], \end{cases} \tag{9.6}$$

then it is easy to show that $y_i = S_i^n \ltimes_{j=1}^{n} y_j = S_i^n \delta_{2^n}^{i_k}$, $i = 1, 2, \cdots, n$. Noticing that $y_i = [x_i, \bar{x}_i]^T$, we need to check whether $y_i = \delta_2^1$ or $x_i = 1$. Set,

$$S(i_k) = \left\{ v_i \,\middle|\, y_i = \delta_2^1,\ 1 \leqslant i \leqslant n \right\}, \tag{9.7}$$

then $S(i_k) \subseteq \mathscr{V}$ is the internally stable set corresponding to $b_{i_k} = 0$, and all the internally stable sets of $\mathscr{G}$ are $\{S(i_k) \mid k = 1, 2, \cdots, m\}$.

4). Let:

$$\alpha_0 = \max_{1 \leqslant k \leqslant m} \left\{ |S(i_k)| \right\}, \quad \mathscr{S} = \left\{ S(i_k) \,\middle|\, |S(i_k)| = \alpha_0 \right\},$$

then α_0 is the internally stable number of $\mathscr{G}$, that is, $\alpha(\mathscr{G}) = \alpha_0$, and $\mathscr{S}$ is the set of all the absolutely maximum internally stable sets of $\mathscr{G}$.

In the above algorithm, with the product $\ltimes_{i=1}^{n} y_i$, each y_i can be uniquely determined (return to its original value) by the formula $y_i = S_i^n \ltimes_{i=1}^{n} y_i$ [5].

Algorithm 9.1 also provides a way to determine the absolutely maximum internally stable sets. In the following, we put forward another method to find all the maximum internally stable sets. To this end, we present a theorem first.

Theorem 9.2

Consider a graph $\mathscr{G}$ with n nodes $\mathscr{V} = \{v_1, v_2, \cdots, v_n\}$. Given a vertex subset $S \subseteq \mathscr{V}$ with its characteristic vector $V_S = [x_1, x_2, \cdots, x_n]$, let $\ltimes_{i=1}^{n} [x_i, \bar{x}_i]^T := \delta_{2^n}^k$, where $1 \leqslant k \leqslant 2^n$ can be uniquely determined. Then, S is an absolutely maximum internally stable set if and only if:

$$b_k = \max_{1 \leqslant i \leqslant 2^n} \{b_i\} \geqslant 0, \tag{9.8}$$

where,

$$[b_1, b_2, \cdots, b_k, \cdots, b_{2^n}] := J_1 \widetilde{M}, \quad \widetilde{M} = \sum_{i=1}^{n} M_i - (n+1) M_S, \tag{9.9}$$

$$M_i = (E_d)^{n-1} W_{[2^i, 2^{n-i}]}, \quad i = 1, 2, \cdots, n, \tag{9.10}$$

M_S and J_1 are the same as in Theorem 9.1. ■

Proof. From the proof of Theorem 9.1 and Definition 9.1, S being an absolutely maximum internally stable set is equivalent to that V_S is a global solution to the following constrained optimization problem:

$$\max \sum_{i=1}^{n} x_i, \tag{9.11}$$

$$\text{s.t.} \quad \sum_{i=1}^{n} \sum_{j \neq i} a_{ij} x_i x_j = 0. \tag{9.12}$$

According to [2], this constrained optimization problem can be changed to find a global maximum point $[x_1^*, \cdots, x_n^*]$ of the following function:

$$f(x_1, x_2, \cdots, x_n) = \sum_{i=1}^{n} x_i - (n+1) \sum_{i=1}^{n} \sum_{j \neq i} a_{ij} x_i x_j$$

such that $f(x_1^*, \cdots, x_n^*) \geqslant 0$.

On the other hand, letting $y_i = [x_i, \bar{x}_i]^T$, $i = 1, 2, \cdots, n$, we have:

$$\begin{aligned} y_i &= (E_d)^{n-1} y_{i+1} \cdots y_n y_1 \cdots y_{i-1} y_i \\ &= (E_d)^i W_{[2^i, 2^{n-i}]} y_1 y_2 \cdots y_i y_{i+1} \cdots y_n \\ &= M_i \ltimes_{j=1}^n y_j, \end{aligned}$$

from which we obtain:

$$x_i = J_1 M_i \ltimes_{j=1}^n y_j, \quad i = 1, 2, \cdots, n. \tag{9.13}$$

Thus, with (9.13) and the proof of Theorem 9.1, the pseudo-Boolean function $f(x_1, x_2, \cdots, x_n)$ can be expressed as:

$$f(x_1, x_2, \cdots, x_n) = J_1 \widetilde{M} \ltimes_{i=1}^n y_i,$$

where $\widetilde{M}$ is given in (9.9).

Based on the above analysis and noticing that $\ltimes_{i=1}^n [x_i, \bar{x}_i]^T = \delta_{2^n}^k \in \Delta_{2^n}$, we conclude that if S is an absolutely maximum internally stable set, then the k-th component of the row vector $J_1 \widetilde{M}$ is the largest among all the non-negative components, which is what (9.8) implies.

Conversely, if (9.8) holds, then the k-th component of $J_1 \widetilde{M}$ is the largest among all the non-negative components, which implies that $V_S = [x_1, \cdots, x_n]$ is the maximum point of $f(x_1, x_2, \cdots, x_n)$ such that $f(x_1, x_2, \cdots, x_n) \geqslant 0$. According to [2], $V_S = [x_1, \cdots, x_n]$ is a global solution to the constrained optimization problem (9.11)–(9.12), which means that S is an absolutely maximum internally stable set, and thus the proof is completed. ∎

The proof of Theorem 9.2 provides us a way to find all the absolutely maximum internally stable sets for any graph.

ALGORITHM 9.2

Given a graph $\mathscr{G}$ with nodes $\mathscr{V} = \{v_1, v_2, \cdots, v_n\}$, for each node v_i we assign it a characteristic logical variable $x_i \in \mathscr{D}$ and let $y_i = [x_i, \bar{x}_i]^T$. To determine all the absolutely maximum internally stable sets of $\mathscr{G}$, we can do by the following steps:

1). Compute the matrix $\widetilde{M}$ given in Theorem 9.2.

2). Extract the first row of $\widetilde{M}$, that is, $J_1 \widetilde{M}$, and denote it by $[b_1, b_2, \cdots, b_{2^n}]$. If for all i, $b_i < 0$, then $\mathscr{G}$ has no absolutely maximum internally stable set (except $\emptyset$) and stop the calculation. Otherwise, find out the maximum components of $[b_1, b_2, \cdots, b_{2^n}]$, and set:

$$K = \left\{ i_k \,\middle|\, b_{i_k} = \max_{1 \leqslant i \leqslant 2^n} \{b_i\},\ k = 1, 2, \cdots, m \right\}.$$

3). For each index $i_k \in K$, $k = 1,2,\cdots,m$, let $\ltimes_{i=1}^{n} y_i = \delta_{2^n}^{i_k}$. Using the formula (9.6), compute $y_i = S_i^n \delta_{2^n}^{i_k}$, $i = 1,2,\cdots,n$. Set:

$$S(i_k) = \left\{ v_i \,\middle|\, y_i = \delta_2^1,\ 1 \leqslant i \leqslant n \right\},$$

then $S(i_k) \subseteq \mathscr{V}$ is the absolutely maximum internally stable set corresponding to b_{i_k}, and $\alpha(\mathscr{G}) = |S(i_k)|$. Thus, all the absolutely maximum internally stable sets of $\mathscr{G}$ are $\{S(i_k) \mid k = 1,2,\cdots,m\}$.

With Theorem 9.2 and Algorithm 9.2, we have the following corollary.

COROLLARY 9.2

Consider a graph $\mathscr{G}$ with n nodes. Then, $\mathscr{G}$ has no internally stable set except $\emptyset$ if and only if $b_i < 0$ holds for all i, $1 \leqslant i \leqslant 2^n$, where b_i is defined by (9.9).

Next, we consider the maximum weight stable set problem of graphs.

We have the following result on determining whether or not a vertex subset is a maximum weight stable set.

Theorem 9.3

Consider a graph $\mathscr{G}$ with n nodes $\mathscr{V} = \{v_1, v_2, \cdots, v_n\}$, and let a non-negative weight function $w : \mathscr{V} \mapsto \mathbb{R}$ be given. Given a vertex subset $S \subseteq \mathscr{V}$ with its characteristic vector $V_S = [x_1, x_2, \cdots, x_n]$, let $\ltimes_{i=1}^{n} [x_i, \bar{x}_i]^T := \delta_{2^n}^{k}$, where $1 \leqslant k \leqslant 2^n$ can be uniquely determined. Then, S is a maximum weight stable set if and only if:

$$\mathrm{col}_k(\rho) = \max_{1 \leqslant i \leqslant 2^n} \left\{ \mathrm{col}_i(\rho) \right\} \geqslant 0,$$

where,

$$\rho = J_1 \widehat{M} \in \mathbb{R}^{2^n}, \quad \widehat{M} = l_0 \sum_{i=1}^{n} w(v_i) M_i - \left(l_0 \sum_{i=1}^{n} w(v_i) + 1 \right) M_S, \tag{9.14}$$

M_i, M_S and J_1 are the same as those in Theorem 9.2, $\mathrm{col}_i(\rho)$ stands for the i-th component of ρ, and l_0 is the smallest positive integer such that all $l_0 w(v_i)$, $i = 1,2,\cdots,n$ are integers. ■

Proof. From the proof of Theorem 9.1 and Definition 9.2, S being a maximum weight stable set is equivalent to that V_S is a global solution to the following constrained optimization problem:

$$\max \sum_{i=1}^{n} w(v_i)x_i,$$

$$\text{s.t.} \qquad \sum_{i=1}^{n}\sum_{j\neq i} a_{ij}x_ix_j = 0.$$

Note that there exists the smallest positive integer L_0 such that all $l_0w(v_i)$, $i = 1,2,\cdots,n$ are positive integers, the above problem can be changed to a constrained optimization problem of an integer-valued pseudo function. Based on this and [2], the above problem can be changed to find a global maximum point $[x_1^*,\cdots,x_n^*]$ of the following function:

$$f(x_1,x_2,\cdots,x_n) = \sum_{i=1}^{n} l_0w(v_i)x_i - \left(\sum_{i=1}^{n} l_0w(v_i)+1\right)\sum_{i=1}^{n}\sum_{j\neq i} a_{ij}x_ix_j \tag{9.15}$$

such that $f(x_1^*,\cdots,x_n^*) \geqslant 0$. Noting that the integer-valued pseudo function (9.15) has a structural matrix of $\widehat{M}$ given in (9.14), similar to the proof of Theorem 9.2, we can prove the theorem. ∎

The proof of Theorem 9.3 provides us an algorithm to find all the maximum weight stable sets for a given graph.

ALGORITHM 9.3

Given a graph $\mathscr{G}$ with nodes $\mathscr{V} = \{v_1,v_2,\cdots,v_n\}$, let $w : \mathscr{V} \mapsto \mathbb{R}$ be a non-negative weight function. For each node v_i, we assign a characteristic logical variable $x_i \in \mathscr{D}$ and set $y_i = [x_i,\bar{x}_i]^T$. To determine all the maximum weight stable sets of $\mathscr{G}$, we can follow the following steps:

1). Find the smallest positive integer l_0 such that all $l_0w(v_i)$, $i = 1,2,\cdots,n$ are positive integers.

2). Compute the matrix $\widehat{M}$ given in Theorem 9.3.

3). Extract the first row of $\widehat{M}$, and denote it by $\rho =: [b_1,b_2,\cdots,b_{2^n}]$. If for all i, $b_i < 0$, then $\mathscr{G}$ has no maximum weight stable set (except $\emptyset$) and stop the calculation. Otherwise, find out all the maximum components of ρ, and set

$$K = \left\{i_k \,\middle|\, b_{i_k} = \max_{1\leqslant i\leqslant 2^n}\{b_i\},\ k = 1,2,\cdots,m\right\}.$$

4). For each $i_k \in K$, $k = 1,2,\cdots,m$, let $\ltimes_{i=1}^{n} y_i = \delta_{2^n}^{i_k}$. Using the formula (9.6), calculate $y_i = S_i^n\delta_{2^n}^{i_k}$, $i = 1,2,\cdots,n$. Set

$$S(i_k) = \left\{v_i \,\middle|\, y_i = \delta_2^1,\ 1\leqslant i\leqslant n\right\},$$

then $S(i_k) \subseteq \mathscr{V}$ is the maximum weight stable set corresponding to b_{i_k}. Thus, all the maximum weight stable sets of $\mathscr{G}$ are $\{S(i_k) \mid k = 1, 2, \cdots, m\}$.

In the following section, we study the vertex coloring problem of graphs. First, we define a new pseudo-logic function.

DEFINITION 9.3

A n-ary k-valued pseudo-logic function $f(x_1, x_2, \cdots, x_n)$ is a mapping from Δ_k^n to $\mathbb{R}^k$.

Assume that $A = [a_{ij}] \in \mathbb{R}^{n\times m}$, $B = [b_{ij}] \in \mathbb{R}^{n\times m}$, then we can define:

$$A \odot B = [c_{ij}], \quad c_{ij} = a_{ij}b_{ij}, \quad i = 1, 2, \cdots, n; \; j = 1, 2, \cdots, m,$$

which is called the Hadamard product of matrices/vectors.

With Lemma 9.1 and the Hadamard product, we have the following result.

Lemma 9.2

Assume that $f(x_1, x_2, \cdots, x_n)$ is an n-ary k-valued pseudo-logic function. Then,

$$f(x_1, x_2, \cdots, x_n) = \gamma_0 + \sum_{k=1}^{m} \gamma_k \prod_{i \in A_k} x_i, \tag{9.16}$$

where $\gamma_i \in \mathbb{R}^k$, $i = 1, 2, \cdots, m$, $A_1, A_2, \cdots, A_m$ are non-empty subsets of $\{1, 2, \cdots, n\}$, and the product $\prod$ is the Hadamard product of vectors. ■

Proof. Notice that each component of $f(x_1, x_2, \cdots, x_n)$ is a pseudo-Boolean function. Applying Lemma 9.1 to each component of $f(x_1, x_2, \cdots, x_n)$ can yield (9.16). ▮

Now we are ready to study the coloring problem. Consider a graph $\mathscr{G} = \{\mathscr{V}, \mathscr{E}\}$ with $\mathscr{V} = \{v_1, v_2, \cdots, v_n\}$, and assume that its adjacency matrix is $A = [a_{ij}]$. Let a coloring mapping $\phi : \mathscr{V} \longmapsto N := \{c_1, \cdots, c_k\}$ be given, where $c_1, \cdots, c_k$ stand for k kinds of different colors. Here, ϕ is not necessarily surjective.

The coloring problem is to find a suitable coloring mapping ϕ such that for any $v_i, v_j \in \mathscr{V}$, if $(v_i, v_j) \in \mathscr{E}$, then $\phi(v_i) \neq \phi(v_j)$.

For each vertex $v_i \in \mathscr{V}$, assign a k-valued characteristic logical variable $x_i \in \Delta_k$ as follows:

$$x_i = \delta_k^j, \text{ if } \phi(v_i) = c_j \in N, \quad i = 1, 2, \cdots, n.$$

Then, we have the following results.

Theorem 9.4

Consider a graph $\mathscr{G}=\{\mathscr{V},\mathscr{E}\}$, and let a coloring mapping $\phi:\mathscr{V}\longmapsto N=\{c_1,\cdots,c_k\}$ be given. Then, the coloring problem is solvable with the given ϕ, if and only if the following n-ary k-valued pseudo-logic equation:

$$\sum_{i=1}^{n}\sum_{j\neq i}a_{ij}x_i\odot x_j=0_k \tag{9.17}$$

is solvable, where 0_k is the k-dimensional zero vector. ■

Proof. ($\Rightarrow$) If the coloring problem is solvable with the given mapping ϕ, then for any $v_i\neq v_j\in\mathscr{V}$, if $(v_i,v_j)\in\mathscr{E}$, then $\phi(v_i)\neq\phi(v_j)$. Thus, if $a_{ij}=1$, then $x_i\neq x_j$, which implies that $x_i\odot x_j=0_k$ or $a_{ij}x_i\odot x_j=0_k$. With this, it is easy to see that:

$$\sum_{i=1}^{n}\sum_{j\neq i}a_{ij}x_i\odot x_j=0_k$$

holds true. In other words, (9.17) is solvable.

($\Leftarrow$) Assume that the equation (9.17) has a solution $(x_1,\cdots,x_n)$. Since $a_{ij}\geqslant 0$ and $x_i\in\Delta_k$, this solution satisfies $a_{ij}x_i\odot x_j=0_k$, $i,j=1,2,\cdots,n$, $i\neq j$. Thus, if $a_{ij}=1$, then $x_i\odot x_j=0_k$, which implies that $x_i\neq x_j$ or $\phi(v_i)\neq\phi(v_j)$. Hence, ϕ is a solution to the coloring problem. ■

Theorem 9.5

Consider a graph $\mathscr{G}=\{\mathscr{V},\mathscr{E}\}$, and let a color set $N=\{c_1,\cdots,c_k\}$ be given. Then, the coloring problem of $\mathscr{G}$ is solvable with a mapping $\phi:\mathscr{V}\longmapsto N$, if and only if:

$$0_k\in\mathrm{Col}(M):=\Big\{\mathrm{Col}_1(M),\ \mathrm{Col}_2(M),\ \cdots,\ \mathrm{Col}_{k^n}(M)\Big\},$$

where,

$$M=\sum_{i=1}^{n}\sum_{j\neq i}a_{ij}M_{ij}^{H}, \tag{9.18}$$

and $M_{ij}^{H}\in\mathbb{R}^{k\times k^n}$, given in (9.19) below, is the structural matrix of the k-valued pseudo-logic function $x_i\odot x_j$. ■

Proof. First, we show that $f(x_1,x_2,\cdots,x_n):=\sum_{i=1}^{n}\sum_{j\neq i}a_{ij}x_i\odot x_j$ given in Theorem 9.4 can be expressed as:

$$f(x_1,x_2,\cdots,x_n)=M\ltimes_{i=1}^{n}x_i.$$

In fact, there exists a unique matrix $M_{ij}^H \in \mathbb{R}^{k\times k^n}$ such that:

$$f_{ij}(x_1,\cdots,x_n) = x_i \odot x_j = M_{ij}^H \ltimes_{i=1}^n x_i.$$

With this, we have:

$$f(x_1,x_2,\cdots,x_n) = \sum_{i=1}^n \sum_{j\neq i} a_{ij} f_{ij}(x_1,x_2,\cdots,x_n) = \sum_{i=1}^n \sum_{j\neq i} a_{ij} M_{ij}^H \ltimes_{i=1}^n x_i = M \ltimes_{i=1}^n x_i.$$

Thus, the equation (9.17) can be rewritten as:

$$M \ltimes_{i=1}^n x_i = 0_k. \tag{9.19}$$

Noticing that $\ltimes_{i=1}^n x_i \in \Delta_{k^n}$, the equation (9.19) is solvable if and only if 0_k is one of columns of M. Thus, the theorem follows from Theorem 9.4. ∎

The structural matrix M_{ij}^H in Theorem 9.5 can be calculated as follows: Set [5]:

$$\begin{cases} S_{1,k}^n = \delta_k[\underbrace{1\cdots 1}_{k^{n-1}}\ \underbrace{2\cdots 2}_{k^{n-1}}\cdots.\underbrace{k\cdots k}_{k^{n-1}}], \\ S_{2,k}^n = \delta_k[\underbrace{\underbrace{\underbrace{1\cdots 1}_{k^{n-2}}\ \underbrace{2\cdots 2}_{k^{n-2}}\cdots\underbrace{k\cdots k}_{k^{n-2}}}\ \cdots\ \underbrace{\underbrace{1\cdots 1}_{k^{n-2}}\ \underbrace{2\cdots 2}_{k^{n-2}}\cdots\underbrace{k\cdots k}_{k^{n-2}}}}_{k}], \\ \vdots \\ S_{n,k}^n = \delta_2[\underbrace{\underbrace{1\ 2\cdots k}\cdots\underbrace{1\ 2\cdots k}}_{k^{n-1}}], \end{cases} \tag{9.20}$$

then,

$$\begin{aligned} f_{ij}(x_1,\cdots,x_n) &= x_i \odot x_j = H_k x_i \ltimes x_j = H_k S_{i,k}^n(\ltimes_{i=1}^n x_i) S_{j,k}^n(\ltimes_{i=1}^n x_i) \\ &= H_k S_{i,k}^n [I_{k^n} \otimes S_{j,k}^n] M_{r,k^n} \ltimes_{i=1}^n x_i, \end{aligned}$$

where $M_{r,k^n} = \mathrm{Diag}\{\delta_{k^n}^1, \delta_{k^n}^2, \cdots, \delta_{k^n}^{k^n}\}$ is the power-reducing matrix, and $H_k = \mathrm{Diag}\{E_1,E_2,\cdots,E_k\}$ with $E_i = [0,\cdots,0,\underbrace{1}_{i-\text{th}},0,\cdots,0]_k = (\delta_k^i)^T$, $i = 1,2,\cdots,k$.

Therefore,

$$M_{ij}^H = H_k S_{i,k}^n [I_{k^n} \otimes S_{j,k}^n] M_{r,k^n}. \tag{9.21}$$

The proof of Theorem 9.5 suggests an algorithm to find out all the coloring schemes with no more than k colors for a given graph.

ALGORITHM 9.4

Assume that $\mathscr{G}$ is a graph with nodes $\mathscr{V} = \{v_1,v_2,\cdots,v_n\}$, and let a color set $N = \{c_1,c_2,\cdots,c_k\}$ be given. For each node v_i, we assign it a k-valued characteristic logical variable $x_i \in \Delta_k$. To find all the k-colorings of $\mathscr{G}$, that is, to find out all the color mappings such that the coloring problem of $\mathscr{G}$ is solved with no more than k different colors, we can do by the following steps:

1). Compute the matrix M given in (9.18).

2). Check whether $0_k \in \mathrm{Col}(M)$. If not, the coloring problem with the given color set has no solution, and stop the computation. Otherwise, label the columns which equal to 0_k and set:

$$K = \left\{ j \,\middle|\, \mathrm{col}_j(M) = 0_k \right\}.$$

3). For each index $j \in K$, let $\ltimes_{i=1}^{n} x_i = \delta_{k^n}^{j}$. Using (9.20), compute $x_i = S_{i,k}^{n} \ltimes_{i=1}^{n} x_i = S_{i,k}^{n} \delta_{k^n}^{j}$, $i = 1, 2, \cdots, n$. With the obtained solution $(x_1, \cdots, x_n)$, define:

$$\begin{aligned} S_{c_1}^{j} &:= \left\{ v_i \mid x_i = \delta_k^1,\ 1 \leqslant i \leqslant n \right\}, \\ S_{c_2}^{j} &:= \left\{ v_i \mid x_i = \delta_k^2,\ 1 \leqslant i \leqslant n \right\}, \\ &\ \ \vdots \\ S_{c_k}^{j} &:= \left\{ v_i \mid x_i = \delta_k^k,\ 1 \leqslant i \leqslant n \right\}, \end{aligned}$$

then a coloring scheme corresponding to index $j \in K$ is given as: All the vertices in $S_{c_i}^{j}$ are colored by color c_i, $i = 1, 2, \cdots, k$.

4). All the coloring schemes are given as follows:

$$\phi_j(v_i) = c_s, \quad \text{if } v_i \in S_{c_s}^{j}, \quad i = 1, 2, \cdots, n;\ s = 1, 2, \cdots, k;\ j \in K.$$

That is,

$$\begin{array}{ccc} \text{Color} & & \text{Vertices} \\ c_1 & \longrightarrow & S_{c_1}^{j} \\ c_2 & \longrightarrow & S_{c_2}^{j} \\ \vdots & & \vdots \\ c_k & \longrightarrow & S_{c_k}^{j}, \end{array}$$

for all $j \in K$. The number of all the coloring schemes is $|K|$.

With $x_i \in \Delta_k$, $i = 1, 2, \cdots, n$, the product $\ltimes_{i=1}^{n} x_i$ is well defined. Conversely, with the product $\ltimes_{i=1}^{n} x_i$, each x_i can be uniquely determined (return to its original value) by the formula $x_i = S_{i,k}^{n} \ltimes_{i=1}^{n} x_i$ [3].

With Theorems 9.4 and 9.5, we have the following result on $S_{c_i}^{j}$, $i = 1, 2, \cdots, k$, $j \in K$ defined in Algorithm 9.4.

PROPOSITION 9.1

Assume that $0_k \in \mathrm{Col}(M)$ holds true. Then,

(1). Each set $S^j_{c_i}$ defined in Algorithm 9.4 is an internally stable set of $\mathscr{G}$, and,
(2). For each index $j \in K$, $\{S^j_{c_1}, S^j_{c_2}, \cdots, S^j_{c_k}\}$ is a coloring partition of $\mathscr{G}$.

Proof. (1). If $S^j_{c_i} = \emptyset$, it is naturally an internally stable set. Otherwise, we choose $v_s, v_t \in S^j_{c_i}$. If $a_{st} = 1$, then it is easy to know from Theorem 9.4 or 9.5 that $a_{st}x_s \odot x_t = 0$, which implies that $x_s \odot x_t = 0$. Noticing that $x_s, x_t \in \Delta_k$, we obtain $x_s \neq x_t$, which means that v_s, v_t belong to two different sets of the form $S^j_{c_i}$. This is a contradiction with $v_s, v_t \in S^j_{c_i}$, and thus $a_{st} = 0$. Therefore, $S^j_{c_i}$ is an internally stable set of $\mathscr{G}$.

(2). Along with (1), to show that $\{S^j_{c_1}, S^j_{c_2}, \cdots, S^j_{c_k}\}$ is a coloring partition, we only need to prove that:
(a) $S^j_{c_i} \cap S^j_{c_s} = \emptyset$, $i \neq s$, and (b) $\bigcup_{i=1}^k S^j_{c_i} = \mathscr{V}$. In fact, if $S^j_{c_i} \cap S^j_{c_s} \neq \emptyset$ $(i \neq s)$, then there exists at least one vertex v_t such that $v_t \in S^j_{c_i}$ and $v_t \in S^j_{c_s}$ hold simultaneously. Thus, from the construction of such sets, $\delta^i_k = x_t = \delta^s_k$, which implies that $i = s$. This is a contradiction, and then (a) holds true. Now, we show that (b) holds, too. Fixing $j \in K$, from the construction of $S^j_{c_i}$, $i = 1, 2, \cdots, k$, it is easy to see that $\bigcup_{i=1}^k S^j_{c_i} \subseteq \mathscr{V}$. On the other hand, for any $v_s \in \mathscr{V}$, its k-valued characteristic logical variable $x_s \in \Delta_k$. Since the coloring problem is solvable, we can let $x_s := \delta^i_k$, from which and the construction of $S^j_{c_i}$ we have $v_s \in S^j_{c_i} \subseteq \bigcup_{i=1}^k S^j_{c_i}$. This implies that $\mathscr{V} \subseteq \bigcup_{i=1}^k S^j_{c_i}$. Thus, (b) follows from the above analysis. ∎

Noticing that ϕ in Theorem 9.4 or 9.5 is not necessarily surjective, it is easy to see that the coloring schemes ϕ_j, $j \in K$ obtained by Algorithm 9.4 contains all the colorings with the number of colors being no more than k. Thus, if the above coloring problem is solvable, the coloring schemes obtained by Algorithm 9.4 contains the minimum coloring mapping, in other words, $\{S^j_{c_1}, S^j_{c_2}, \cdots, S^j_{c_k}\}$ (all $j \in K$) contain the minimum coloring partition.

When the above coloring problem is solvable, the minimum coloring partition can be given as follows. Let:

$$N_j \quad = \quad \left\{ i \mid S^j_{c_i} = \emptyset, \ 1 \leqslant i \leqslant k \right\}, \quad |N_{j_0}| = \max_{j \in K} \left\{ |N_j| \right\},$$

then the minimum coloring partition is:

$$\left\{ S^{j_0}_{c_1}, S^{j_0}_{c_2}, \cdots, S^{j_0}_{c_k} \right\} = \left\{ S^{j_0}_{c_1}, S^{j_0}_{c_2}, \cdots, S^{j_0}_{c_k} \right\} \setminus \left\{ S^{j_0}_{c_i} \mid S^{j_0}_{c_i} = \emptyset \right\},$$

and the chromatic number of $\mathscr{G}$ is:

$$\gamma(\mathscr{G}) = k - |N_{j_0}|.$$

In the next, we use the above results to investigate the group consensus problem of multi-agent systems, and present a new control protocol design procedure for a class of multi-agent systems.

Consider the following multi-agent system:

$$\begin{cases} \dot{x}_i = v_i, \\ \dot{v}_i = u_i, \quad i = 1, 2, \cdots, n, \end{cases} \tag{9.22}$$

where $x_i \in \mathbb{R}$, $v_i \in \mathbb{R}$ and $u_i \in \mathbb{R}$ are the position, velocity and control input of agent i, respectively.

Assume that the information topology of the system (9.22) is described by a directed or undirected graph $\mathscr{G} = \{\mathscr{V}, \mathscr{E}\}$, where $\mathscr{V} = \{1, 2, \cdots, n\}$ consists of all the agents. It is noted here that $\mathscr{G}$ is not necessarily a balanced graph when $\mathscr{G}$ is directed.

Applying Algorithm 9.4 with $k \, (\geqslant \gamma(\mathscr{G}))$ colors to the graph $\mathscr{G}$, we can obtain a minimum coloring partition of $\mathscr{V}$, denoted by:

$$S_1, \; S_2, \; \cdots, \; S_\gamma,$$

where $\gamma = \gamma(\mathscr{G})$ is the chromatic number of $\mathscr{G}$. From Proposition 9.1, each S_i is an internally stable set, based on which the graph $\mathscr{G}$ can be equivalently changed as Figure 9.1.

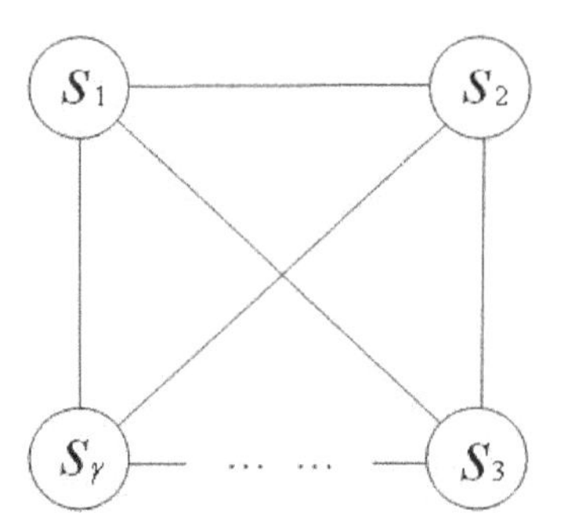

Figure 9.1 A coloring partition of $\mathscr{V}$

For each agent i, define:

$$N_i^F := S_j, \;\; \text{if } i \in S_j, \;\; \text{and} \;\; N_i^E := \{j \mid (j, i) \in \mathscr{E}, \; i \neq j\},$$

which are called the "Friends" and "Enemies" neighborhoods of agent i, respectively. Obviously, the agents belonging to the same internally stable set have the same "Friends" neighborhood, i.e., the internally stable set itself; and for agent $i \in S_j = N_i^F$, its enemies disperse in S_k, $k \neq j$.

The objective of this part is as follows: Design a distributed protocol based on the information from enemies such that all the agents are attracted to friends and avoid enemies' and maintain the given distance from enemies. This can be described as the *group consensus problem*. Given a required distance d_{ij} for each pair (S_i, S_j), $1 \leqslant i < j \leqslant \gamma$, design a linear control protocol:

$$u_i = u_i(x_i, x_{j_1}, \cdots, x_{j_{i_0}}),\ \ j_k \in N_i^E,\ \ k = 1, 2, \cdots, j_{i_0},\ \ i = 1, 2, \cdots, n, \tag{9.23}$$

such that for $\forall s \in S_i$ and $\forall k \in S_j$ $(i \leqslant j)$:

$$\lim_{t \to \infty} |x_s(t) - x_k(t)| = \begin{cases} d_{ij}, & i \neq j; \\ 0, & i = j. \end{cases} \quad \text{and} \quad \lim_{t \to \infty} |v_s(t) - v_k(t)| = 0.$$

Consider the distance d_{ij}. Without loss of generality, we suppose:

$$d_{12} \leqslant d_{13} \leqslant \cdots \leqslant d_{1\gamma}.$$

Then, we have the following results on the group consensus problem.

Theorem 9.6

Consider the multi-agent system (9.22) with its topology of graph $\mathscr{G}$. Assume that the required distances satisfy:

$$d_{ij} = d_{1j} - d_{1i},\ \ 1 \leqslant i < j \leqslant \gamma, \tag{9.24}$$

where $d_{11} := 0$. Then, the group consensus problem is solvable if and only if $\mathscr{G}$ has a directed spanning tree (or $\mathscr{G}$ is connected when $\mathscr{G}$ is an undirected graph). ■

Proof. For each agent $i \in S_k$, let,

$$y_i = x_i - d_{1k},\ \ i = 1, 2, \cdots, |S_k|,\ \ k = 1, 2, \cdots, \gamma, \tag{9.25}$$

then the system (9.22) can be expressed as:

$$\begin{cases} \dot{y}_i = v_i, \\ \dot{v}_i = u_i, \ \ i = 1, 2, \cdots, n. \end{cases} \tag{9.26}$$

On the other hand, $\forall s \in S_i$ and $\forall k \in S_j$, we obtain from (9.24) and (9.25) that:

$$\begin{aligned} \lim_{t \to \infty} |y_s - y_k| &= \lim_{t \to \infty} |x_s - d_{1i} - (x_k - d_{1j})| \\ &= \lim_{t \to \infty} |x_s - x_k - (d_{1i} - d_{1j})| \\ &= \begin{cases} \lim_{t \to \infty} |-(x_k - x_s) + d_{ij}|, & i < j; \\ \lim_{t \to \infty} |(x_s - x_k) - d_{ji}|, & i \geqslant j. \end{cases} \end{aligned}$$

where we define $d_{ii} := 0,\ 1 < i \leqslant \gamma$. Thus,

$$\lim_{t \to \infty} |y_s - y_k| = 0 \Leftrightarrow \lim_{t \to \infty} |x_s(t) - x_k(t)| = \begin{cases} d_{ij}, & i < j; \\ 0, & i = j. \end{cases}$$

which implies that the group consensus problem is equivalent to the state consensus of the system (9.26). From [4], the group consensus problem is solvable if and only if $\mathscr{G}$ has a directed spanning tree (or $\mathscr{G}$ is connected when $\mathscr{G}$ is an undirected graph). ∎

The condition (9.24) is a natural one for the position requirement of S_i. If the dimension of agents' positions is large than one, the condition (9.24) will be in the form of inequalities. The results of this section can be generalized to high-order multi-agent systems.

Theorem 9.7

Consider the multi-agent system (9.22) with its topology of graph $\mathscr{G}$. Assume that $\mathscr{G}$ has a directed spanning tree (or $\mathscr{G}$ is a connected undirected graph) and the required distances satisfy (9.24). Then, the group consensus protocol (9.23) can be designed as:

$$u_i = \sum_{j\in N_i^E} a(v_j - v_i) + \sum_{\substack{s=1 \\ s\neq k}}^{\gamma} \sum_{j\in N_i^E\cap S_s} b(x_j - d_{1s} - x_i + d_{1k}), \ \ \forall i \in S_k, \ \ k = 1,2,\cdots,\gamma, \tag{9.27}$$

where $a,b > 0$ are real numbers satisfying $a^2 > c_0 b$, and $c_0 > 0$ is a sufficiently large real number. ∎

Proof. Applying (9.25) to each agent, the multi-agent system can be expressed as (9.26). On the other hand, by (9.25), the control (9.27) can be rewritten as:

$$\begin{aligned} u_i &= \sum_{j\in N_i^E} a(v_j - v_i) + \sum_{\substack{s=1 \\ s\neq k}}^{\gamma} \sum_{j\in N_i^E\cap S_s} b(y_j - y_i) \\ &= \sum_{j\in N_i^E} \Big[a(v_j - v_i) + b(y_j - y_i)\Big], \quad i \in S_k, \ \ k = 1,2,\cdots,\gamma. \end{aligned} \tag{9.28}$$

From [4], the system (9.26) can reach the state consensus under the control (9.28), which implies, in the proof of Theorem 9.6, the group consensus of the system (9.22) is solved by the control protocol (9.27). Thus, the proof is completed. ∎

Finally, we give three examples to illustrate the effectiveness of the results/algorithms obtained in this section.

Example 9.1:

Consider the graph $G = \{\mathscr{V}, \mathscr{E}\}$ shown in Figure 9.2. We use Algorithm 9.2 to find all the absolutely maximum stable sets of the graph.

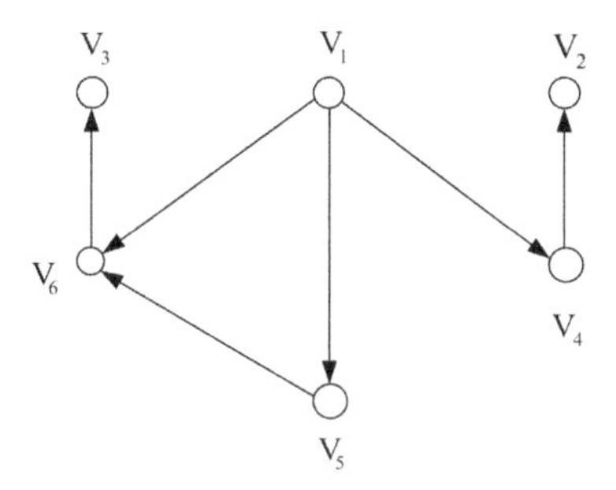

Figure 9.2 A directed graph

The adjacency matrix of this graph is as follows (see (9.1)):

$$A = \begin{bmatrix} 0 & 0 & 0 & 0 & 0 & 0 \\ 0 & 0 & 0 & 1 & 0 & 0 \\ 0 & 0 & 0 & 0 & 0 & 1 \\ 1 & 0 & 0 & 0 & 0 & 0 \\ 1 & 0 & 0 & 0 & 0 & 0 \\ 1 & 0 & 0 & 0 & 1 & 0 \end{bmatrix}.$$

For each node i, we assign a characteristic logical variable $x_i \in \mathscr{D}$ and let $y_i = [x_i, \bar{x}_i]^T$, $i = 1, 2, \cdots, 6$. Since there are many '0' in A, we only need to calculate the structural matrices of x_i and $a_{ij}x_ix_j$ with $a_{ij} = 1$. By (9.3) and (9.10), and using the MATLAB toolbox provided by D. Cheng and his co-workers, we easily obtain:

$$\begin{aligned}
M_{24} = \delta_2[\ & 1\,1\,1\,1\,2\,2\,2\,2\,1\,1\,1\,1\,2\,2\,2\,2\,2\,2\,2\,2\,2\,2\,2\,2\,2\,2\,2\,2\,2\,2\,2\,2 \\
& 1\,1\,1\,1\,2\,2\,2\,2\,1\,1\,1\,1\,2\,2\,2\,2\,2\,2\,2\,2\,2\,2\,2\,2\,2\,2\,2\,2\,2\,2\,2\,2],
\end{aligned}$$

$$\begin{aligned}
M_{36} = \delta_2[\ & 1\,2\,1\,2\,1\,2\,1\,2\,2\,2\,2\,2\,2\,2\,2\,2\,1\,2\,1\,2\,1\,2\,1\,2\,2\,2\,2\,2\,2\,2\,2\,2 \\
& 1\,2\,1\,2\,1\,2\,1\,2\,2\,2\,2\,2\,2\,2\,2\,2\,1\,2\,1\,2\,1\,2\,1\,2\,2\,2\,2\,2\,2\,2\,2\,2],
\end{aligned}$$

$$\begin{aligned}
M_{41} = \delta_2[\ & 1\,1\,1\,1\,2\,2\,2\,2\,1\,1\,1\,1\,2\,2\,2\,2\,1\,1\,1\,1\,2\,2\,2\,2\,1\,1\,1\,1\,2\,2\,2\,2 \\
& 2\,2\,2\,2\,2\,2\,2\,2\,2\,2\,2\,2\,2\,2\,2\,2\,2\,2\,2\,2\,2\,2\,2\,2\,2\,2\,2\,2\,2\,2\,2\,2],
\end{aligned}$$

$$\begin{aligned}
M_{51} = \delta_2[\ & 1\,1\,2\,2\,1\,1\,2\,2\,1\,1\,2\,2\,1\,1\,2\,2\,1\,1\,2\,2\,1\,1\,2\,2\,1\,1\,2\,2\,1\,1\,2\,2 \\
& 2\,2\,2\,2\,2\,2\,2\,2\,2\,2\,2\,2\,2\,2\,2\,2\,2\,2\,2\,2\,2\,2\,2\,2\,2\,2\,2\,2\,2\,2\,2\,2],
\end{aligned}$$

$$\begin{aligned}
M_{61} = \delta_2[\ & 1\,2\,1\,2\,1\,2\,1\,2\,1\,2\,1\,2\,1\,2\,1\,2\,1\,2\,1\,2\,1\,2\,1\,2\,1\,2\,1\,2\,1\,2\,1\,2 \\
& 2\,2\,2\,2\,2\,2\,2\,2\,2\,2\,2\,2\,2\,2\,2\,2\,2\,2\,2\,2\,2\,2\,2\,2\,2\,2\,2\,2\,2\,2\,2\,2],
\end{aligned}$$

$$\begin{aligned}
M_{65} = \delta_2[\ & 1\,2\,2\,2\,1\,2\,2\,2\,1\,2\,2\,2\,1\,2\,2\,2\,1\,2\,2\,2\,1\,2\,2\,2\,1\,2\,2\,2\,1\,2\,2\,2 \\
& 1\,2\,2\,2\,1\,2\,2\,2\,1\,2\,2\,2\,1\,2\,2\,2\,1\,2\,2\,2\,1\,2\,2\,2\,1\,2\,2\,2\,1\,2\,2\,2];
\end{aligned}$$

$$
\begin{aligned}
M_1 = \delta_2[\;& 1\ 1\ 1\ 1\ 1\ 1\ 1\ 1\ 1\ 1\ 1\ 1\ 1\ 1\ 1\ 1\ 1\ 1\ 1\ 1\ 1\ 1\ 1\ 1\ 1\ 1\ 1\ 1\ 1\ 1\ 1\ 1\\
& 2\ 2\ 2\ 2\ 2\ 2\ 2\ 2\ 2\ 2\ 2\ 2\ 2\ 2\ 2\ 2\ 2\ 2\ 2\ 2\ 2\ 2\ 2\ 2\ 2\ 2\ 2\ 2\ 2\ 2\ 2\ 2],\\
M_2 = \delta_2[\;& 1\ 1\ 1\ 1\ 1\ 1\ 1\ 1\ 1\ 1\ 1\ 1\ 1\ 1\ 1\ 1\ 2\ 2\ 2\ 2\ 2\ 2\ 2\ 2\ 2\ 2\ 2\ 2\ 2\ 2\ 2\ 2\\
& 1\ 1\ 1\ 1\ 1\ 1\ 1\ 1\ 1\ 1\ 1\ 1\ 1\ 1\ 1\ 1\ 2\ 2\ 2\ 2\ 2\ 2\ 2\ 2\ 2\ 2\ 2\ 2\ 2\ 2\ 2\ 2],\\
M_3 = \delta_2[\;& 1\ 1\ 1\ 1\ 1\ 1\ 1\ 1\ 2\ 2\ 2\ 2\ 2\ 2\ 2\ 2\ 1\ 1\ 1\ 1\ 1\ 1\ 1\ 1\ 2\ 2\ 2\ 2\ 2\ 2\ 2\ 2\\
& 1\ 1\ 1\ 1\ 1\ 1\ 1\ 1\ 2\ 2\ 2\ 2\ 2\ 2\ 2\ 2\ 1\ 1\ 1\ 1\ 1\ 1\ 1\ 1\ 2\ 2\ 2\ 2\ 2\ 2\ 2\ 2],\\
M_4 = \delta_2[\;& 1\ 1\ 1\ 1\ 2\ 2\ 2\ 2\ 1\ 1\ 1\ 1\ 2\ 2\ 2\ 2\ 1\ 1\ 1\ 1\ 2\ 2\ 2\ 2\ 1\ 1\ 1\ 1\ 2\ 2\ 2\ 2\\
& 1\ 1\ 1\ 1\ 2\ 2\ 2\ 2\ 1\ 1\ 1\ 1\ 2\ 2\ 2\ 2\ 1\ 1\ 1\ 1\ 2\ 2\ 2\ 2\ 1\ 1\ 1\ 1\ 2\ 2\ 2\ 2],\\
M_5 = \delta_2[\;& 1\ 1\ 2\ 2\ 1\ 1\ 2\ 2\ 1\ 1\ 2\ 2\ 1\ 1\ 2\ 2\ 1\ 1\ 2\ 2\ 1\ 1\ 2\ 2\ 1\ 1\ 2\ 2\ 1\ 1\ 2\ 2\\
& 1\ 1\ 2\ 2\ 1\ 1\ 2\ 2\ 1\ 1\ 2\ 2\ 1\ 1\ 2\ 2\ 1\ 1\ 2\ 2\ 1\ 1\ 2\ 2\ 1\ 1\ 2\ 2\ 1\ 1\ 2\ 2],\\
M_6 = \delta_2[\;& 1\ 2\ 1\ 2\ 1\ 2\ 1\ 2\ 1\ 2\ 1\ 2\ 1\ 2\ 1\ 2\ 1\ 2\ 1\ 2\ 1\ 2\ 1\ 2\ 1\ 2\ 1\ 2\ 1\ 2\ 1\ 2\\
& 1\ 2\ 1\ 2\ 1\ 2\ 1\ 2\ 1\ 2\ 1\ 2\ 1\ 2\ 1\ 2\ 1\ 2\ 1\ 2\ 1\ 2\ 1\ 2\ 1\ 2\ 1\ 2\ 1\ 2\ 1\ 2].
\end{aligned}
$$

Thus,

$$
\begin{aligned}
[b_1\ b_2\ \cdots\ b_{64}] &= J_1\widetilde{M} = J_1\left[\sum_{i=1}^{6} M_i - 7(M_{24}+M_{36}+M_{41}+M_{51}+M_{61}+M_{65})\right]\\
&= [\begin{array}{rrrrrrrr}
-36 & -16 & -23 & -10 & -23 & -3 & -10 & 3\\
-30 & -17 & -17 & -11 & -17 & -4 & -4 & 2\\
-30 & -10 & -17 & -4 & -24 & -4 & -11 & 2\\
-24 & -11 & -11 & -5 & -18 & -5 & -5 & 1\\
-16 & -3 & -10 & -4 & -10 & 3 & -4 & 2\\
-10 & -4 & -4 & -5 & -4 & 2 & 2 & 1\\
-10 & 3 & -4 & 2 & -11 & 2 & -5 & 1\\
-4 & 2 & 2 & 1 & -5 & 1 & 1 & 0
\end{array}],
\end{aligned}
$$

from which it is easy to see that:

$$\max_{1\le i\le 64}\{b_i\} = 3$$

and the corresponding column index set is:

$$K = \{i_k \mid b_{i_k} = 3\} = \{8,\ 38,\ 50\}.$$

For each index $i_k \in K$, let $\ltimes_{i=1}^{6} y_i = \delta_{2^6}^{i_k}$. By computing $y_i = S_i^6 \delta_{2^6}^{i_k}$, $i = 1, 2, \cdots, 6$ (see (9.6)), we have:

$$
\begin{aligned}
i_1 &= 8 \sim (x_1,x_2,x_3,x_4,x_5,x_6) = (1,\ 1,\ 1,\ 0,\ 0,\ 0),\\
i_2 &= 38 \sim (x_1,x_2,x_3,x_4,x_5,x_6) = (0,\ 1,\ 1,\ 0,\ 1,\ 0),\\
i_3 &= 50 \sim (x_1,x_2,x_3,x_4,x_5,x_6) = (0,\ 0,\ 1,\ 1,\ 1,\ 0).
\end{aligned}
$$

Thus, all the absolutely maximum stable sets are as follows:

$$S(i_1) = \{v_1, v_2, v_3\}, \quad S(i_2) = \{v_2, v_3, v_5\}, \quad S(i_3) = \{v_3, v_4, v_5\}.$$

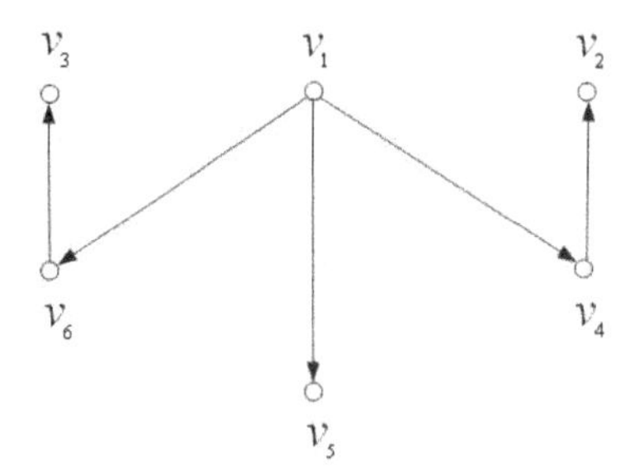

Figure 9.3 A directed graph

Example 9.2:

Consider the graph $G = \{\mathscr{V}, \mathscr{E}\}$ shown in Figure 9.3. Letting a two-color set $N = \{c_1 = \text{Red},\ c_2 = \text{Blue}\}$ be given, we use Algorithm 9.4 to find out all the coloring schemes for $\mathscr{G}$.

For each node i, we assign a characteristic logical variable $x_i \in \Delta$, $i = 1, 2, \cdots, 6$. The adjacency matrix of this graph is as follows:

$$A = \begin{bmatrix} 0 & 0 & 0 & 0 & 0 & 0 \\ 0 & 0 & 0 & 1 & 0 & 0 \\ 0 & 0 & 0 & 0 & 0 & 1 \\ 1 & 0 & 0 & 0 & 0 & 0 \\ 1 & 0 & 0 & 0 & 0 & 0 \\ 1 & 0 & 0 & 0 & 0 & 0 \end{bmatrix},$$

and by (9.21), (9.18) and the MATLAB toolbox, it is easy to obtain:

$$\begin{array}{rcl} M & = & \sum_{i=1}^{6} \sum_{j \neq i} a_{ij} M_{ij}^{H} \\ & = & \left[\begin{array}{l} 5\,3\,4\,2\,3\,1\,2\,0\,4\,3\,3\,2\,2\,1\,1\,0\,4\,2\,3\,1\,3\,1\,2\,0\,3\,2\,2\,1\,2\,1\,1\,0 \\ 0\,0\,0\,0\,0\,0\,0\,0\,0\,1\,0\,1\,0\,1\,0\,1\,0\,0\,0\,0\,1\,1\,1\,1\,0\,1\,0\,1\,1\,2\,1\,2 \end{array} \right. \\ & & \left. \begin{array}{l} 2\,1\,2\,1\,1\,0\,1\,0\,1\,1\,1\,1\,0\,0\,0\,0\,1\,0\,1\,0\,1\,0\,1\,0\,0\,0\,0\,0\,0\,0\,0\,0 \\ 0\,1\,1\,2\,1\,2\,2\,3\,0\,2\,1\,3\,1\,3\,2\,4\,0\,1\,1\,2\,2\,3\,3\,4\,0\,2\,1\,3\,2\,4\,3\,5 \end{array} \right]. \end{array}$$

It is observed that:

$$\mathrm{col}_8(M) = \mathrm{col}_{57}(M) = 0_2,$$

and the corresponding column index set is:

$$K = \{8,\ 57\}.$$

For each index $j \in K$, let $\ltimes_{i=1}^{6} x_i = \delta_{2^6}^{j}$. By computing $x_i = S_{i,2}^{6}\delta_{2^6}^{j}$, $i = 1,2,\cdots,6$ (see (9.20)), we obtain:

$$\begin{aligned}\delta_{2^6}^{8} &\sim [x_1,x_2,x_3,x_4,x_5,x_6] = \begin{bmatrix} 1 & 1 & 1 & 0 & 0 & 0 \\ 0 & 0 & 0 & 1 & 1 & 1 \end{bmatrix},\\ \delta_{2^6}^{57} &\sim [x_1,x_2,x_3,x_4,x_5,x_6] = \begin{bmatrix} 0 & 0 & 0 & 1 & 1 & 1 \\ 1 & 1 & 1 & 0 & 0 & 0 \end{bmatrix},\end{aligned}$$

from which we obtain the following two coloring schemes:

$$S_{c_1}^{8} = \{v_1,\ v_2,\ v_3\}\ (\text{Red}),\quad S_{c_2}^{8} = \{v_4,\ v_5,\ v_6\}\ (\text{Blue})$$

and,

$$S_{c_1}^{57} = \{v_4,\ v_5,\ v_6\}\ (\text{Red}),\quad S_{c_2}^{57} = \{v_1,\ v_2,\ v_3\}\ (\text{Blue}).$$

Example 9.3:

Consider the following multi-agent system:

$$\begin{cases} \dot{x}_i = v_i, \\ \dot{v}_i = u_i, \quad i = 1,2,\cdots,6, \end{cases}$$

where $x_i \in \mathbb{R}$, $v_i \in \mathbb{R}$ and $u_i \in \mathbb{R}$ are the position, velocity and control input of agent i, respectively.

Assume that the information topology of the system is given by the graph $\mathscr{G}$ shown in Figure 9.2. From Example 9.2, we know that the minimum coloring partition of $\mathscr{G}$ is as follows: $S_1 = \{1,2,3\}$, $S_2 = \{4,5,6\}$. Given a required distance $d = 2$ between the two sets, for each agent i we design a control u_i such that the whole system reaches group consensus, that is, for all $i,j = 1,2,\cdots,6$,

$$\lim_{t\to\infty} |x_i(t) - x_j(t)| = \begin{cases} d, & i \in S_1 \text{ and } j \in S_2; \\ 0, & \text{both } i,j \in S_1 \text{ or both } i,j \in S_2, \end{cases}$$

and,

$$\lim_{t\to\infty} |v_i(t) - v_j(t)| = 0.$$

It is easy to check that all the conditions of Theorem 9.6 are satisfied, and thus the group consensus problem is solvable. By Theorem 9.7, the desired control protocol is designed as:

$$\begin{cases} u_1 = 0, \\ u_2 = a(v_4 - v_2) + b(x_4 - x_2 - d), \\ u_3 = a(v_6 - v_3) + b(x_6 - x_3 - d), \\ u_4 = a(v_1 - v_4) + b(x_1 - x_4 + d), \\ u_5 = a(v_1 - v_5) + b(x_1 - x_5 + d), \\ u_6 = a(v_1 - v_6) + b(x_1 - x_6 + d), \end{cases} \tag{9.29}$$

where $a,b > 0$ are real numbers, and a is sufficiently large.

To show the effectiveness of the control (9.29), we carry out some simulation results with the following choices. Initial Condition: $[x_1(0), x_2(0), x_3(0), x_4(0), x_5(0), x_6(0)] = [1, -1, 5, -2, 3, 4]$ and $[v_1(0), v_2(0), v_3(0), v_4(0), v_5(0), v_6(0)] = [0, 0.5, 1, 0.5, 0.2, 0.5]$; Parameters: $a = 4$ and $b = 1$. The simulation results are shown in Figures 5.3 and 5.4, which are the responses of the agents' positions and velocities, respectively.

It is observed from Figures 9.4 and 9.5 that the group consensus with the given distance between the two groups is reached under the protocol (9.29). Simulation shows that the control design method given in this paper is very effective.

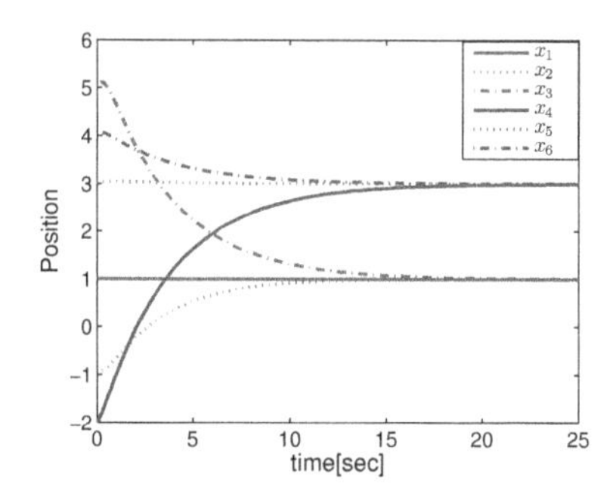

Figure 9.4 Response of the positions

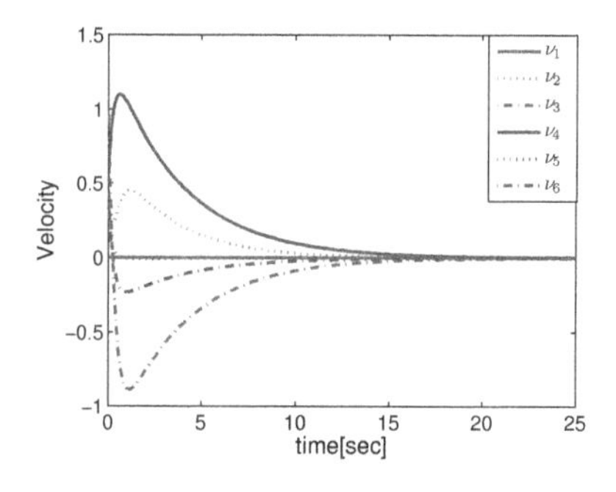

Figure 9.5 Responses of the velocities

9.3 ROBUST GRAPH COLORING PROBLEM

This section investigates the robust graph coloring problem, which is derived from coloring problem. Thus, firstly, we recall the concept of the coloring problem.

The coloring problem: Given a graph $\mu G = \{\mu V, E\}$ with $\mu V = \{v_1, v_2, \ldots, v_n\}$, let $\phi : \mu V \longmapsto N := \{c_1, \ldots, c_k\}$ be a mapping, where $c_1, \ldots, c_k$ stand for k kinds of different colors. The coloring problem is to find a suitable color mapping ϕ such that for any $v_i, v_j \in \mu V$, if $(v_i, v_j) \in \mu E$, then $\phi(v_i) \neq \phi(v_j)$.

For each vertex $v_i \in \mu V$, assign a k-valued characteristic logical variable $x_i \in \Delta_k$ as follows [6]:

$$x_i = \delta_k^j, \ if \ \phi(v_i) = c_j \in N, \ i = 1,2,\ldots,n.$$

Then, the following results were obtained in [6].

Lemma 9.3: see [6]

Consider a graph $\mu G = \{\mu V, \mu E\}$, and let a color mapping $\phi : \mu V \longmapsto N = \{c_1,\ldots,c_k\}$ be given. Then the coloring problem is solvable with the given ϕ, if and only if the following n-ary k-valued pseudo-logic equation:

$$\sum_{i=1}^{n} \sum_{j \neq i} a_{ij} x_i \odot x_j = 0_k$$

is solvable, where 0_k is the k-dimensional zero vector, and $\odot$ is the Hadamard product of matrices/vectors. ■

Lemma 9.4: see [6]

Consider a graph $\mu G = \{\mu V, \mu E\}$, and let a color set $N = \{c_1,\ldots,c_k\}$ be given. Then the coloring problem of μG is solvable with a mapping $\phi : \mu V \longmapsto N$, if and only if:

$$0_k \in Col(M) : \{Col_1(M), Col_2(M), \ldots, Col_{k^n}(M)\},$$

where $M = \sum_{i=1}^{n} \sum_{j \neq i} a_{ij} M_{ij}^H$, and M_{ij}^H is the structural matrix of the k-valued pseudo-logic function $x_i \odot x_j$. ■

In this section, we investigate the robust k-coloring problem by the semi-tensor product method, and present the main results of this paper.

Consider a simple graph $\mu G = (\mu V, \mu E)$ with n vertices $\mu V = \{v_1, v_2, \ldots, v_n\}$, and assume that its adjacency matrix is $A = [a_{ij}]$. Let $\mu \bar{G} = (\mu V, \mu \bar{E})$ be the complementary graph of μG, and for each complementary edge $(v_i, v_j) \in \mu \bar{E}$, we assign a positive number p_{ij} to represent the penalty for v_i and v_j. Let $P = \{p_{ij}\}$ be the penalty set, and $\phi : \mu V \mapsto \mu N := \{c_1,\ldots,c_k\}$ be a k-coloring mapping, where $c_1,\ldots,c_k$ stand for k kinds of different colors. The rigidity level of the coloring mapping ϕ is defined as:

$$R(\phi) = \sum_{\substack{(v_i,v_j) \in \mu \bar{E} \\ \phi(v_i) = \phi(v_j)}} p_{ij}.$$

The robust graph coloring problem (RGCP) is to find a suitable k-coloring mapping ϕ such that the k-coloring problem is not only solvable with ϕ but also,

$R(\phi)$ is minimized, that is,

$$\min R(\phi) = \sum_{\substack{(v_i,v_j)\in\mu\bar{E},\\ \phi(v_i)=\phi(v_j)}} p_{ij}.$$

Note that:

1. If the complementary edge $(v_i,v_j) \in \mu\bar{E}$, whose endpoints share the same color, is added to the graph μG, then the given coloring mapping ϕ will be invalid. In this case, $p_{ij} > 0$ will sign the invalidness of $(v_i,v_j) \in \mu\bar{E}$, that is, the complementary edge (v_i,v_j) cannot be added for the given coloring mapping.
2. We can assign different penalty for each complementary edge according to different application and physical meaning in practice.
3. The rigidity level is a measurement of coloring robustness of a given coloring mapping ϕ. Obviously, the lower rigidity level represents greater robustness of the coloring.

In order to investigate the RGCP, we define a new pseudo-logic function first.

DEFINITION 9.4

A n-ary k-valued pseudo-logic function $g(x_1,x_2,\ldots,x_n)$ is a mapping from Δ_k^n to $\mathbb{R}$, where $\Delta_k^n : \underbrace{\Delta_k \times \Delta_k \times \ldots \times \Delta_k}_{n}$ and $x_i \in \Delta_k,\ \ i = 1,2,\cdots,n$.

A point $(x_1^*,x_2^*,\ldots,x_n^*) \in \Delta_k^n$ is said to be a global minimum point of g, if $g(x_1^*,x_2^*,\ldots,x_n^*) \le g(x_1,x_2,\ldots,x_n)$ for any $(x_1,x_2,\ldots,x_n) \in \Delta_k^n$, and $g(x_1^*,x_2^*,\ldots,x_n^*)$ is called the global minimum.

For each vertex $v_i \in \mu V$, we assign a k-valued characteristic logical variable $x_i \in \Delta_k$ as follows:

$$x_i = \delta_k^j,\ \ if\ \ \phi(v_i) = c_j \in \mu N,\ i = 1,2,\ldots,n.$$

Let $V_\phi := [x_1,x_2,\ldots,x_n]$, and define:

$$x_i \star x_j = \begin{cases} 1, & x_i = x_j, \\ 0, & \text{otherwise.} \end{cases}$$

Then, the rigidity level of the coloring mapping ϕ can be expressed as:

$$R(x_1,\ldots,x_n) = \sum_{i=1}^{n}\sum_{j\neq i}(1-a_{ij})p_{ij}x_i \star x_j, \tag{9.30}$$

which is a n-ary k-valued pseudo-logic function.

Moreover, by Lemma 9.3 and the above analysis, it is easy to see that the robust coloring problem is equivalent to the following constrained optimization problem:

$$\min R(\phi) = \sum_{i=1}^{n} \sum_{j \neq i} (1 - a_{ij}) p_{ij} x_i \star x_j,$$

$$s.t. \quad \sum_{i=1}^{n} \sum_{j \neq i} a_{ij} x_i \odot x_j = 0_k.$$

It is worth noting that:

1. If all $p_{ij} = 1$, then the RGCP is to find a suitable coloring mapping ϕ such that the number of complementary edges whose vertices share the same color is minimum. Or equivalently, to find a suitable coloring mapping ϕ such that the number of possible complementary edges are maximum, where a possible complementary edge is one that if it is added to the original graph, the coloring mapping ϕ is still valid.

2. If all $p_{ij} = c > 0$, where c is a constant, the RGCP is the same as that in the case of all $p_{ij} = 1$. Other case is different.

Based on the above discussion, our study is divided into the following two cases: (I) The RGCP with all p_{ij} the same, say, $p_{ij} = 1$, and (II) The RGCP with different p_{ij}.

Case I: The RGCP with all $p_{ij} = 1$

In this case, using the semi-tensor product, we express the rigidity level of a given coloring into a matrix algebraic form via the structural matrix first, which leads to the following result.

PROPOSITION 9.2

For the rigidity level (9.30), there exists a unique matrix $\tilde{M} \in \mathbb{R}^{1 \times k^n}$ such that:

$$R(\phi) = R(x_1, x_2, \ldots, x_n) = \tilde{M} \ltimes_{i=1}^{n} x_i, \; x_i \in \Delta_k, \tag{9.31}$$

where,

$$\tilde{M} = \sum_{i=1}^{n} \sum_{j \neq i} (1 - a_{ij}) M_{ij}, \quad M_{ij} = L_k S_{i,k}^n \left[I_{k^n} \otimes S_{j,k}^n \right] M_{r_{k^n}},$$

$L_k = [E_1, E_2, \ldots, E_k]$, and,

$$S_{i,k}^n = \mathbf{1}_{k^{i-1}} \otimes I_k \otimes \mathbf{1}_{k^{n-i}}, i = 1, 2, \cdots, n \tag{9.32}$$

are given in [6].

Proof. With (9.32), we have:

$$\begin{aligned} R_{ij}(x_1,x_2,\ldots,x_n) &:= x_i \star x_j = L_k x_i \ltimes x_j \\ &= L_k S^n_{i,k}(\ltimes^n_{i=1} x_i) S^n_{j,k}(\ltimes^n_{i=1} x_i) \\ &= L_k S^n_{i,k}\left[I_{k^n} \otimes S^n_{j,k}\right] M_{r_{k^n}} \ltimes^n_{i=1} x_i \\ &= M_{ij} \ltimes^n_{i=1} x_i. \end{aligned}$$

Based on this, we obtain:

$$\begin{aligned} R(x_1,x_2,\ldots,x_n) &= \sum_{i=1}^{n}\sum_{j\neq i}(1-a_{ij}) x_i \star x_j \\ &= \sum_{i=1}^{n}\sum_{j\neq i}(1-a_{ij}) M_{ij} \ltimes^n_{i=1} x_i \\ &:= \tilde{M} \ltimes^n_{i=1} x_i. \end{aligned}$$

Next, we prove that $\tilde{M}$ is unique.

In fact, if there exists another $\tilde{M}' \in \mathbb{R}^{1\times k^n}$ such that (9.31) holds, then, for any $(x_1,x_2,\cdots,x_n) \in \Delta^n_k$ with $\ltimes^n_{i=1} x_i = \delta^j_{k^n}$, on one hand, $R(x_1,x_2,\ldots,x_n) = \tilde{M}\delta^j_{k^n} = Col_j(\tilde{M})$, and on the other hand, $R(x_1,x_2,\ldots,x_n) = \tilde{M}'\delta^j_{k^n} = Col_j(\tilde{M}')$, where $Col_j(.)$ stands for the j-th column of a matrix. Thus, $Col_j(\tilde{M}) = Col_j(\tilde{M}'), j = 1,2,\cdots,k^n$, which implies that $\tilde{M} = \tilde{M}'$. ∎

$\tilde{M}$ is the structural matrix of the pseudo-logical function $R(x_1,x_2,\ldots,x_n)$, and the minimum component of $\tilde{M}$ is the global minimum of the rigidity level $R(x_1,x_2,\ldots,x_n)$.

According to Proposition 9.2 and Lemma 9.4, the RGCP can be expressed in an algebraic form of matrices as:

$$\begin{aligned} &\min R(\phi) = \tilde{M} \ltimes^n_{i=1} x_i, \\ &s.t. \quad M \ltimes^n_{i=1} x_i = 0_k, \end{aligned} \tag{9.33}$$

where $M = \sum^n_{i=1}\sum_{j\neq i} a_{ij} M^H_{ij}$, $M^H_{ij} = H_k S^n_{i,k}\left[I_{k^n} \otimes S^n_{j,k}\right] M_{r,k^n}$, $H_k = Diag\{E_1,E_2,\ldots,E_k\}$, and $\tilde{M}$ is the same as that in Proposition 9.2.

With above analysis, we can establish the following algorithm to obtain the most robust coloring scheme for any simple graph.

ALGORITHM 9.5

Given a simple graph μG with n vertices $\mu V = \{v_1,v_2,\ldots,v_n\}$, let a color set $\mu N = \{c_1,c_2,\ldots,c_k\}$ be given. For each vertex v_i, we assign a k-valued characteristic logical variable $x_i \in \Delta_k$. We can obtain the most robust coloring scheme of μG in the following steps:

S1: Compute the matrix M and the row vector $\tilde{M} = (r_1, r_2, \ldots, r_{k^n})$ given in (9.33).

S2: Check whether $0_k \in Col(M)$ or not. If $0_k \notin Col(M)$, the coloring problem with the given color set has no solution, which implies that the RGCP is not solvable, and the algorithm is ended. Otherwise, label the columns which equal 0_k and set:

$$K = \left\{ j \mid Col_j(M) = 0_k \right\}.$$

S3: Calculate $r^* = \min\left\{ r_j \mid j \in K \right\}$, and the corresponding column index j^*.

S4: Let $\ltimes_{i=1}^{n} x_i = \delta_{k^n}^{j^*}$. Using (9.32), compute:

$$x_i = S_{i,k}^{n} \ltimes_{i=1}^{n} x_i = S_{i,k}^{n} \delta_{k^n}^{j^*}, \quad i = 1, 2, \ldots, n,$$

and the most robust coloring scheme is given as:

$$\phi_{j^*}(v_i) = c_m, \; if \; x_i = \delta_k^m, \; i = 1, 2, \ldots, n; \; m = 1, 2, \ldots, k.$$

Next, we present a theorem, which provides us with more practical method to solve the RGCP.

Theorem 9.8

A point $(x_1^*, x_2^*, \ldots, x_n^*) \in \Delta_k^n$ is a global solution to the following optimization problem:

$$\min R(x_1, x_2, \ldots, x_n) = \sum_{i=1}^{n} \sum_{j \neq i} (1 - a_{ij}) x_i \star x_j, \tag{9.34}$$

$$s.t. \quad \sum_{i=1}^{n} \sum_{j \neq i} a_{ij} x_i \odot x_j = 0_k, \tag{9.35}$$

if and only if $(x_1^*, x_2^*, \ldots, x_n^*)$ is a global minimum point of the function:

$$F(x_1, x_2, \ldots, x_n) := R(x_1, x_2, \ldots, x_n) + (\nu + 1) \mathbf{1}_k \sum_{i=1}^{n} \sum_{j \neq i} a_{ij} x_i \odot x_j$$

such that $F(x_1^*, x_2^*, \ldots, x_n^*) \leq \nu$, where $F(x_1, x_2, \ldots, x_n)$ is a n-ary k-valued pseudo-logic function and $\nu = \sum_{i=1}^{n} \sum_{j \neq i} (1 - a_{ij})$. ■

Proof. Obviously, $0 \leq R(x_1, x_2, \ldots, x_n) \leq \nu$.

($\Rightarrow$) Assume that $(x_1^*, x_2^*, \ldots, x_n^*)$ is a global solution to the optimization problem:

$$\min R(x_1, x_2, \ldots, x_n),$$

$$s.t. \quad \sum_{i=1}^{n}\sum_{j\neq i} a_{ij}x_i \odot x_j = 0_k.$$

We show that $(x_1^*, x_2^*, \ldots, x_n^*)$ is a global minimum point of the function $F(x_1, x_2, \ldots, x_n)$.

In fact, if there exists $(y_1^*, y_2^*, \ldots, y_n^*) \in \Delta_k^n$ such that:

$$F(y_1^*, y_2^*, \ldots, y_n^*) < F(x_1^*, \ldots, x_n^*). \tag{9.36}$$

Now, we prove that $(y_1^*, y_2^*, \ldots, y_n^*)$ satisfies (9.35). If not, there must exist $1 \leq i_0 \leq n, 1 \leq j_0 \leq n$, and $j_0 \neq i_0$, such that $a_{i_0 j_0} y_{i_0}^* \odot y_{j_0}^* \neq 0_k$, which implies that $y_{i_0}^* \odot y_{j_0}^* \neq 0_k$. Then $y_{i_0}^* \neq y_{j_0}^*$, and $\mathbf{1}_k y_{i_0}^* \odot y_{j_0}^* = 1$. With this, we have:

$$\mathbf{1}_k \sum_{i=1}^{n}\sum_{j\neq i} a_{ij} y_i^* \odot y_j^* \geq 1.$$

Thus,

$$\begin{aligned} F(y_1^*, y_2^*, \ldots, y_n^*) &= R(y_1^*, y_2^*, \ldots, y_n^*) \\ &\quad + (\nu+1)\mathbf{1}_k \sum_{i=1}^{n}\sum_{j\neq i} a_{ij} y_i^* \odot y_j^* \\ &\geq R(y_1^*, y_2^*, \ldots, y_n^*) + (\nu+1) \\ &\geq \nu + 1. \end{aligned}$$

On the other hand, since $(x_1^*, x_2^*, \ldots, x_n^*)$ satisfies (9.35), we have:

$$\begin{aligned} F(x_1^*, x_2^*, \ldots, x_n^*) &= R(x_1^*, x_2^*, \ldots, x_n^*) \\ &\quad + (\nu+1)\mathbf{1}_k \sum_{i=1}^{n}\sum_{j\neq i} a_{ij} x_i^* \odot x_j^* \\ &= R(x_1^*, x_2^*, \ldots, x_n^*) \leq \nu. \end{aligned} \tag{9.37}$$

Hence,

$$F(x_1^*, x_2^*, \ldots, x_n^*) < F(y_1^*, y_2^*, \ldots, y_n^*),$$

which is a contradiction with (9.36). Therefore $(y_1^*, y_2^*, \ldots, y_n^*)$ satisfies (9.35).

From above, it is easy to know that:

$$F(y_1^*, y_2^*, \ldots, y_n^*) = R(y_1^*, y_2^*, \ldots, y_n^*),$$

$$F(x_1^*, x_2^*, \ldots, x_n^*) = R(x_1^*, x_2^*, \ldots, x_n^*).$$

Thus, the inequality (9.36) can be expressed as:

$$R(y_1^*, y_2^*, \ldots, y_n^*) < R(x_1^*, x_2^*, \ldots, x_n^*),$$

which is in contradiction with:

$$R(x_1^*, x_2^*, \ldots, x_n^*) = \min R(x_1, x_2, \ldots, x_n).$$

Hence,

$$F(x_1^*,\ldots,x_n^*) = \min F(x_1,x_2,\ldots,x_n),$$

which implies that $(x_1^*,x_2^*,\ldots,x_n^*)$ is a global minimum point of the function $F(x_1,x_2,\ldots,x_n)$.

Moreover, it is easy to see from (9.37) that $F(x_1^*,x_2^*,\ldots,x_n^*) \leq \nu$.

($\Leftarrow$) Assume that $(x_1^*,x_2^*,\ldots,x_n^*)$ is a global minimum point of the function $F(x_1,x_2,\ldots,x_n)$, such that $F(x_1^*,\ldots,x_n^*) \leq \nu$.

We show that $(x_1^*,\ldots,x_n^*)$ satisfies (9.35).

In fact, if not, we have:

$$\begin{aligned} F(x_1^*,\ldots,x_n^*) &= R(x_1^*,\ldots,x_n^*) + (\nu+1)\mathbf{1}_k \sum_{i=1}^{n}\sum_{j\neq i} a_{ij}x_i^* \odot x_j^* \\ &\geq R(x_1^*,\ldots,x_n^*) + (\nu+1) \geq \nu+1 > \nu. \end{aligned}$$

This is a contradiction. Hence, $(x_1^*,\ldots,x_n^*)$ satisfies (9.35), and $F(x_1^*,\ldots,x_n^*) = R(x_1^*,\ldots,x_n^*)$.

Furthermore, for every $(x_1,x_2,\ldots,x_n) \in \Delta_k^n$, which satisfies (9.35), we have:

$$F(x_1,x_2,\ldots,x_n) = R(x_1,x_2,\ldots,x_n).$$

Hence, $(x_1^*,x_2^*,\ldots,x_n^*)$ is a global minimum point of the function $R(x_1,x_2,\ldots,x_n)$, that is:

$$R(x_1^*,\ldots,x_n^*) = \min R(x_1,x_2,\ldots,x_n).$$

Thus, the proof is completed. ■

According to Theorem 9.8, we have the following result to solve the RGCP.

Theorem 9.9

Consider a simple graph $\mu G = \{\mu V, \mu E\}$ with n vertices $\mu V = \{v_1,v_2,\ldots,v_n\}$, and let a coloring mapping $\phi : \mu V \longmapsto \mu N = \{c_1,\ldots,c_k\}$, with $V_\phi = [x_1,x_2,\ldots,x_n]$, be given. Set $\ltimes_{i=1}^{n} x_i := \delta_{k^n}^{s}$, where $1 \leq s \leq k^n$ can be uniquely determined. Then the coloring ϕ is the most robust coloring if and only if:

$$b_s = \min_{1\leq i\leq k^n} \{b_i\} \leq \nu,$$

where,

$$[b_1,b_2,\ldots,b_{k^n}] = \hat{M}, \quad \hat{M} = \tilde{M} + (\nu+1)\mathbf{1}_k M,$$

and $\tilde{M}$, M and ν are the same as those in (9.33) and Theorem 9.8. ■

Proof. According to Theorem 9.8 and the above analysis of the RGCP, the RGCP is equivalent to finding a minimum point $(x_1^*, \dots, x_n^*) \in \Delta_k^n$ of the following function:

$$\begin{aligned} F(x_1,x_2,\dots,x_n) &= \sum_{i=1}^{n}\sum_{j\neq i}(1-a_{ij})x_i \star x_j \\ &\quad +(\nu+1)\mathbf{1}_k \sum_{i=1}^{n}\sum_{j\neq i} a_{ij}x_i \odot x_j, \end{aligned}$$

such that $F(x_1^*, x_2^*, \dots, x_n^*) \leq \nu$.

On the other hand, by Lemma 9.4 and Proposition 9.2, we have:

$$\begin{aligned} F(x_1,x_2,\dots,x_n) &= [\tilde{M} + (\nu+1)\mathbf{1}_k M] \ltimes_{i=1}^{n} x_i \\ &= \hat{M} \ltimes_{i=1}^{n} x_i. \end{aligned}$$

Let $\ltimes_{i=1}^{n} x_i = \delta_{k^n}^s$, then it is easy to see that:

$$\hat{M} \ltimes_{i=1}^{n} x_i = Col_s(\hat{M}) = b_s.$$

Hence, if the coloring with the given ϕ is the most robust coloring, then the s-th component of the row vector $\hat{M}$ is the minimum among all the components, which are less than ν. That is, $b_s = \min_{1\leq i\leq k^n}\{b_i\} \leq \nu$.

On the contrary, if $b_s = \min_{1\leq i\leq k^n}\{b_i\} \leq \nu$, then the s-th component of the row vector $\hat{M}$ is the minimum among all the components, which are less than ν. Hence, $(x_1,x_2,\dots,x_n)$, which contents $\ltimes_{i=1}^{n} x_i = \delta_{k^n}^s$, is the minimum point of $F(x_1,x_2,\dots,x_n)$ such that $F(x_1,x_2,\dots,x_n) \leq \nu$. According to Theorem 9.8, $(x_1,x_2,\dots,x_n)$ is a global optimal solution to the constrained optimization problem (9.34)–(9.35), which implies that $(x_1,x_2,\dots,x_n)$ is the most robust coloring scheme. ∎

Based on the proof of Theorem 9.9, we present an algorithm to find all the most robust coloring schemes for any simple graph.

ALGORITHM 9.6

Given a simple graph μG with n vertices $\mu V = \{v_1, v_2, \dots, v_n\}$, let a color set $\mu N = \{c_1, c_2, \dots, c_k\}$ be given. For each vertex v_i, we assign a k-valued characteristic logical variable $x_i \in \Delta_k$. To find the most robust coloring schemes of μG, we take the following steps:

S1: Compute the row vector $\hat{M}$ given in Theorem 9.9, and denote it by $[b_1, b_2, \dots, b_{k^n}]$. If for every $1 \leq i \leq k^n$, $b_i > \nu$, then the RGCP has no solution and the algorithm is ended. Otherwise, go to the next step.

S2: Calculate $b_s = \min_{1\leq i\leq k^n}\{b_i\} \leq \nu$, and set:

$$K = \left\{ s \;\middle|\; b_s = \min_{1\leq i\leq k^n}\{b_i\} \leq \nu \right\}.$$

S3: For each index $s \in K$, let $\ltimes_{i=1}^{n} x_i = \delta_{k^n}^{s}$. Using (9.32), computer $x_i = S_{i,k}^{n} \ltimes_{i=1}^{n} x_i = S_{i,k}^{n} \delta_{k^n}^{s}$, $i = 1,2,\ldots,n$, and the corresponding most robust coloring scheme is given as:

$$\phi_s(v_i) = c_m, \; if \; x_i = \delta_k^m, \; i = 1,2,\ldots,n; \; m = 1,2,\ldots,k;$$

$s \in K$.

Case II: The RGCP with different p_{ij}

By a similar argument to Case I, we have the following result to determine whether a coloring is the most robust coloring or not.

Theorem 9.10

Consider a simple graph $\mu G = \mu V, \mu E$ with n vertices $\mu V = v_1, v_2, \cdots, v_n$. Its complementary graph is given by $\bar{\mu G} = \mu V, \bar{\mu E}$. Let a penalty set $P = \{p_{ij}\}$ and a coloring mapping $\phi : \mu V \longmapsto \mu N = \{c_1,\ldots,c_k\}$ with $V_\phi = [x_1,x_2,\ldots,x_n]$, be given. Set $\ltimes_{i=1}^{n} x_i := \delta_{k^n}^{s}$, where $1 \le s \le k^n$ can be uniquely determined. Then, the coloring mapping ϕ is the most robust coloring if and only if:

$$b_s = \min_{1 \le i \le k^n} \{b_i\} \le \sum_{i=1}^{n} \sum_{j \ne i} (1 - a_{ij}) p_{ij},$$

where,

$$[b_1, b_2, \ldots, b_{k^n}] = \widehat{M},$$

$$\widehat{M} = \sum_{i=1}^{n} \sum_{j \ne i} (1 - a_{ij}) p_{ij} M_{ij} + \left[\sum_{i=1}^{n} \sum_{j \ne i} (1 - a_{ij}) p_{ij} + 1 \right] \mathbf{1}_k M,$$

and M_{ij} and M are the same as those in Proposition 9.2 and (9.33). ■

Proof. The proof is similar to that of Theorem 9.9, and thus it is omitted. ▮

Theorem 9.10 can also provide us with an effective algorithm to find the most robust coloring schemes for any simple graph, which is similar to Algorithm 9.6 and thus omitted.

In the following section, we would apply the above results to a significant administrative issue: examination timetabling.

Examination timetabling is a significant administrative issue that arises in academic institutions. In an examination timetabling problem, number of examinations are allocated into a given number of time slots subject to constrains, which are usually divided into two independent categories: hard and soft constraints [7, 8]. Hard constraints need to be satisfied under any circumstances, while soft constraints are desirable to be satisfied, but they are not essential. For a real-world university timetabling problem, it is usually impossible to satisfy all the soft constraints. Based on this, an examination timetabling is to find a feasible timetable, which satisfies all of the hard constraints and the violation of the soft constraints is minimal.

Consider an examination timetabling of n exams with k available time slots. Assume that there are so-called "standard students" and "non-standard students". Standard students are those who study according to basic studies plans and non-standard ones are those who repeated courses. Besides, the hard constraints and soft ones, considered in this paper, are listed as follows.

The hard constraints are:

(1) All exams must be scheduled, and each exam must be scheduled only once.

(2) No standard student can take two exams concurrently.

The soft constraints are:

(1) No non-standard student can take two exams concurrently.

(2) The exams should be arranged as evenly as possible for all time slots.

Taking account of the hard constraints, we can obtain a feasible examination timetable by solving the k-coloring of graph $\mu G = \{\mu V, \mu E\}$, where $\mu V = \{v_1, v_2, \ldots, v_n\}$ represents the set of exam courses and the edge $(v_i, v_j) \in \mu E$ exists when the examination courses v_i and v_j share at least one standard student.

In the following section, we consider the violation of the soft constraints for the feasible timetable.

For each examination course $v_i \in \mu V$, we assign a k-valued characteristic logical variable $x_i \in \Delta_k$ as follows:

$$x_i = \delta_k^s, \text{ if } v_i \text{ is assigned to time slot } s.$$

Assume that $D = [d_{ij}]_{n \times n}$ is a conflict matrix, where d_{ij} is the proportion of non-standard students taking both exams v_i and v_j, $i, j = 1, 2, \ldots, n$. Then the violation of the soft constraint (1) can be quantified as:

$$f_1(x_1, x_2, \ldots, x_n) = \sum_{i=1}^{n} \sum_{j \neq i} (1 - a_{ij}) \bar{p}_{ij} x_i \star x_j,$$

where $\bar{p}_{ij}$ denotes the penalty for the exams v_i and v_j, depending on d_{ij}.

If $a_{ij} = 1$, the value of d_{ij} does not affect the violation of the soft constraint (1). In this case, for the simplicity, we denote $d_{ij} = 0$, and similarly $d_{ii} = 0$.

In the same way, the violation degree of the soft constraint (2) is formulated as:

$$f_2(x_1, x_2, \ldots, x_n) = \sum_{i=1}^{n} \sum_{j \neq i} (1 - a_{ij}) x_i \star x_j,$$

which implies that the violation of the soft constraint (2) will be decreased as the number of exams arranged on the same time slot tends to be uniform for all the time slots.

Consider both the soft constraints and assuming each one with a weight w_k, $k = 1, 2$, the violation of the soft constraints for the feasible timetable can be defined as follows:

$$\begin{aligned} H &= w_1 f_1(x_1,x_2,\ldots,x_n) + w_2 f_2(x_1,x_2,\ldots,x_n) \\ &= \sum_{i=1}^{n}\sum_{j\neq i}(1-a_{ij})(w_1\bar{p}_{ij}+w_2)x_i \star x_j. \end{aligned}$$

Thus, the examination timetabling problem is to find a suitable timetable for the following optimization problem:

$$\min H = \sum_{i=1}^{n}\sum_{j\neq i}(1-a_{ij})(w_1\bar{p}_{ij}+w_2)x_i \star x_j.$$

Applying the results of Theorem 9.10, we have the following result on the examination timetabling problem.

PROPOSITION 9.3

Consider n examinations $V = \{v_1, v_2, \ldots, v_n\}$ with its topology of graph $\mu G = \{\mu V, \mu E\}$ for the standard student's courses incompatibilities, and let k available time slots $\mu N = \{c_1, \ldots, c_k\}$, a penalty set $P = \{\bar{p}_{ij}\}$ for the soft constraint (1) and an examination timetable mapping $\phi : \mu V \longmapsto \mu N$ with $V_\phi = [x_1, x_2, \ldots, x_n]$, be given. Set $\ltimes_{i=1}^{n} x_i := \delta_{k^n}^{s}$, where $1 \leq s \leq k^n$ can be uniquely determined. Then, ϕ is the most feasible timetable mapping for the soft constraints (1) and (2) if and only if:

$$b_s = \min_{1\leq i\leq k^n}\{b_i\} \leq \sum_{i=1}^{n}\sum_{j\neq i}(1-a_{ij})(w_1\bar{p}_{ij}+w_2),$$

where,

$$\begin{aligned} &[\ b_1, b_2, \ldots, b_{k^n}] \\ = &\sum_{i=1}^{n}\sum_{j\neq i}(1-a_{ij})(w_1\bar{p}_{ij}+w_2)M_{ij} \\ &+ \left[\sum_{i=1}^{n}\sum_{j\neq i}(1-a_{ij})(w_1\bar{p}_{ij}+w_2)+1\right]\mathbf{1}_k \sum_{i=1}^{n}\sum_{j\neq i} a_{ij} M_{ij}^{H}, \end{aligned}$$

w_1 and w_2 are the weights of the soft constraints (1) and (2), respectively, and M_{ij} and M_{ij}^{H} are the same as those in Proposition 9.2 and (9.33).

Finally, we give two examples to illustrate the effectiveness of the results obtained in this paper.

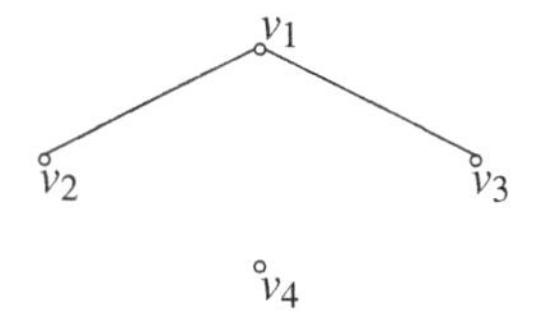

Figure 9.6 Graph $\mu G = \{V, E\}$

Example 9.4:

Consider the graph $\mu G = \{V, E\}$ shown in Figure 9.3. Letting a two-color set $N = \{C_1 = red, C_2 = blue\}$ be given, we use Algorithm 9.5 to find out the most robust coloring schemes for μG.

For each vertex v_i, we assign a characteristic logical variable $x_i \in \Delta, i = 1,2,3,4$. The adjacency matrix of this graph is as follows:

$$A = [a_{ij}]_{4\times 4} = \begin{bmatrix} 0 & 1 & 1 & 0 \\ 1 & 0 & 0 & 0 \\ 1 & 0 & 0 & 0 \\ 0 & 0 & 0 & 0 \end{bmatrix}.$$

By (9.6) and the MATLAB toolbox which is provided by D. Cheng and his co-workers, we can easily obtain:

$$\begin{aligned} \tilde{M} &= (r_1, r_2, \ldots, r_{16}) \\ &= [8\,2\,4\,2\,4\,2\,4\,6\,6\,4\,2\,4\,2\,4\,2\,8], \end{aligned}$$

$$M = \begin{bmatrix} 4\,4\,2\,2\,2\,2\,0\,0\,0\,0\,0\,0\,0\,0\,0\,0 \\ 0\,0\,0\,0\,0\,0\,0\,0\,0\,0\,2\,2\,2\,2\,4\,4 \end{bmatrix}.$$

It is observed that:

$$Col_7(M) = Col_8(M) = Col_9(M) = Col_{10}(M) = 0_2,$$

and the corresponding column index set is:

$$K = \{7, 8, 9, 10\}.$$

Calculate:

$$r^* = \min\{r_j \mid j \in K\} = 4,$$

and the corresponding column index $j^* = 7, 10$.

By computing $x_i = S^4_{i,2}\delta^{j^*}_{2^4}$, we have:

$$\delta^7_{2^4} \sim [x_1, x_2, x_3, x_4] = \begin{bmatrix} 1 & 0 & 0 & 1 \\ 0 & 1 & 1 & 0 \end{bmatrix},$$

$$\delta_{2^4}^{10} \sim [x_1,x_2,x_3,x_4] = \begin{bmatrix} 0 & 1 & 1 & 0 \\ 1 & 0 & 0 & 1 \end{bmatrix}.$$

Thus, all the most robust coloring schemes are as follows:

$$S_{c_1}^{7} = \{v_1,v_4\}\,(red),\ S_{c_2}^{7} = \{v_2,v_3\}\,(blue).$$

and,

$$S_{c_1}^{10} = \{v_1,v_4\}\,(blue),\ S_{c_2}^{10} = \{v_2,v_3\}\,(red).$$

Example 9.5:

Consider an examination timetabling of 4 exams with 2 available time slots. Assume that the information topology of the examinations for the standard students is given by a graph $\mu G = \{\mu V, \mu E\}$ with $\mu V = \{v_1,v_2,v_3,v_4\}$ shown in Figure 1, where v_1,v_2,v_3,v_4 stand for the 4 different examinations. Moreover, let the matrix $D = [d_{ij}]_{4\times 4}$, where d_{ij} is the proportion of non-standard students taking both the examinations v_i and v_j, be given as:

$$D = [d_{ij}]_{4\times 4} = \begin{bmatrix} 0 & 0 & 0 & 0.04 \\ 0 & 0 & 0.01 & 0.02 \\ 0 & 0.01 & 0 & 0.03 \\ 0.04 & 0.02 & 0.03 & 0 \end{bmatrix}.$$

In this example, we assume that the weight of the soft constraints (1) and (2) in Proposition 9.3 is $w_1 = 1$ and $w_2 = 0$. Then, we apply Proposition 9.3 to find all the feasible examination timetables to ensure that no standard student takes two examinations concurrently and make the examinations sharing more non-standard students be scheduled at different time slots as much as possible.

For each exam v_i, we assign it a 2-valued characteristic logical variable $x_i \in \Delta_2$, $i = 1,2,3,4$.

In order to obtain a feasible timetable, the examination which shares more non-standard students being scheduled the different time slot has a lower violation than another one with the same time slot, we let the penalty for the examinations v_i and v_j be defined as $p_{ij} = 4^{100s}$, if $d_{ij} = s$, where 4 denotes the number of the complement edges for the graph μG.

Using Proposition 9.3, we obtain:

$$\begin{aligned} \widehat{M} &= [b_1,b_2,\ldots,b_{16}] \\ &= [3404\ 2732\ 1906\ 1490\ 2006\ 1394\ 520\ 168 \\ &\quad\ 168\ 520\ 1394\ 2042\ 1490\ 1906\ 2732\ 3404]. \end{aligned}$$

It is easy to know that

$$\begin{aligned} b_s &= \min_{1\le i\le 16} \{b_i\} \le 681 \\ &= 168, \end{aligned}$$

and the corresponding column index $s = 8,9$.

By computing $x_i = S^4_{i,2}\delta^s_{2^4}$, we have:

$$\delta^8_{2^4} \sim [x_1, x_2, x_3, x_4] = \begin{bmatrix} 1 & 0 & 0 & 0 \\ 0 & 1 & 1 & 1 \end{bmatrix},$$

$$\delta^9_{2^4} \sim [x_1, x_2, x_3, x_4] = \begin{bmatrix} 0 & 1 & 1 & 1 \\ 1 & 0 & 0 & 0 \end{bmatrix}.$$

Thus, we can obtain all the most feasible time schemes as:

$$S^8_{T_1} = \{v_1\} \text{(time slot 1)}, \ S^8_{T_2} = \{v_2, v_3, v_4\} \text{(time slot 2)}.$$

and,

$$S^9_{T_1} = \{v_2, v_3, v_4\} \text{(time slot 1)}, \ S^9_{T_2} = \{v_1\} \text{(time slot 2)}.$$

9.4 T-COLORING AND LIST COLORING PROBLEMS

In this section, we shall introduce T-coloring and minimum coloring problems in sequence.

9.4.1 T-COLORING PROBLEM

First, we present some results of the T-colorings of graphs by the semi-tensor product method.

Consider a simple graph $\mu G = \{\mu V, \mu E\}$ with n vertices $\mu V = \{v_1, v_2, \ldots, v_n\}$, and assume that its adjacency matrix is $A = [a_{ij}]$. Let T be a finite non-negative integer set satisfying $0 \in T$, $\phi : \mu V \mapsto \mu N := \{c_1, \ldots, c_k\}$ be a k coloring mapping, where $c_1, \ldots, c_k$ stand for k different consecutive positive integers with $c_1 < c_2 < \ldots < c_k$.

The T-coloring problem is to find a suitable k-coloring mapping ϕ, such that for any $v_i, v_j \in \mu V$, if $(v_i, v_j) \in \mu E$, then:

$$|\phi(v_i) - \phi(v_j)| \notin T.$$

If we assume that $T = \{t_1, t_2, \cdots, t_m\}$, where t_i is non-negative integer, it is easy to see that $|\phi(v_i) - \phi(v_j)| \notin T$ is equivalent to:

$$|\phi(v_i) - \phi(v_j)| \neq t_s, s = 1, 2, \cdots, m.$$

For each vertex $v_i \in \mu V$, we assign a k-valued characteristic logical variable $x_i \in \Delta_k$ as follows:

$$x_i = \delta^j_k, \ if \ \phi(v_i) = c_j \in \mu N, \ i = 1, 2, \ldots, n,$$

and define.

$$x_i \star x_j = \begin{cases} 1, & x_i = x_j; \\ 0, & \text{otherwise.} \end{cases}$$

Then, we have the following result to determine whether or not coloring mapping ϕ is feasible.

Theorem 9.11

Consider a graph $\mu G = \{\mu V, \mu E\}$, let a set $T = \{t_1, t_2, \cdots, t_m\}$ and a coloring mapping $\phi : \mu V \longmapsto N = \{c_1, \ldots, c_k\}$ be given. Then the T-coloring problem is solvable with the given ϕ, if and only if the following n-ary k-valued pseudo-logic equation:

$$\sum_{i=1}^{n}\sum_{j\neq i} a_{ij} \sum_{s=1}^{m} x_i \star (\oslash^{t_s} x_j) = 0 \tag{9.38}$$

is solvable. ■

Proof. ($\Rightarrow$) If the T-coloring problem is solvable with the given mapping ϕ, then for any $v_i \neq v_j \mu V$, if $(v_i, v_j) \in \mu E$, then $|\phi(v_i) - \phi(v_j)| \notin T$, which implies that $|\phi(v_i) - \phi(v_j)| \neq t_s, s = 1, 2, \cdots, m$. Without loss of generality, we assume that $\phi(v_i) - \phi(v_j) \neq t_s, s = 1, 2, \cdots, m$. Thus, if $a_{ij} = 1$, then $x_i \star (\oslash^{t_s} x_j) = 0, s = 1, 2, \cdots, m$. So, $a_{ij} \sum_{s=1}^{m} x_i \star (\oslash^{t_s} x_j) = 0$. With this, we have:

$$\sum_{i=1}^{n}\sum_{j\neq i} a_{ij} \sum_{s=1}^{m} x_i \star (\oslash^{t_s} x_j) = 0.$$

In other words, Eq. (9.38) is solvable.

($\Leftarrow$) Assume that $(x_1, x_2, \cdots, x_n)$ is a solution of Eq. (9.38). Since $a_{ij} \geq 0$ and $x_i \in \Delta_k$, then $a_{ij} \sum_{s=1}^{m} x_i \star (\oslash^{t_s} x_j) = 0$. If $a_{ij} = 1$, then $\sum_{s=1}^{m} x_i \star (\oslash^{t_s} x_j) = 0$. With $x_i \star (\oslash^{t_s} x_j) \geq 0$, we have $x_i \star (\oslash^{t_s} x_j) = 0, s = 1, 2, \cdots, m$, which means that $\phi(v_i) - \phi(v_j) \neq t_s, s = 1, 2, \cdots, m$. Hence, ϕ is a solution of the T-coloring problem. ■

Next, using the semi-tensor product, we express Eq. (9.38) in a matrix algebraic form via the structural matrix, and have the following result.

Theorem 9.12

Consider a graph $\mu G = \{\mu V, \mu E\}$, let a set $T = \{t_1, t_2, \cdots, t_m\}$ and a coloring mapping $\phi : \mu V \longmapsto N = \{c_1, \ldots, c_k\}$ be given. Then the T-coloring problem is solvable with the given ϕ, if and only if:

$$0 \in Col(M) := \{Col_1(M), Col_2(M), \cdots, Col_{k^n}(M)\},$$

where,

$$M = \sum_{i=1}^{n}\sum_{j\neq i} a_{ij}M_{ij},$$

$$M_{ij} = \sum_{s=1}^{m} L_k(I_k \otimes M_{o,k}^{t_s})S_{i,k}^n \left[I_{k^n} \otimes S_{j,k}^n\right] M_{r,k^n},$$

$L_k = [E_1, E_2, \ldots, E_k]$, and,

$$S_{i,k}^n = \mathbf{1}_{k^{i-1}} \otimes I_k \otimes \mathbf{1}_{k^{n-i}}, i = 1,2,\cdots,n \tag{9.39}$$

are given as [5]. ■

Proof. First, we prove that:

$$f(x_1,x_2,\cdots,x_n) = \sum_{i=1}^{n}\sum_{j\neq i} a_{ij} \sum_{s=1}^{m} x_i \star (\oslash^{t_s} x_j)$$

can be expressed as

$$f(x_1,x_2,\cdots,x_n) = M \ltimes_{i=1}^{n} x_i.$$

With L_k and $S_{i,k}^n$, we have:

$$\begin{aligned} f_{ij}(x_1,x_2,\ldots,x_n) &:= \sum_{s=1}^{m} x_i \star (\oslash^{t_s} x_j) \\ &= \sum_{s=1}^{m} L_k(I_k \otimes M_{o,k}^{t_s}) x_i \ltimes x_j \\ &= \sum_{s=1}^{m} L_k(I_k \otimes M_{o,k}^{t_s}) S_{i,k}^n (\ltimes_{i=1}^{n} x_i) S_{j,k}^n (\ltimes_{i=1}^{n} x_i) \\ &= \sum_{s=1}^{m} L_k(I_k \otimes M_{o,k}^{t_s}) S_{i,k}^n \left[I_{k^n} \otimes S_{j,k}^n\right] M_{r,k^n} \ltimes_{i=1}^{n} x_i \\ &= M_{ij} \ltimes_{i=1}^{n} x_i. \end{aligned}$$

Based on this, we obtain:

$$\begin{aligned} f(x_1,x_2,\ldots,x_n) &= \sum_{i=1}^{n}\sum_{j\neq i} a_{ij} \sum_{s=1}^{m} x_i \star (\oslash^{t_s} x_j) \\ &= \sum_{i=1}^{n}\sum_{j\neq i} a_{ij} M_{ij} \ltimes_{i=1}^{n} x_i \\ &= M \ltimes_{i=1}^{n} x_i. \end{aligned}$$

Thus, Eq. (9.38) can be rewritten as

$$M \ltimes_{i=1}^{n} x_i = 0.$$

Noticing that $\ltimes_{i=1}^{n} x_i \in \Delta_{k^n}$, Eq. (9.38) is solvable if and only if 0 is one of columns of M. Thus, the theorem follows from Theorem 9.11. ∎

COROLLARY 9.3

Consider the graph $\mu G = \{\mu V, \mu E\}$, set $T = \{t_1, t_2, \cdots, t_m\}$ and $N = \{c_1, \ldots, c_k\}$ in Theorem 9.11. Then the T-coloring problem has no solution solvable if and only if:

$$Col_i(M) \neq 0, \ \forall\, i = 1, 2, \cdots, n.$$

With above analysis, we can establish the following algorithm to obtain all the T-coloring schemes for any simple graph.

ALGORITHM 9.7

Given a simple graph μG with n vertices $\mu V = \{v_1, v_2, \ldots, v_n\}$, let a color set $\mu N = \{c_1, c_2, \ldots, c_k\}$ and a set $T = \{t_1, t_2, \cdots, t_m\}$ be given. For each vertex v_i, we assign a k-valued characteristic logical variable $x_i \in \Delta_k$. We can obtain all the T-coloring schemes of μG by taking the following steps:

S1: Compute the matrix M given in Theorem 9.12.

S2: Check whether $0 \in Col(M)$ or not. If $0 \notin Col(M)$, the T-coloring problem with the given color set and T set has no solution, and the algorithm ends. Otherwise, label the columns which equal 0 and set:

$$K = \left\{ j \,\middle|\, Col_j(M) = 0 \right\}.$$

S3: For each index $j \in K$, let $\ltimes_{i=1}^{n} x_i = \delta_{k^n}^{j}$. Using (9.39), compute $x_i = S_{i,k}^{n} \ltimes_{i=1}^{n} x_i = S_{i,k}^{n} \delta_{k^n}^{j}$, $i = 1, 2, \ldots, n$, and the corresponding T-coloring scheme is given as:

$$\phi_j(v_i) = c_m, \ if\ x_i = \delta_k^m, \ i = 1, 2, \ldots, n; \ m = 1, 2, \ldots, k.$$

S4: Set,

$$S_1^j := \{v_i \mid \phi_j(v_i) = c_1, 1 \leq i \leq m\},$$
$$S_2^j := \{v_i \mid \phi_j(v_i) = c_2, 1 \leq i \leq m\},$$
$$\vdots$$
$$S_k^j := \{v_i \mid \phi_j(v_i) = c_k, 1 \leq i \leq m\}.$$

and let,

$$N_j = \{i \mid S_i^j = \emptyset, 1 \leq i \leq k\}, \ \left| N_{j_0} \right| = \max_{j \in K} \{ \left| N_j \right| \}.$$

Then the minimum T-coloring partition is:

$$\{S_1^{j_0}, S_2^{j_0}, \ldots, S_k^{j_0}\} \backslash \{S_i^{j_0} \mid S_i^{j_0} = \emptyset\},$$

and the T-chromatic number of μG is:

$$\chi_T(\mu G) = k - \mid N_{j_0} \mid .$$

S5: Let,

$$\bar{N_j} = N \backslash N_j, \;\; d_j = \max \bar{N_j} - \min \bar{N_j},$$

then the minimum span among all T-colorings of μG is:

$$sp_T(\mu G) = \min_{j \in K}\{d_j\}.$$

In this section, as an application, we use the above results to investigate adjacent-frequency constraint assignment problem.

Consider a wireless communication network with n transmitters, μV, respected by $\mu V = \{v_1, v_2, \cdots, v_n\}$, where $v_1, v_2, \cdots, v_n$ stand for n different transmitters. Assume $N = \{1, \ldots, k\}$ is k available frequencies, $d(v_i, v_j)$ denotes the distance between transmitters v_i and v_j, and d_0, d_1 represent two different frequency constraint distance, satisfying $d_0 > d_1 > 0$. Let a frequency assignment mapping $\phi : \mu V \longmapsto N = \{1, \ldots, k\}$ be given.

The frequency assignment problem with adjacent-channel constraint is to find a suitable frequency assignment mapping ϕ such that for any $v_i, v_j \in \mu V$, if $d(v_i, v_j) \leq d_s$, where $d(v_i, v_j)$ denotes the distance between cells v_i and v_j, then:

$$|\phi(v_i) - \phi(v_j)| \neq s, s = 0, 1.$$

For each vertex $v_i \in \mu V$, assign a k-valued characteristic logical variable $x_i \in \Delta_k$ as follows [6]:

$$x_i = \delta_k^j, \;\; if \;\; \phi(v_i) = j \in N, \; i = 1, 2, \ldots, n.$$

The adjacent transmitters matrix can be modeled by $B = [b_{ij}]$, such that:

$$b_{ij} = \begin{cases} \delta_3^1, & d(v_i, v_j) \geq d_0; \\ \delta_3^2, & d_1 \leq d(v_i, v_j) < d_0; \\ \delta_3^3, & \text{otherwise.} \end{cases}$$

9.4.2 LIST COLORING PROBLEM

In this section, we investigate the list coloring problem by the semi-tensor product method.

Consider a simple graph $\mu G = (\mu V, \mu E)$ with n vertices $\mu V = \{v_1, v_2, \ldots, v_n\}$, and assume that its adjacency matrix is $A = [a_{ij}]$. Let $\mu C := \{c_1, c_2, \cdots, c_k\}$ be a set of colors, where $c_1, c_2, \cdots, c_k$ stand for k kinds of different colors.

Let L be list assignment for μG, i.e., a function that assigns nonempty finite subsets of μC to vertices of μG.

The list L coloring problem is to find a suitable coloring mapping $\phi : \mu V \longmapsto \mu C$ such that:

(1) $\phi(v_i) \neq \phi(v_j)$, if $(v_i, v_j) \in \mu E$;

(2) $\phi(v_i) \in L(v_i)$ for every vertex $v_i \in \mu V$.

The colorings that satisfy only the first of the above conditions, are called k colorings of μG.

If μG has a list L coloring (k coloring), we say that the graph μG is list L colorable (k colorable).

The coloring that only satisfies the second of the above conditions, are called coloring respects L of μG.

For the list $L(v_i)$, one can define a $k \times k$ matrix $\hat{L}_i = [L_{1i}, L_{2i}, \cdots, L_{ki}]$, called the characteristic logical matrix of $L(v_i)$, as follows:

$$\hat{L}_{ij} = \begin{cases} \delta_k^j, & c_j \in L(v_i); \\ 0_k, & c_j \notin L(v_i). \end{cases}$$

For each vertex $v_i \in \mu V$, we assign a k-valued characteristic logical variable $x_i \in \Delta_k$ as follows:

$$x_i = \delta_k^j, \ if \ \phi(v_i) = c_j \in \mu C := \{c_1, c_2, \cdots, c_k\}, \ i = 1, 2, \ldots, n.$$

and let $x = (x_1, x_2, \cdots, x_n)$.

Then we have the following result.

Theorem 9.13

$\phi(v_i) \in L(v_i)$ holds, if and only if:

$$\hat{L}_i \ltimes x_i = x_i, \qquad i = 1, 2, \cdots, n. \tag{9.40}$$

■

Proof. ($\Rightarrow$) Assume that $\phi(v_i) = c_j$, then $x_i = \delta_k^j$. if $\phi(v_i) \in L(v_i)$, which implies that $c_j \in L(v_i)$, then $\hat{L}_{ij} = \delta_k^j$, thus $\hat{L}_i \ltimes x_i = \hat{L}_{ij} = \delta_k^j = x_i$.

($\Leftarrow$) Assume that $\hat{L}_i \ltimes x_i = x_i$, $i = 1, 2, \cdots, n$, then, $\hat{L}_{ij} = \delta_k^j$, which implies that $c_j \in L(v_i)$. So $\phi(v_i) \in L(v_i)$. ■

Define $x = x_1 \ltimes x_2 \ltimes \cdots \ltimes x_n$. By the pseudo-commutative property of semi-tensor product, one can convert the equation (9.40) into the following algebraic form:

$$\begin{aligned} x &= (\hat{L}_1 \ltimes x_1) \ltimes (\hat{L}_2 \ltimes x_2) \ltimes \cdots \ltimes (\hat{L}_n x_n) \\ &= \hat{L}_1 \ltimes (I_k \otimes \hat{L}_2) \ltimes \cdots \ltimes (I_{k^{n-1}} \otimes \hat{L}_n) \ltimes_{i=1}^n x_i \\ &:= \daleth \ltimes x, \end{aligned} \tag{9.41}$$

where $\urcorner = \hat{L}_1 \ltimes (I_k \otimes \hat{L}_2) \ltimes \cdots \ltimes (I_{k^{n-1}} \otimes \hat{L}_n)$.

Then we have:

$$(\urcorner - I_{k^n}) \ltimes x = 0_{k^n}.$$

Based on the above analysis, we have the following results about the coloring respects L.

Theorem 9.14

Consider a graph $\mu G = \{\mu V, \mu E\}$ with a k list assignment L, and let a coloring mapping $\phi : \mu V \longmapsto \mu C = \{c_1, c_2, \cdots, c_k\}$ with $x = (x_1, x_2, \ldots, x_n)$ be given. Set $\ltimes_{i=1}^{n} x_i := \delta_{k^n}^{s}$, where $1 \le s \le k^n$ can be uniquely determined. Then the coloring ϕ is a coloring respects L if and only if:

$$Col_s(\urcorner - I_{k^n}) = 0_{k^n}. \tag{9.42}$$

■

Proof. ($\Rightarrow$) Suppose that the coloring ϕ with $\ltimes_{j=1}^{n} x_j = \delta_{k^n}^{s}$ is a coloring respects L, then $\ltimes_{j=1}^{n} x_j$ is a solution to Equation (9.41).

That is

$$(\urcorner - I_{k^n}) \ltimes_{j=1}^{n} x_j = Col_s(\urcorner - I_{k^n}) = 0_{k^n}.$$

($\Leftarrow$) Suppose (9.42) is satisfied. Then,

$$(\urcorner - I_{k^n}) \delta_{k^n}^{s} = Col_s(\urcorner - I_{k^n}) = 0_{k^n}.$$

That is to say the coloring ϕ with $\ltimes_{j=1}^{n} x_j = \delta_{k^n}^{s}$ is a coloring respects L. ■

From the above proof, we have the following corollary.

COROLLARY 9.4

Consider the graph $\mu G = \{\mu V, \mu E\}$ in Theorem 2. Then, the graph μG has a coloring respects L if and only if the equation:

$$(\urcorner - I_{k^n}) \ltimes_{j=1}^{n} x_j = 0_k$$

is solvable. Furthermore, the number of zero components in $\urcorner - I_{k^n}$ are just the number of all the coloring respects L schemes.

According to Lemma 9.4 and Theorem 2, we have the following result about the list L coloring problem.

Theorem 9.15

Consider a graph $\mu G = \{\mu V, \mu E\}$ with a list assignment L, and let a coloring mapping $\phi : \mu V \longmapsto \mu C = \{c_1, \ldots, c_k\}$ be given. Then the coloring with the given ϕ with its vector form $\ltimes_{j=1}^{n} x_j = \delta_{k^n}^{s}$ is a list L coloring if and only if:

$$s \in P \cap Q,$$

where $P = \{j | Col_j(\urcorner - I_{k^n}) = 0_{k^n}\}$, $Q = \{j | Col_j \bar{M} = 0_k\}$ and,

$$\bar{M} = \sum_{i=1}^{n} \sum_{j \neq i} a_{ij} H_k S_{i,k}^{n} [I_{k^n} \otimes S_{j,k}^{n}] M_{r_k}.$$

■

Based on above analysis, we can establish the following algorithm to obtain all the list coloring schemes with the color list L for a given graph.

ALGORITHM 9.8

Given a simple graph μG with n vertices $\mu V = \{v_1, v_2, \ldots, v_n\}$, let $\mu C := \{c_1, c_2, \cdots, c_k\}$ be a set of colors, and L be a list assignment of the graph μG with $L(v_i) \subseteq \mu C$. For each vertex v_i, we assign a k-valued characteristic logical variable $x_i \in \Delta_k$. We can obtain all the list L coloring schemes of μG by taking the following steps:

S1: Compute the matrix P and Q given in Theorem 9.15.

S2: Check whether $P \cap Q = \emptyset$. If $J \cap Q = \emptyset$, the coloring problem with the given color list L has no solution, and the algorithm ends. Otherwise, set,

$$K = \{j \mid j \in P \ and \ j \in Q\}.$$

S3: For each element $s \in K$, let $\ltimes_{i=1}^{n} x_i = \delta_{k^n}^{s}$. Compute $x_i = S_{i,k}^{n} \ltimes_{i=1}^{n} x_i = S_{i,k}^{n} \delta_{k^n}^{s}$, $i = 1, 2, \ldots, n$, and the corresponding list L coloring scheme is given as:

$$\phi_s(v_i) = c_m, \ if \ x_i = \delta_k^{m}, \ i = 1, 2, \ldots, n; \ m = 1, 2, \ldots, k.$$

S4: Set,

$$S_1^s := \{v_i \mid \phi_j(v_i) = c_1, 1 \leq i \leq m\},$$
$$S_2^s := \{v_i \mid \phi_j(v_i) = c_2, 1 \leq i \leq m\},$$
$$\vdots$$
$$S_k^s := \{v_i \mid \phi_j(v_i) = c_k, 1 \leq i \leq m\},$$

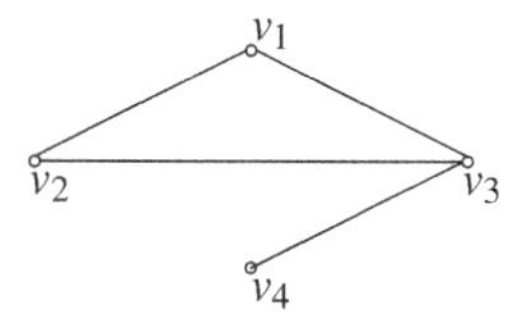

Figure 9.7 Graph μG

then a list L coloring schemes corresponding to element $s \in K$ is given as: all the vertices in S_i^s are colored with color $c_i, i = 1, 2, \cdots, k$. The number of list L coloring schemes is $|K|$.

In this subsection, we give one example to illustrate the effectiveness of the results obtained in this subsection.

Example 9.6:

Consider the graph $\mu G = \{V, E\}$ shown in Figure 9.6. $\mu C := \{C_1 = red,\ C_2 = blue,\ C_3 = yellow\}$ is a set of colors. Let a color assignment L of the graph μG be given as $L(v_1) = \{C_1 = red,\ C_2 = blue,\ C_3 = yellow\}$, $L(v_2) = \{C_1 = red,\ C_3 = yellow\}$, $L(v_3) = \{C_2 = blue,\ C_3 = yellow\}$, $L(v_4) = \{C_1 = red,\ C_3 = yellow\}$, we use Algorithm 9.8 to find out all the list L coloring schemes for μG.

For each vertex v_i, we assign a characteristic logical variable $x_i \in \Delta, i = 1, 2, \cdots, 4$. The adjacency matrix of this graph is as follows:

$$\begin{bmatrix} 0 & 1 & 1 & 0 \\ 1 & 0 & 1 & 0 \\ 1 & 1 & 0 & 1 \\ 0 & 0 & 1 & 0 \end{bmatrix}.$$

By the MATLAB toolbox which is provided by D. Cheng and his co-workers, we can easily obtain:

$$P = \{4, 6, 7, 9, 22, 24, 25, 27, 31, 33, 34, 36, 49, 51, 52, 54,$$

$$58, 60, 61, 63, 76, 78, 79, 81\}.$$

$$Q = \{16, 17, 22, 24, 34, 35, 47, 48, 58, 60, 65, 66\}.$$

It is observed that:

$$K = \{22, 24, 34, 58, 60\}.$$

For each element $s \in K$, let $\ltimes_{i=1}^4 x_i = \delta_{3^4}^s$. By computing $x_i = S_{i,3}^4 \delta_{3^4}^s$, we have

$$\delta_{3^4}^{22} \sim [x_1, x_2, x_3, x_4] = \begin{bmatrix} 1 & 0 & 0 & 1 \\ 0 & 0 & 1 & 0 \\ 0 & 1 & 0 & 0 \end{bmatrix},$$

$$\delta_{3^4}^{24} \sim [x_1, x_2, x_3, x_4] = \begin{bmatrix} 1 & 0 & 0 & 0 \\ 0 & 0 & 1 & 0 \\ 0 & 1 & 0 & 1 \end{bmatrix},$$

$$\delta_{34}^{34} \sim [x_1,x_2,x_3,x_4] = \begin{bmatrix} 0 & 1 & 0 & 1 \\ 1 & 0 & 0 & 0 \\ 0 & 0 & 1 & 0 \end{bmatrix},$$

$$\delta_{34}^{58} \sim [x_1,x_2,x_3,x_4] = \begin{bmatrix} 0 & 1 & 0 & 1 \\ 0 & 0 & 1 & 0 \\ 1 & 0 & 0 & 0 \end{bmatrix},$$

$$\delta_{34}^{60} \sim [x_1,x_2,x_3,x_4] = \begin{bmatrix} 0 & 1 & 0 & 0 \\ 0 & 0 & 1 & 0 \\ 1 & 0 & 0 & 1 \end{bmatrix}.$$

Thus, we obtain the following five coloring schemes:

$$S_1^{22} = \{v_1,v_4\}\,(red),\ S_2^{22} = \{v_3\}\,(blue),\ S_3^{22} = \{v_2\}\,(yellow),$$

$$S_1^{24} = \{v_1\}\,(red),\ S_2^{22} = \{v_3\}\,(blue),\ S_3^{22} = \{v_2,v_4\}\,(yellow),$$

$$S_1^{34} = \{v_2,v_4\}\,(red),\ S_2^{22} = \{v_1\}\,(blue),\ S_3^{22} = \{v_3\}\,(yellow),$$

$$S_1^{58} = \{v_2,v_4\}\,(red),\ S_2^{22} = \{v_3\}\,(blue),\ S_3^{22} = \{v_1\}\,(yellow).$$

and,

$$S_1^{60} = \{v_2\}\,(red),\ S_2^{22} = \{v_3\}\,(blue),\ S_3^{22} = \{v_1,v_4\}\,(yellow).$$

REFERENCES

1. Minty, G. J. (1980). On maximal independent sets of vertices in claw-free graphs. J. of Combin. Theory, 28: 284–304.
2. Liu, Y. C. and Zhang, W. (1993). Boolean Methodology. Shanghai Technology Literature Press.
3. Cheng, D. and Qi, H. (2010). A linear representation of dynamics of Boolean networks. IEEE Trans. Aut. Contr., 55(10): 2251–2258.
4. Zhu, J., Tian, Y. and Kuang, J. (2009). On the general consensus protocol of multi-agent systems with double-integrator dynamics. Linear Algebra and Its Applications, 431(5-7): 701–715.
5. Cheng, D. and Qi, H. (2007). Semi-Tensor Product of Matrices—Theory and Applications. Beijing, Science Press.
6. Wang, Y., Zhang, C. and Liu, Z. (2012). A matrix approach to graph maximum stable set and coloring problems with application to multi-agent systems. Automatica, 48(7): 1227–1236.
7. Carter, M., Laporte, G. and Lee, S. (1996). Examination timetabling: algorithmic strategies and applications. Journal of the Operational Resaerch Society, 47(3): 373–383.
8. Carter, M. (1986). A survey of practical applications of examination timetabling. Operations Research, 34(2): 193–202.

10 Finite-Field Networks

10.1 INTRODUCTION TO FINITE-FIELD NETWORKS

Multi-agent systems are made up of a group of nodes or agents that communicate with each other locally through a graph, which can be used to solve problems that are difficult or impossible for a monolithic system or an individual agent to solve [1–3]. It is worth pointing out that consensus of real-valued multi-agent systems often requires infinite memory and communication resources, which is rather expensive, in reality. Moreover, real-valued consensus algorithms often converge in infinite time, and thus are not suitable for capacity and memory constrained networks with time constraints. Considering this, F. Pasqualetti et al. [4] proposed the use of finite fields to design consensus algorithms for networks of cooperative agents. The main advantages of finite-field consensus networks are that they require finite memory, computation and communication resources, and converge in finite time. Hence, finite-field consensus algorithms are suitable for capacity and memory constrained networks, and for applications subject to time constraints [5, 6].

A finite-field network consists of four basic components:

- the finite field $\mathbb{F}_p = \{0, 1, \cdots, p-1\}$ with characteristic p a prime number;
- a set of n agents, and each agent takes value from $\mathbb{F}_p$;
- a directed graph $\mathscr{G} = (\mathscr{V}, \mathscr{E})$, where $i \in \mathscr{V}$ denotes the i-th agent, with $\mathscr{V} = \{1, \cdots, n\}$, and $(i, j) \in \mathscr{E}$ if there exists a directed edge from agent j to i;
- a linear distributed protocol in which each agent i is associated with a state $x_i \in \mathbb{F}_p$ and each agent updates its state as a weighted combination of the states of its in-neighbors $\mathscr{N}_i^{in}$. The evolution of the network state $x = (x_1, \cdots, x_n)^T$ is described as:

$$x(t+1) = Ax(t), \tag{10.1}$$

where $A = (a_{ij}) \in \mathbb{F}_p^{n \times n}$ is the weighted adjacency matrix of $\mathscr{G}$, $a_{ij} \in \mathbb{F}_p$ is the weight associated with the edge (i, j), $a_{ij} = 0$ whenever $(i, j) \notin \mathscr{E}$, and all the operations in (10.1) are performed in the field $\mathbb{F}_p$.

Given a finite field $\mathbb{F}_p = \{0, 1, \cdots, p-1\}$ with characteristic p a prime number, the addition operator "$+_p$" and the multiplication operator "$\times_p$" are defined as in modular arithmetic, that is, by performing the operation in the set of integers $\mathbb{Z}$, dividing by p, and taking the remainder. The structural matrix of "$+_p$" is:

$$M_{+,p} = \delta_p[U_1\ U_2\ \cdots\ U_p], \tag{10.2}$$

where $U_1 = (1, \cdots, p)$ and $U_s = (s, \cdots, p, 1, \cdots, s-1), s = 2, \cdots, p$. The structural matrix of "$\times_p$" is:

$$M_{\times,p} = \delta_p[V_1\ V_2\ \cdots\ V_p], \tag{10.3}$$

where $V_s = \Big((0 \times s)mod(p) + 1,\ (1 \times s)mod(p) + 1,\ \cdots,\ ((p-1) \times s)mod(p) + 1\Big), s = 1, \cdots, p.$

10.2 CONSENSUS OF FINITE-FIELD NETWORKS WITH SWITCHING TOPOLOGIES AND LINEAR PROTOCOLS

In this section, we assume that the finite-field network has w different directed graphs $\mathscr{G}_r, r = 1, \cdots, w$. Correspondingly, there are w different weighted adjacency matrices, denoted by $A_r = (a_{ij}^r) \in \mathbb{F}_p^{n\times n}, r = 1, \cdots, w$. We assume that each A_r is row-stochastic. At each time instant, the network chooses a deterministic directed graph, to evolve its state. Then, the evolution of the finite-field network with switching topologies and linear protocols can be described as the following switched system:

$$x(t+1) = A_{\sigma(t)}x(t), \tag{10.4}$$

where $\sigma : \mathbb{N} \mapsto \{1, \cdots, w\}$ is the switching signal, and all the operations are performed in the field $\mathbb{F}_p$.

The iteration (10.4) is said to achieve consensus under arbitrary switching signal, if for all initial states $x(0) \in \mathbb{F}_p^n$ and any switching signal $\sigma : \mathbb{N} \mapsto \{1, \cdots, w\}$, there exists a finite time $T \in \mathbb{N}$ and an integer $\alpha \in \mathbb{F}_p$ such that $x(t) = \alpha \mathbf{1}_n$ holds for all $t \geq T$.

Noting that the iteration (10.4) with $\sigma(t) \equiv r$ achieves consensus if and only if $\mathscr{G}_{A_r}$ contains exactly p cycles, corresponding to the unit cycles around the vertices $\alpha\mathbf{1}_n, \alpha \in \mathbb{F}_p$, we make a basic assumption.

Assumption 1 For each A_r, $\mathscr{G}_{A_r}$ contains exactly p cycles, corresponding to the unit cycles around the vertices $\alpha\mathbf{1}_n, \alpha \in \mathbb{F}_p$.

We convert the iteration (10.4) into an algebraic form via the semi-tensor product method.

For the r-th subnetwork of (10.4), identifying $a_{ij}^r, x_i(t) = k \sim \delta_p^{k+1}$ and setting $x(t) = \ltimes_{i=1}^n x_i(t) \in \Delta_{p^n}$, we have:

$$\begin{cases} x_1(t+1) = S_1^r x(t), \\ \quad\vdots \\ x_n(t+1) = S_n^r x(t), \end{cases} \tag{10.5}$$

where,

$$S_i^r = (M_{+,p})^{n-1} \ltimes_{k=1}^n \left[I_{p^{k-1}} \otimes (M_{\times,p} \ltimes a_{ik}^r)\right] \in \mathscr{L}_{p\times p^n}, \tag{10.6}$$

and $M_{+,p}$ and $M_{\times,p}$ are given in (10.2) and (10.3), respectively. Multiplying the equations in (10.5) together yields the following algebraic form:

$$x(t+1) = L_r x(t), \tag{10.7}$$

where $L_r \in \mathscr{L}_{p^n \times p^n}$ and $Col_i(L_r) = \ltimes_{j=1}^{n} Col_i(S_j^r)$, $i = 1, \cdots, p^n$.

Finally, we obtain the algebraic form of the iteration (10.4) as:

$$x(t+1) = L_{\sigma(t)} x(t). \tag{10.8}$$

Moreover, $\alpha \mathbf{1}_n \sim \delta_{p^n}^{c(\alpha)}$ with $c(\alpha) = \alpha \frac{p^n - 1}{p-1} + 1$, $\alpha \in \mathbb{F}_p$.

In the following section, we analyze the consensus of the iteration (10.4) by using the algebraic form (10.8). Before that, we state some useful results on the switching point reachability.

DEFINITION 10.1

Consider system (10.8). Let $x_0 \in \Delta_{p^n}$. Then, a point $x \in \Delta_{p^n}$ is said to be switching reachable from x_0, if one can find an integer $k > 0$ and a switching signal σ, such that under the switching signal, the trajectory of system (10.8) starting from x_0 reaches x at time k.

PROPOSITION 10.1

Consider system (10.8). Then,

1). $x = \delta_{p^n}^r$ is switching reachable from $x_0 = \delta_{p^n}^s$ at time k, if and only if:

$$\left(M^k\right)_{rs} > 0, \tag{10.9}$$

where,

$$M = \sum_{i=1}^{w} L_i, \tag{10.10}$$

and $\left(M^k\right)_{rs}$ denotes the (r,s)-th element of M^k;

2). $x = \delta_{p^n}^r$ is switching reachable from $x_0 = \delta_{p^n}^s$, if and only if:

$$\mathscr{C}_{rs} > 0, \tag{10.11}$$

where,

$$\mathscr{C} = \sum_{k=1}^{p^n} M^k, \tag{10.12}$$

and the operation is performed in $\mathbb{R}$.

Now, based on Proposition 10.1, we have the following result.

Theorem 10.1

Consider the iteration (10.4). Suppose that Assumption 1 holds. Then, the iteration (10.4) achieves consensus under arbitrary switching signal, if and only if there exists a positive integer $\tau \leq p^n$ such that:

$$Row_i(M^\tau)\mathbf{1}_{p^n} = 0 \tag{10.13}$$

holds for any $i \in \{1, \cdots, p^n\}/\{c(\alpha) : \alpha \in \mathbb{F}_p\}$. ■

Proof. Noting that (10.13) is equivalent to:

$$Row_i(M^\tau) = [\underbrace{0 \cdots 0}_{p^n}], \tag{10.14}$$

we prove that the iteration (10.4) achieves consensus under arbitrary switching signal, if and only if there exists a positive integer $\tau \leq p^n$ such that (10.14) holds for any $i \in \{1, \cdots, p^n\}/\{c(\alpha) : \alpha \in \mathbb{F}_p\}$.

(Sufficiency) Suppose that (10.14) holds for any $i \in \{1, \cdots, p^n\}/\{c(\alpha) : \alpha \in \mathbb{F}_p\}$. We prove that the iteration (10.4) achieves consensus under arbitrary switching signal.

In fact, since (10.8) is equivalent to (10.4), we consider (10.8). Starting from any initial state x_0 and under any switching signal $\sigma : \mathbb{N} \mapsto \{1, \cdots, w\}$, from Proposition 10.1 and (10.14), it is easy to see that:

$$x(\tau; x_0, \sigma) \in \{\delta_{p^n}^{c(\alpha)} : \alpha \in \mathbb{F}_p\}.$$

On the other hand, $\delta_{p^n}^{c(\alpha)}$, $\alpha \in \mathbb{F}_p$ are fixed points of $x(t+1) = L_r x(t), r = 1, \cdots, w$. Thus, there exists $\alpha \in \mathbb{F}_p$ such that $x(\tau; x_0, \sigma) = \delta_{p^n}^{c(\alpha)}$.

Next, we prove that $x(t; x_0, \sigma) = \delta_{p^n}^{c(\alpha)}$ holds for all $t \geq \tau$ by induction.

Assume that $x(t; x_0, \sigma) = \delta_{p^n}^{c(\alpha)}$ holds for $t = k \geq \tau$, we consider the case of $k+1$. In this case, we have:

$$x(k+1) = L_{\sigma(k)} x(k; x_0, \sigma) = L_{\sigma(k)} \delta_{p^n}^{c(\alpha)} = \delta_{p^n}^{c(\alpha)}.$$

Thus, $x(t; x_0, \sigma) = \delta_{p^n}^{c(\alpha)}$ holds for $t = k+1$.

By induction, $x(t; x_0, \sigma) = \delta_{p^n}^{c(\alpha)}$ holds for all $t \geq \tau$.

From the arbitrariness of x_0 and σ, one can see that the iteration (10.4) achieves consensus under arbitrary switching signal.

(Necessity) Assume that the iteration (10.4) achieves consensus under arbitrary switching signal. Then, for the system (10.8), starting from any initial state x_0 and under any switching signal $\sigma : \mathbb{N} \mapsto \{1, \cdots, w\}$, there exist an integer $T(x_0, \sigma) \in \mathbb{N}$ and $\alpha \in \mathbb{F}_p$ such that:

$$x(t; x_0, \sigma) = \delta_{p^n}^{c(\alpha)}$$

holds for all $t \geq T(x_0,\sigma)$, and,

$$x(t;x_0,\sigma) \neq \delta_{p^n}^{c(\alpha)}, \ \forall\, 0 \leq t \leq T(x_0,\sigma)-1.$$

In the following section, we prove that $T(x_0,\sigma) \leq p^n$ holds for all $x_0 \in \Delta_{p^n}$ and all $\sigma : \mathbb{N} \mapsto \{1,\cdots,w\}$.

In fact, if the conclusion is not true, then one can find an initial state x_0 and a switching signal σ, such that $T(x_0,\sigma) > p^n$. Under σ, we have $x(t;x_0,\sigma) \neq \delta_{p^n}^{c(\alpha)}$, $\forall\, 0 \leq t \leq T(x_0,\sigma)-1$, and $x(t;x_0,\sigma) = \delta_{p^n}^{c(\alpha)}$ holds for $t = T(x_0,\sigma)$. Since the number of different states for the system (10.8) is p^n, there must exist two integers $0 \leq t_1 < t_2 \leq T(x_0,\sigma)-1$ such that:

$$x(t_1;x_0,\sigma) = x(t_2;x_0,\sigma) \neq \delta_{p^n}^{c(\alpha)}.$$

Now, for the initial state $\widehat{x}_0 = x(t_1;x_0,\sigma)$, we set $\widehat{\sigma}(t) = \sigma(t_1+t)$, $t = 0,1,\cdots,t_2-t_1-1$. Then, under the switching signal $\widehat{\sigma}(t)$, one can see that $\widehat{x}(t_2-t_1;\widehat{x}_0,\widehat{\sigma}) = x(t_2;x_0,\sigma) = x(t_1;x_0,\sigma)$. Generally, for $k \in \mathbb{N}$, we construct the following switching signal:

$$\widehat{\sigma}(t) = \begin{cases} \sigma(t_1), \quad t = k(t_2-t_1), \\ \sigma(t_1+1),\ t = k(t_2-t_1)+1, \\ \vdots \\ \sigma(t_2-1),\ t = (k+1)(t_2-t_1)-1, \end{cases} \tag{10.15}$$

which is periodic. Then, under the switching signal $\widehat{\sigma}(t)$, the trajectory of the system (10.8) starting from the initial state $\widehat{x}(0) = x(t_1;x_0,\sigma)$ forms a cycle:

$$\begin{aligned} &\{x(t_1;x_0,\sigma), x(t_1+1;x_0,\sigma),\cdots,x(t_2-1;x_0,\sigma); \\ &x(t_1;x_0,\sigma), x(t_1+1;x_0,\sigma),\cdots,x(t_2-1;x_0,\sigma); \\ &\cdots\}, \end{aligned}$$

which is a contradiction to the fact that the iteration (10.4) achieves consensus under arbitrary switching signal. Thus, $T(x_0,\sigma) \leq p^n$ holds for all $x_0 \in \Delta_{p^n}$ and all $\sigma : \mathbb{N} \mapsto \{1,\cdots,w\}$.

Set $\tau = \max\limits_{x_0,\sigma} T(x_0,\sigma) \leq p^n$. One can see that for the system (10.8), starting from any initial state $x_0 = \delta_{p^n}^r$ and under any switching signal $\sigma : \mathbb{N} \mapsto \{1,\cdots,w\}$, there exists $\alpha \in \mathbb{F}_p$ such that $x(\tau;x_0,\sigma) = \delta_{p^n}^{c(\alpha)}$. That is, $\left(M^\tau\right)_{r,i} = 0$ holds for all $r = 1,\cdots,p^n$ and $i \in \{1,\cdots,p^n\}/\{c(\alpha) : \alpha \in \mathbb{F}_p\}$.

Therefore, (10.14) holds for any $i \in \{1,\cdots,p^n\}/\{c(\alpha) : \alpha \in \mathbb{F}_p\}$.

We give an example to illustrate the above results.

Example 10.1:

Consider a fully connected network with three agents over the field $\mathbb{F}_3$. Assume that the network has two weighted adjacency matrices:

$$A_1 = \begin{bmatrix} 2 & 1 & 1 \\ 2 & 1 & 1 \\ 2 & 1 & 1 \end{bmatrix},$$

and

$$A_2 = \begin{bmatrix} 1 & 1 & 2 \\ 1 & 1 & 2 \\ 1 & 1 & 2 \end{bmatrix}.$$

At each time instant, the network chooses one of A_1 and A_2 to evolve its state. Then, the evolution of the network with switching topologies and linear protocols can be described as the following switched system:

$$x(t+1) = A_{\sigma(t)}x(t), \tag{10.16}$$

where $\sigma : \mathbb{N} \mapsto \{1,2\}$ is the switching signal, and all the operations are performed in the field $\mathbb{F}_3$. The objective of this example is to verify whether or not the iteration (10.16) achieves consensus under arbitrary switching signal.

First, we convert the iteration (10.16) into the algebraic form.

Identifying $0 \sim \delta_3^1$, $1 \sim \delta_3^2$ and $2 \sim \delta_3^3$, and setting $x(t) = \ltimes_{i=1}^3 x_i(t)$, for the iteration (10.16) with $\sigma(t) \equiv 1$, we have the following algebraic form:

$$x(t+1) = L_1 x(t), \tag{10.17}$$

where,

$$\begin{aligned} L_1 \quad = \quad & \delta_{27}[1\ 14\ 27\ 14\ 27\ 1\ 27\ 1\ 14 \\ & 27\ 1\ 14\ 1\ 14\ 27\ 14\ 27\ 1 \\ & 14\ 27\ 1\ 27\ 1\ 14\ 1\ 14\ 27]. \end{aligned}$$

For the iteration (10.16) with $\sigma(t) \equiv 2$, we have the following algebraic form:

$$x(t+1) = L_2 x(t), \tag{10.18}$$

where:

$$\begin{aligned} L_2 \quad = \quad & \delta_{27}[1\ 27\ 14\ 14\ 1\ 27\ 27\ 14\ 1 \\ & 14\ 1\ 27\ 27\ 14\ 1\ 1\ 27\ 14 \\ & 27\ 14\ 1\ 1\ 27\ 14\ 14\ 1\ 27]. \end{aligned}$$

Thus, the iteration (10.16) has the following algebraic form:

$$x(t+1) = L_{\sigma(t)}x(t), \tag{10.19}$$

where $\sigma : \mathbb{N} \mapsto \{1,2\}$ is the switching signal. Moreover, $c(0) = 1$, $c(1) = 14$, and $c(2) = 27$.

Second, based on Theorem 10.1, we check whether or not the iteration (10.16) achieves consensus under arbitrary switching signal.

It is easy to see that A_1 and A_2 are row-stochastic. Moreover, for each $i \in \{1,2\}$, one can see from L_i that the system $x(t+1) = L_i x(t)$ has exactly 3 attractors, corresponding to the fixed points δ_{27}^{1}, δ_{27}^{14} and δ_{27}^{27}. Thus, Assumption 1 holds.

Set $M = L_1 + L_2$. A simple computation gives:

$$
\begin{aligned}
M\mathbf{1}_{27} = & [18\,0\,0\,0\,0\,0\,0\,0\,0 \\
& 0\,0\,0\,0\,18\,0\,0\,0\,0 \\
& 0\,0\,0\,0\,0\,0\,0\,0\,18]^T,
\end{aligned}
$$

which implies that $Row_i(M)\mathbf{1}_{27} = 0,\ \forall\ i \in \{1,\cdots,27\}/\{1,14,27\}$. Therefore, by Theorem 10.1, the iteration (10.16) achieves consensus under arbitrary switching signal.

10.3 LEADER-FOLLOWER CONSENSUS OF MULTI-AGENT SYSTEMS OVER FINITE FIELDS

In this section, we study the consensus of leader-follower multi-agent systems with time-delay over finite fields.

Consider a leader-follower multi-agent system over $\mathbb{F}_p := \{0,\cdots,p-1\}$ which consists of one leader and N followers, where p is a prime number. The dynamics of the leader is the following modular system:

$$
x_0(t+1) = Ax_0(t-\tau_{0j}), \tag{10.20}
$$

where $x_0(t) = (x_0^1(t),\cdots,x_0^n(t))^T \in \mathbb{F}_p^n$ and $A \in \mathbb{F}_p^{n\times n}$. The dynamics of the i-th follower ($i \in \{1,\cdots,N\}$) is given as follows:

$$
x_i(t+1) = A_i x_i(t-\tau_{ij}), \tag{10.21}
$$

where $x_i(t) = (x_i^1(t),\cdots,x_i^n(t))^T \in \mathbb{F}_p^n$, $A_i \in \mathbb{F}_p^{n\times n}$.

In the following, we convert the leader-follower multi-agent system over $\mathbb{F}_p$ into an algebraic form.

Using the vector form of values in $\mathbb{F}_p$ and setting $x_0(t) = \ltimes_{i=1}^{n} x_0^i(t) \in \Delta_{p^n}$. For the dynamics of the leader (10.20), we have:

$$
\begin{aligned}
x_0^i(t+1) &= a_{i1}^0 \times_p x_0^1(t-\tau_{01}) +_p \cdots +_p \\
& \quad a_{in}^0 \times_p x_0^n(t-\tau_{0n}) \\
&= \sum_{k=\tau}^{0} \{\sum_{j=1}^{n} {}^k c_{ij}^0 x_0^j(t-k)\},
\end{aligned} \tag{10.22}
$$

where:

$$
{}^k c_{ij}^0 = \begin{cases} a_{ij}^0, \text{ if } j \in \mathcal{N}_i^{in}(t-k) \cup \{i\}, \\ 0, \text{ otherwise.} \end{cases} \tag{10.23}
$$

Using the canonical vector form of variables in $\mathbb{F}_p$ and setting $x(t) = \ltimes_{i=1}^{N} x_i(t)$, we have:

$$\begin{aligned}
& \sum_{j=1}^{n} {}^{k}c_{ij}^{0} x_0^j(t-k) \\
= \; & (M_{+,p})^{n-1} M_{\times,p} \ltimes {}^{k}c_{i1}^{0} x_0^1(t-k) M_{\times,p} \ltimes \\
& {}^{k}c_{i2}^{0} x_0^2(t-k) \cdots M_{\times,p} \ltimes {}^{k}c_{in}^{0} x_0^n(t-k) \\
= \; & (M_{+,p})^{n-1} \ltimes_{j=1}^{n} [I_{p^{j-1}} \otimes (M_{\times,p} \ltimes {}^{k}c_{ij}^{0})] \\
& x_0^1(t-k) \cdots x_0^n(t-k) \\
:= \; & M_{i,k} x_0(t-k),
\end{aligned} \tag{10.24}$$

where $M_{i,k} = (M_{+,p})^{n-1} \ltimes_{j=1}^{n} [I_{p^{j-1}} \otimes (M_{\times,p} \ltimes {}^{k}c_{ij}^{0})]$.

Thus,

$$\begin{aligned}
x_0^i(t+1) \; = \; & (M_{+,p})^{\tau} M_{i,\tau} x_0(t-\tau) M_{i,\tau-1} \ltimes \\
& x_0(t-\tau+1) \cdots M_{i,0} x_0(t) \\
= \; & (M_{+,p})^{\tau} \ltimes_{j=0}^{\tau} (I_{p^{nj}} \otimes M_{i,\tau-j}) \ltimes \\
& x_0(t-\tau) \cdots x_0(t) \\
:= \; & M_0^i z_0(t),
\end{aligned} \tag{10.25}$$

where $M_0^i = (M_{+,p})^{\tau} \ltimes_{j=0}^{\tau} (I_{p^{nj}} \otimes M_{i,\tau-j}) \in \mathscr{L}_{p \times p^{n(\tau+1)}}$ and $z_0(t) = \ltimes_{i=t-\tau}^{t} x_0(i)$.

Hence, we obtain that:

$$\begin{cases}
x_0^1(t+1) = M_0^1 z_0(t), \\
x_0^2(t+1) = M_0^2 z_0(t), \\
\vdots \\
x_0^n(t+1) = M_0^n z_0(t).
\end{cases} \tag{10.26}$$

By the Khatri-Rao matrix product, we can present the following algebraic form of (10.20):

$$x_0(t+1) = M_0 z_0(t) \tag{10.27}$$

where $M_0 = M_0^1 * M_0^2 * \cdots * M_0^n \in \mathscr{L}_{p^n \times p^{n(\tau+1)}}$.

Similarly, for the dynamics of the i-th follower (10.21), we have:

$$\begin{cases}
x_1(t+1) = M_1 z_1(t), \\
x_2(t+1) = M_2 z_2(t), \\
\vdots \\
x_N(t+1) = M_N z_N(t).
\end{cases} \tag{10.28}$$

Multiplying the $N+1$ equations of (10.27) and (10.28) by the Khatri-Rao matrix product, we can convert the leader follower system into the following algebraic form:

$$\begin{aligned} & x(t+1) = x_0(t+1) \ltimes x_1(t+1) \ltimes \cdots \ltimes x_N(t+1) \\ = & M_0 z_0(t) M_1 z_1(t) \cdots M_N z_N(t) \\ = & M_0 \ltimes \prod_{i=1}^{N} (I_{p^{n(\tau+1)i}} \otimes M_i) z_0(t) z_1(t) \cdots z_N(t) \\ := & Mz(t), \end{aligned} \tag{10.29}$$

where $M = M_0 \ltimes \prod_{i=1}^{N} (I_{p^{n(\tau+1)i}} \otimes M_i) \in \mathscr{L}_{p^{n(N+1)} \times p^{n(N+1)(\tau+1)}}$, and $z(t) = z_0(t) \ltimes \cdots \ltimes z_N(t)$.

Set,

$$z(t+1) = \ltimes_{i=t-\tau+1}^{t+1} x(i).$$

We have:

$$\begin{aligned} z(t+1) &= \ltimes_{i=t-\tau+1}^{t} x(i) \ltimes x(t+1) \\ &= \ltimes_{i=t-\tau+1}^{t} x(i) \ltimes Mz(t) \\ &= (I_{p^{n(N+1)\tau}} \otimes M) \ltimes_{i=t-\tau+1}^{t} x(i) z(t) \\ &= (I_{p^{n(N+1)\tau}} \otimes M) \ltimes_{i=t-\tau+1}^{t} x(i) x(t-\tau) \\ &\quad \ltimes_{i=t-\tau+1}^{t} x(i) \\ &= (I_{p^{n(N+1)\tau}} \otimes M) W_{[p^{n(N+1)}, p^{n(N+1)\tau}]} \\ &\quad x(t-\tau) [\ltimes_{i=t-\tau+1}^{t} x(i)]^2 \\ &= (I_{p^{n(N+1)\tau}} \otimes M) W_{[p^{n(N+1)}, p^{n(N+1)\tau}]} \\ &\quad (I_{p^{n(N+1)}} \otimes M_{r, p^{n(N+1)(\tau+1)}}) z(t) \\ &:= \hat{M} z(t) \end{aligned} \tag{10.30}$$

where $\hat{M} = (I_{p^{n(N+1)\tau}} \otimes M) W_{[p^{n(N+1)}, p^{n(N+1)\tau}]} (I_{p^{n(N+1)}} \otimes M_{r, p^{n(N+1)(\tau+1)}})$.

The leader-follower consensus problem of the iteration (10.20) and (10.21) is defined as follows.

DEFINITION 10.2

The follower (10.21) achieves (finite-time) consensus with the leader (10.20) in $\mathbb{F}_p$, if there exists an integer $T \in \mathbb{Z}_+$ such that:

$$x_i^s(t) = x_0^s(t), i = 1, \cdots, N \tag{10.31}$$

holds for any initial condition, any $s = 1, 2, \cdots, n$ and any integer $t \geq T$.

Define,

$$
\begin{aligned}
A &= \{(\delta_{p^n}^i)^{N+1} : i = 1, \cdots, p^n\} && (10.32)\\
&:= \{\delta_{p^n(N+1)}^{i_1}, \delta_{p^n(N+1)}^{i_2}, \cdots, \delta_{p^n(N+1)}^{i_{p^n}}\} && (10.33)
\end{aligned}
$$

and

$$
\begin{aligned}
B &= \{\delta_{p^n(N+1)(\tau+1)}^{j} = \delta_{p^n(N+1)}^{j_1} \ltimes \cdots \ltimes \delta_{p^n(N+1)}^{j_{\tau+1}}\}\\
&:= \{\delta_{p^n(N+1)(\tau+1)}^{l_1}, \cdots, \delta_{p^n(N+1)(\tau+1)}^{l_{p^n(\tau+1)}}\}, && (10.34)
\end{aligned}
$$

where $i_1 < \cdots < i_{p^n}$, $l_1 < \cdots < l_{p^n(\tau+1)}$.

DEFINITION 10.3

System (10.30) is said to be stable at B with respect to trajectory, if for any initial trajectory $z_0 \in \Delta_{p^n(N+1)(\tau+1)}$, there exists a positive integer T such that:

$$z(t; z_0) \in B \tag{10.35}$$

holds for any integer $t \geq T$.

Theorem 10.2

The follower (10.21) achieves (finite-time) consensus with the leader (10.20), if and only if the algebraic form (10.30) is stable at B with respect to trajectory. ■

Proof. (Necessity) Suppose that the follower (10.21) achieves (finite-time) consensus with the leader (10.20). By Definition 10.2, for any initial state $x_i(0) \in \Delta_{p^n}, i = 1, 2, \cdots, N$, there exists an integer $T \in \mathbb{Z}_+$ such that $x_i^s(t) = x_0^s(t)$ holds for any $s = 1, 2, \cdots, n$ and any integer $t \geq T$. Hence, we have that $x(t) \in A$ holds for any integer $t \geq T$.

Set $\hat{T} = T + \tau$, then:

$$z(t) = \ltimes_{i=t-\tau}^{t} x(i) \in B$$

holds for any integer $t \geq \hat{T}$. By Definition 10.3, system (10.30) is stable at B with respect to trajectory.

(Sufficiency) Suppose that system (10.30) is stable at B with respect to trajectory. By Definition 10.3, for any $z_0 \in \Delta_{p^n(N+1)(\tau+1)}$, there exists an integer $T \in \mathbb{Z}_+$ such that:

$$z(t; z_0) \in B$$

holds for any integer $t \geq T$. Considering the form of B and the unique factorization formula of $z(t)$, one can see that $x(i) \in A$ holds for any $i \in \{t-\tau,\cdots,t\}$ and any integer $t \geq T$.

Based on the above analysis, we have the following result.

Theorem 10.3

System (10.30) is stable at B with respect to trajectory, if and only if there exists a positive integer $\mu \leq p^{n(N+1)(\tau+1)}$, such that:

$$\sum_{c\in\Gamma} Row_c(\hat{M}^{\mu}) = \mathbf{0}^T_{p^{n(N+1)(\tau+1)}}, \tag{10.36}$$

where $\Gamma = \{1,\cdots,p^{n(N+1)(\tau+1)}\} \setminus \{l_1,\cdots,l_{p^{n(\tau+1)}}\}$. ■

Proof. (Sufficiency) We firstly prove that (10.36) holds for any integer $t \geq \mu$ by induction.

Obviously, (10.36) holds for $t=\mu$. Assuming that (10.36) holds for $t=\gamma\geq\mu$, we prove the case of $t=\gamma+1$. In fact, $Row_c(\hat{M}^{\gamma+1}) = Row_c(\hat{M}^{\gamma})\hat{M} = \mathbf{0}^T_{p^{n(N+1)(\tau+1)}}$ holds for any $c \in \{1,\cdots,p^{n(N+1)(\tau+1)}\} \setminus \{l_1,\cdots,l_{p^{n(\tau+1)}}\}$. Which implies that (10.36) holds for $t=\gamma+1$. By induction, (10.36) holds for any integer $t \geq \mu$.

Thus, for any $z(0) \in \Delta_{p^{n(N+1)(\tau+1)}}$, we have:

$$z(t) = (\hat{M})^t z(0) \in B$$

holds for any integer $t \geq \mu$.

By Definition 10.3, system (10.30) is stable at B with respect to trajectory.

(Necessity) Suppose that system (10.30) is stable at B with respect to trajectory. Then, for any $z(0) \in \Delta_{p^{n(N+1)(\tau+1)}}$, there exists an integer $T(z(0)) \in \mathbb{Z}_+(\leq p^{n(N+1)(\tau+1)})$ such that:

$$z(t) \in B$$

holds for any integer $t \geq T(z(0))$.

Let $\mu = \max_{z(0)\in\Delta_{p^{n(N+1)(\tau+1)}}} \{T(z(0))\}$. One can see that:

$$z(t) = \hat{M}^t z(0) \in B$$

holds for any integer $t \geq \mu$ and any $z(0) \in \Delta_{p^{n(N+1)(\tau+1)}}$, which implies that:

$$Row_c(\hat{M}^{\mu}) = \mathbf{0}^T_{p^{n(n+1)(\tau+1)}}$$

holds for any $c \in \{1,\cdots,p^{n(N+1)(\tau+1)}\} \setminus \{l_1,\cdots,l_{p^{n(\tau+1)}}\}$.

Therefore,

$$\sum_{c\in\Gamma} Row_c(\hat{M}^{\mu}) = \mathbf{0}^T_{p^{n(N+1)(\tau+1)}}.$$

Finally, we apply the obtained new results to the consensus of leader follower systems with control.

The dynamics of the leader to follow a modular system:

$$x_0(t+1) = Ax_0(t-\tau_{0j}), \tag{10.37}$$

where $x_0(t) = (x_0^1(t),\cdots,x_0^n(t))^T \in \mathbb{F}_p^n$ and $A \in \mathbb{F}_p^{n\times n}$. The dynamics of the i-th follower ($i \in \{1,\cdots,N\}$) is given as follows:

$$x_i(t+1) = A_i x_i(t-\tau_{ij}) + b_i u_i(t), \tag{10.38}$$

where $x_i(t) = (x_i^1(t),\cdots,x_i^n(t))^T \in \mathbb{F}_p^n$, $A_i \in \mathbb{F}_p^{n\times n}$, $b_i = (b_i^1,\cdots,b_i^n)^T \in \mathbb{F}_p^n$, and $u_i(t) \in \mathbb{F}_p$ is the control input.

According to the conversion process of (10.20) and (10.21), we have

$$\begin{aligned}
& x_i^j(t+1) = a_{j1}^i \times_p x_i^1(t-\tau_{i1}) +_p \cdots +_p \\
& a_{jn}^i \times_p x_i^n(t-\tau_{in}) +_p b_i^j u_i(t) \\
= \; & b_i^j u_i(t) +_p \sum_{k=\tau}^{0} \{ \sum_{l=1}^{n} {}^k c_{jl}^i x_i^l(t-k) \} \\
= \; & b_i^j u_i(t) +_p \sum_{k=\tau}^{0} M_{i,k} x_i(t-k) \\
= \; & b_i^j u_i(t) +_p M_{i,\tau} x_i(t-\tau) +_p M_{i,\tau-1} x_i(t-\tau+1) \\
& +_p \cdots +_p M_{i,0} x_i(t) \\
= \; & (M_{+,p})^{\tau+1} M_{\times,p} b_i^j (I_{p^n} \otimes M_{i,\tau}) \prod_{j=1}^{\tau} (I_{p^{n(j+1)}} \otimes \\
& M_{i,\tau-j}) u_i(t) x_i(t-\tau) x_i(t-\tau+1) \cdots x_i(t) \\
:= \; & \hat{M}_i^j u_i(t) x_i(t),
\end{aligned}$$

where $\hat{M}_i^j = (M_{+,p})^{\tau+1} M_{\times,p} b_i^j (I_{p^n} \otimes M_{i,\tau}) \prod_{j=1}^{\tau} (I_{p^{n(j+1)}} \otimes M_{i,\tau-j})$. Hence, we have:

$$\begin{cases} x_i^1(t+1) = \hat{M}_i^1 u_i(t) z_i(t), \\ \vdots \\ x_i^n(t+1) = \hat{M}_i^n u_i(t) z_i(t). \end{cases} \tag{10.39}$$

By the Khatri-Rao matrix product, we can present the following algebraic form of (10.38):

$$x_i(t+1) = \hat{M}_i u_i(t) z_i(t) \tag{10.40}$$

where $\hat{M}_i = \hat{M}_i^1 * \hat{M}_i^2 * \cdots * \hat{M}_i^n$.

Set,

$$H(t+1) = x_1(t+1) \ltimes x_2(t+1) \ltimes \cdots \ltimes x_N(t+1).$$

We have:

$$\begin{aligned} H(t+1) &= \hat{M}_1 u_1(t) z_1(t) \ltimes \cdots \ltimes \hat{M}_N u_N(t) z_N(t) \\ &= \hat{M}_1 \prod_{i=2}^{N} [I_{p^{(n(\tau+1)+1)(i-1)}} \otimes \hat{M}_i \\ &\quad \ltimes (I_{p^{i-1}} \otimes W_{[p,p^{(i-1)n(\tau+1)}]})] \\ &\quad u(t) z_1(t) \cdots z_N(t) \\ &:= Lu(t) z_1(t) \cdots z_N(t), \end{aligned} \tag{10.41}$$

where $L = \hat{M}_1 \prod_{i=2}^{N} [I_{p^{(n(\tau+1)+1)(i-1)}} \otimes \hat{M}_i \ltimes (I_{p^{i-1}} \otimes W_{[p,p^{(i-1)n(\tau+1)}]})]$.

Using the semi-tensor product of matrices, we can present the algebraic form of (10.37) and (10.38) as follow:

$$\begin{aligned} & x(t+1) = x_0(t+1) \ltimes x_1(t+1) \ltimes \cdots \ltimes x_N(t+1) \\ = \;& M_0 z_0(t) \ltimes H(t+1) \\ = \;& M_0 (I_{p^{n(\tau+1)}} \otimes L) W_{[p^N, p^{n(\tau+1)}]} u(t) z(t) \\ := \;& \hat{L} u(t) z(t), \end{aligned} \tag{10.42}$$

where $M = \hat{L} = M_0 (I_{p^{n(\tau+1)}} \otimes L) W_{[p^N, p^{n(\tau+1)}]}$.

Suppose that,

$$\begin{aligned} v_i(t) &= \sum_{j=0}^{N} a_{ij}(x_j(t) + x_i(t)) \\ &= \begin{bmatrix} \sum_{j=0}^{N} a_{ij}(x_j^1(t) + x_i^1(t)) \\ \sum_{j=0}^{N} a_{ij}(x_j^2(t) + x_i^2(t)) \\ \vdots \\ \sum_{j=0}^{N} a_{ij}(x_j^n(t) + x_i^n(t)) \end{bmatrix} \\ &:= \begin{bmatrix} v_i^1(t) \\ v_i^2(t) \\ \vdots \\ v_i^n(t) \end{bmatrix}. \end{aligned} \tag{10.43}$$

For any $l = 1, 2, \cdots, n$, we have,

$$
\begin{aligned}
v_i^l(t) &= \sum_{j=0}^{N} a_{ij}(x_j^l(t) + x_i^l(t)) \\
&= (M_{+,p})^N M_{\times,p} a_{i0} M_{+,p} x_0^l(t) x_i^l(t) M_{\times,p} a_{i1} M_{+,p} \\
&\quad x_1^l(t) x_i^l(t) \cdots M_{\times,p} a_{iN} M_{+,p} x_N^l(t) x_i^l(t) \\
&= (M_{+,p})^N \ltimes_{s=0}^{N} [I_{p^{2s}} \otimes (M_{\times,p} a_{is} M_{+,p})] \\
&\quad x_0^l(t) x_i^l(t) x_1^l(t) x_i^l(t) \cdots x_N^l(t) x_i^l(t) \\
&= (M_{+,p})^N \ltimes_{s=0}^{N} [I_{p^{2s}} \otimes (M_{\times,p} a_{is} M_{+,p})] \\
&\quad \ltimes_{s=1}^{N} [(I_{p^s} \otimes W_{[p,p]})(I_{p^{s+1}} \otimes M_{r,p})] \\
&\quad x_0^l(t) x_1^l(t) \cdots x_N^l(t) x_i^l(t) \\
&= (M_{+,p})^N \ltimes_{s=0}^{N} [I_{p^{2s}} \otimes (M_{\times,p} a_{is} M_{+,p})] \\
&\quad \ltimes_{s=1}^{N} [(I_{p^s} \otimes W_{[p,p]})(I_{p^{s+1}} \otimes M_{r,p})] \\
&\quad [I_{p^i} \otimes (W_{[p,p^{N-i+1}]} M_{r,p})] \\
&\quad x_0^l(t) x_1^l(t) \cdots x_N^l(t) \\
&:= Q_i^l x_0^l(t) x_1^l(t) \cdots x_N^l(t),
\end{aligned} \tag{10.44}
$$

where $Q_i^l = (M_{+,p})^N \ltimes_{s=0}^{N} [I_{p^{2s}} \otimes (M_{\times,p} a_{is} M_{+,p})] \ltimes_{s=1}^{N} [(I_{p^s} \otimes W_{[p,p]})(I_{p^{s+1}} \otimes M_{r,p})][I_{p^i} \otimes (W_{[p,p^{N-i+1}]} M_{r,p})]$.

Thus,

$$
\begin{aligned}
v_i(t) &= \ltimes_{l=1}^{n} v_i^l(t) \\
&= Q_i^1 x_0^1(t) x_1^1(t) \cdots x_N^1(t) Q_i^2 x_0^2(t) x_1^2(t) \cdots x_N^2(t) \\
&\quad \cdots Q_i^n x_0^n(t) x_1^n(t) \cdots x_N^n(t) \\
&= \ltimes_{s=1}^{n} (I_{p^{(N+1)(s-1)}} \otimes Q_i^s) x_0^1(t) x_1^1(t) \cdots x_N^1(t) \\
&\quad x_0^2(t) x_1^2(t) \cdots x_N^2(t) \cdots x_0^n(t) x_1^n(t) \cdots x_N^n(t) \\
&= \ltimes_{s=1}^{n} (I_{p^{(N+1)(s-1)}} \otimes Q_i^s) \ltimes_{r=1}^{N} \{I_{p^{n(r-1)}} \otimes \\
&\quad [\ltimes_{s=1}^{n-1} (I_{p^s} \otimes W_{[p,p^{s(N-r+1)}]})]\} x_0(t) \cdots x_N(t) \\
&:= Q_i x_0(t) \cdots x_N(t),
\end{aligned} \tag{10.45}
$$

where $Q_i = \ltimes_{s=1}^{n} (I_{p^{(N+1)(s-1)}} \otimes Q_i^s) \ltimes_{r=1}^{N} \{I_{p^{n(r-1)}} \otimes [\ltimes_{s=1}^{n-1} (I_{p^s} \otimes W_{[p,p^{s(N-r+1)}]})]\}$.

Hence, we obtain that:

$$\begin{aligned} & v_i(t) = Q_i x_0(t) \cdots x_N(t) \\ = & Q_i D_r[p^{n\tau}, p^n] x_0(t-\tau) x_0(t-\tau+1) \cdots x_0(t) \\ & D_r[p^{n\tau}, p^n] x_1(t-\tau) x_1(t-\tau+1) \cdots x_1(t) \cdots \\ & D_r[p^{n\tau}, p^n] x_N(t-\tau) x_N(t-\tau+1) \cdots x_N(t) \\ = & Q_i D_r[p^{n\tau}, p^n] z_0(t) \cdots D_r[p^{n\tau}, p^n] z_N(t) \\ = & Q_i \ltimes_{j=0}^{N} (I_{p^{n(N+1)j}} \otimes D_r[p^{n\tau}, p^n]) z(t) \\ := & \hat{Q}_i z(t), \end{aligned} \tag{10.46}$$

where $\hat{Q}_i = Q_i \ltimes_{j=0}^{N} (I_{p^{n(N+1)j}} \otimes D_r[p^{n\tau}, p^n])$.

Based on the above analysis, we suppose that the distributed control input in (10.38) has the following form:

$$u_i(t) = K_i v_i(t), \quad i = 1, \cdots, N, \tag{10.47}$$

where $K_i \in \mathscr{L}_{p \times p^n}$ is the distributed state feedback gain matrix.

Therefore,

$$\begin{cases} u_1(t) = K_1 \hat{Q}_1 z(t), \\ \vdots \\ u_N(t) = K_N \hat{Q}_N z(t), \end{cases} \tag{10.48}$$

which shows that:

$$u(t) = Qz(t), \tag{10.49}$$

where $Q = (K_1 \hat{Q}_1) * (K_2 \hat{Q}_2) * \cdots * (K_N \hat{Q}_N)$.

Combining (10.42) and (10.49), one can obtain the following algebraic form of the leader-follower multi-agent system over $\mathbb{F}_p$ with the distributed control (10.43):

$$\begin{aligned} z(t+1) &= \hat{L} Q z^2(k) \\ &= \hat{L} Q M_{r, p^{n(N+1)(\tau+1)}} z(t) \\ &:= \widetilde{L} z(t), \end{aligned} \tag{10.50}$$

where $\widetilde{L} = \hat{L} Q M_{r, p^{n(N+1)(\tau+1)}}$.

Based on the above analysis, we have the following result.

Theorem 10.4

The follower (10.38) achieves (finite-time) consensus with the leader (10.37) under the distributed control (10.43), if and only if there exists a positive integer

$\mu \leq p^{n(N+1)(\tau+1)}$, such that

$$\sum_{c \in \Gamma} Row_c(\widetilde{L}^{\mu}) = \mathbf{0}^T_{p^{n(N+1)(\tau+1)}}. \tag{10.51}$$

■

REFERENCES

1. Li, T., Fu, M., Xie, L. and Zhang, J. (2011). Distributed consensus with limited communication data rate. IEEE Trans. Aut. Contr., 56(2): 279–292.
2. Liu, B., Chu, T., Wang, L. and Xie, G. (2008). Controllability of a leader-follower dynamic network with switching topology. IEEE Trans. Aut. Contr., 53(4): 1009–1013.
3. Li, Z., Ren, W., Liu, X. and Xie, L. (2013). Distributed consensus of linear multi-agent systems with adaptive dynamic protocols. Automatica, 49(7): 1986–1995.
4. Pasqualetti, F., Borra, D. and Bullo, F. (2014). Consensus networks over finite fields. Automatica, 50(2): 349–358.
5. Fagiolini, A. and Bicchi, A. (2013). On the robust synthesis of logical consensus algorithms for distributed intrusion detection. Automatica, 49(8): 2339–2350.
6. Xu, X. and Hong, Y. (2014). Leader-following consensus of multi-agent systems over finite fields. IEEE 53rd Annual Conference on Decision and Control, 2999–3004.

Index